MODERN COSMOLOGY IN RETROSPECT

MODERN COSMOLOGY IN RETROSPECT

Edited by

B. Bertotti, R. Balbinot, S. Bergia and A. Messina

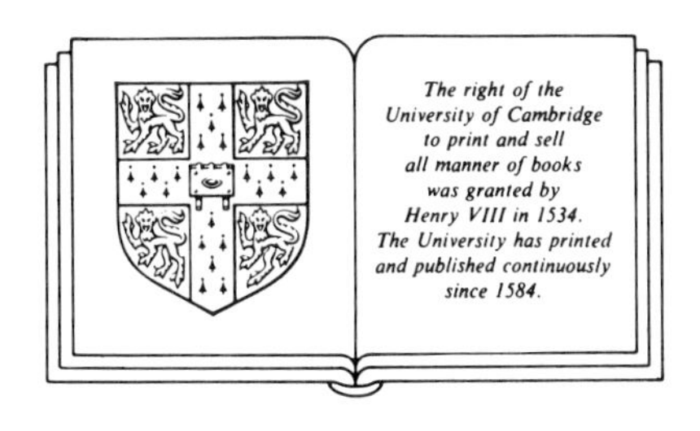

CAMBRIDGE UNIVERSITY PRESS

Cambridge

New York Port Chester Melbourne Sydney

Published by the Press Syndicate of the University of Cambridge
The Pitt Building, Trumpington Street, Cambridge CB2 1RP
40 West 20th Street, New York, NY 10011, USA
10 Stamford Road, Oakleigh, Melbourne 3166, Australia

First published 1990

Printed in Great Britain at the University Press, Cambridge

British Library cataloguing in publication data

Modern cosmology in retrospect.
1. Astronomy. Cosmology, history
I. Bertotti, B. *1930–*
523.109

Library of Congress cataloguing in publication data available

ISBN 0 521 37213 5 hardback

Contents

List of contributors *page* ix
Preface xi

Part I The general framework

1 Cosmology, a peculiar science 3
Bruno Bertotti

2 The early years 11
J. D. North

Part II Riddles of and clues to cosmology

3 Olbers' Paradox in recent times 33
Edward Harrison

4 The part played by Mach's Principle in the genesis of relativistic cosmology 47
J. B. Barbour

5 The mysterious lore of Large Numbers 67
John D. Barrow

Part III Geometrical and physical cosmology

6 Innovation, resistance and change: the transition to the expanding universe 97
G. F. R. Ellis

7 Early inhomogeneous cosmological models in Einstein's theory 115
Andrzej Krasiński

8 Early work on 'big-bang' cosmology and the cosmic blackbody radiation — 129
Ralph A. Alpher and Robert Herman

9 Deciphering the nuclear ashes of the early universe: a personal perspective — 159
Robert V. Wagoner

Part IV The great cosmological debates

10 The cosmological scene 1945–1952 — 189
Hermann Bondi

11 Personal recollections: some lessons for the future — 197
William McCrea

12 An assessment of the evidence against the steady-state theory — 221
F. Hoyle

13 Steady-state cosmology, the arrow of time, and Hoyle and Narlikar's theories — 233
José M. Sánchez-Ron

Part V Cosmological observations and discoveries

14 The observational approach to cosmology: U.S. Observatories pre-World War II — 247
Donald E. Osterbrock

15 Discovery of the cosmic microwave background — 291
Robert M. Wilson

16 The entry of radio astronomy into cosmology: radio stars and Martin Ryle's 2C survey — 309
Woodruff T. Sullivan, III

17 Radio source counts — 331
Peter Scheuer

18 The discovery of quasars — 347
Maarten Schmidt

19 History of dark matter in the universe (1922–1974) — 355
Virginia Trimble

Part VI Dramatis personae

20 Carl Wilhelm Wirtz – a pioneer in observational cosmology — 365
Waltraut C. Seitter and Hilmar W. Duerbeck

21 Cosmic rays and cosmological speculations in the 1920s: the debate between Jeans and Millikan 401
M. De Maria and A. Russo

22 Sinclair Smith (1899–1938) 411
Virginia Trimble

23 Revisiting Fritz Zwicky 415
Alessandro Braccesi

Index 424

Contributors

R. A. Alpher, Department of Physics, Union College, Schenectady, New York 12308, USA.
R. Balbinot, Dipartimento di Fisica, Università degli Studi di Bologna, Via Irnerio 46, 40126 Bologna, Italy.
J. B. Barbour, College Farm, South Newington, Banbury, Oxfordshire OX15 4JH, United Kingdom.
J. D. Barrow, Astronomy Centre, University of Sussex, Brighton BN1 9QH, United Kingdom.
S. Bergia, Dipartimento di Fisica, Università degli Studi di Bologna, Via Irnerio 46, 40126 Bologna, Italy.
B. Bertotti, Dipartimento di Fisica Nucleare e Teorica, Università degli Studi di Pavia, Via Bassi 6, 27100 Pavia, Italy.
Sir Hermann Bondi, The Master's Lodge, Churchill College, Cambridge CB3 0DS, United Kingdom.
A. Braccesi, Dipartimento di Astronomia, Università degli Studi di Bologna, Via Zamboni, 33, 40126 Bologna, Italy.
M. De Maria, Dipartimento di Fisica, Università di Roma 'La Sapienza', Piazzale Aldo Moro, 5, 00185 Roma, Italy.
Hilmar W. Duerbeck, Astronomisches Institut, Westfälische Wilhelms-Universität, Wilhelm-Klemm-Strasse 10, 44 Münster, Federal Republic of Germany.
G. Ellis, SISSA, Strada Costiera 11, 34014 Trieste-Miramare, Italy.
E. Harrison, Department of Physics and Astronomy, University of Massachusetts at Amherst, MA 01003, USA.
R. Herman, College of Engineering, University of Texas, Austin, Texas 78712, USA.
A. Krasiński, Polska Akademia Nauk, Centrum Astronomiczne im. M. Kopernika, ul. Bartycka 18, 00–716 Warszawa, Poland.
Sir William McCrea, Astronomy Centre, University of Sussex, Brighton BN1 9QH, United Kingdom.
A. Messina, Dipartimento di Astronomia, Università degli Studi di Bologna, Via Zamboni 33, 40126 Bologna, Italy.
J. D. North, Kamperfoelieweg 25, 9765H5 Paterswolde, The Netherlands: or

Filosofisch Instituut der Rijksuniversiteit, Westersingel 19, 9718 CA Groningen, The Netherlands.

D. E. Osterbrock, Lick Observatory, University of California, Santa Cruz, CA 95064, USA.

A. Russo, Dipartimento di Fisica, Università di Palermo, Via Archirafi 36, 90123 Palermo, Italy.

J. M. Sánchez-Ron, Departamento de Fisica Teorica, Universidad Autonoma de Madrid, Cantoblanco, 28049 Madrid, Spain.

P. G. Scheuer, Cavendish Laboratory, University of Cambridge, Cambridge CB3 0HE, United Kingdom.

M. Schmidt, California Institute of Technology, 1210 East California Bld., Pasadena, CA 91125, USA.

W. Seitter, Astronomisches Institut, Westfälische Wilhelms-Universität, Wilhelm-Klemm-Strasse 10, 44 Münster, Federal Republic of Germany.

W. T. Sullivan, III, Department of Astronomy FM-20, University of Washington, Seattle, WA 98195, USA.

V. Trimble, Department of Physics, University of California, Irvine, CA 92717, USA.

R. Wagoner, Department of Physics, Stanford University, Stanford, CA 94305–4060, USA.

R. H. Wilson, AT and T Bell Laboratories, Mail Stop HOH-L239, PO Box 400, Holmdel, NJ 07733, USA.

Sir Fred Hoyle, 102 Admirals Walk, West Cliff Road, Bournemouth, Dorset BH2 5HF, United Kingdom.

Preface

Introduction

This book is the outcome of a meeting which was held in Bologna, Italy, from 20 May to 23 May 1988, in connection with the Third ESO–CERN Symposium on Astronomy, Cosmology and Fundamental Physics. The Scientific Committee, formed by Silvio Bergia, Bruno Bertotti, Juergen Ehlers, Max Jammer, John D. North and Dennis W. Sciama, succeeded in obtaining the distinguished participation documented by the names of the contributors to this volume. The book, however, should not be simply taken as the proceedings of the meeting: indeed, Simon Mitton, editorial director of Cambridge University Press, while enthusiastically agreeing to the idea of a book on the subject, suggested from the start that the book should be based on the lectures given at the conference rather than being a mere reprinting of papers read at the conference. We have solicited the authors' help in this respect. At the same time, we felt that some of the discussions should be included to renew for the readers the lively atmosphere of the meeting. They are, we believe, an important part of the book.

Bologna and astronomical studies

The main reason for choosing Bologna as the site of the meeting was that the local university, the Alma Mater Studiorum, the oldest university in Europe, celebrated its ninth centenary in 1988. The meeting was held in celebration of this event.

Although twentieth-century cosmology is a collective enterprise to which Italian science has started contributing only fairly recently, Italian universities and cultural centres can claim an ancient tradition in astronomical studies which helped prepare the ground for the modern investigation of the universe. As far as Bologna is concerned, the liberal arts of quadrivium, which included astronomy, were taught there in capitular schools as early as the eleventh century, and regular lectures in astronomy are documented at the Universitas since about 1280 (Braccesi 1987).

Copernicus spent three and half years at Bologna between 1496 and 1499. There, according to Armitage (1972), while his professed subject of study was canon law, 'his wide range of interests covered many other branches of knowledge as well – particularly mathematics and astronomy. In fact, the most important influence upon him at Bologna was that of the Professor of Astronomy there, one Domenico Maria da Novara (1454–1504).'

In Bologna, Domenico da Novara (or Domenico Novara) and Copernicus made an observation of the lunar parallax, reported by the latter in his 'De revolutionibus' (1543). The use of an eccentric, in Ptolemaic astronomy, to account for the irregularities of the lunar motion, produces, he wrote much later, 'two evident errors': on the one hand, 'the moon ... should appear about four times larger at quadratures ... than when it is new or full'; at the same time, 'its parallax ... should also rise considerably at quadratures'.[1] It seems very likely that da Novara and Copernicus aimed through an astronomical observation to clarify these matters, and, although viewing this episode as a step towards the mature Copernican refutation of Ptolemaic astronomy would certainly be going too far, it seems fair to conclude that problems about the lunar motions, as traditionally accounted for, worried Copernicus even at this early stage.

Copernicus' association with Domenico da Novara is interesting also from another viewpoint. According to Armitage (1972),

> Domenico da Novara was one of the leading spirits in the great revival of free studies which was just then sweeping through Italy and spreading into northern Europe ... The ideas behind this movement largely went back to Plato and even to Pythagoras. On the scientific side, they led men *to try to picture the Universe by means of simple geometrical figures, or relations between numbers* [our emphasis – we note parenthetically that this tendency is recurrent in the history of cosmology]. And his friendship with one of the leaders of this movement must have encouraged Copernicus to go ahead with plans for reforming astronomy along these lines.

In an essay on the early acceptance of Kepler's laws, Russell (1964) stresses that of the seven Italians mentioned by him as having 'either accepted or at least carefully studied Kepler's ideas', four – Magini, Cavalieri, Riccioli and Cassini – were professors at Bologna. Indeed, in 1615 Giovanni Magini, professor of mathematics, published his Supplementum Ephemeridum, 'in which he used Kepler's laws in calculating ephemerides for Mars'. Bonaventura Cavalieri, one of the leading mathematicians of his time, best known for his pioneering work on infinitesimals, and a friend and disciple of Galileo, was also professor of mathematics at Bologna. Russell notes that Cavalieri made 'a careful study of Kepler's works' in his *Directorium Generale* of 1632. 'An

[1] From Copernicus' *Commentariolus* (Italian edn, 1984), p. 57.

important contribution to the spread of Kepler's theories' was also made by the Jesuit astronomer G. B. Riccioli in his *Almagestum Novum* (Bologna, two volumes, 1651).

The last figure mentioned by Russell – Gian Domenico Cassini, one of the leading Italian[2] astronomers – is best known for his discoveries of the 'Cassini division', a major dark separation within Saturn's rings, of Jupiter's Great Red Spot and the planet's rotation about its axis, made possible by the use of the objectives which had spread the renown of Campani (1635–1715) throughout Europe (Braccesi 1987). Cassini is the builder of the sundial which can still be admired in the Bolognese Cathedral of San Petronio. Immediately upon its completion in 1656, he used the new instrument to make combined observations of variations in the Sun's diameter and of the speed of its apparent motion along the ecliptic; his conclusion was that the correlations observed could not be accounted for in terms of an eccentric position of the Earth with respect to an orb carrying along the Sun in a uniform motion. He was thus able to write that 'a real variability of the solar motion had been for the first time revealed by the observations' (as quoted in Braccesi 1987), a deadly blow to the understanding of the heavens as advanced by Aristotelian physics. In addition it should be recalled that Cassini 'used ellipses in his contributions to the planetary tables of Count Malvasia (1662)' (Russell 1964). Russell's conclusion is that 'evidently there was a more or less continuous Keplerian tradition at this university during the whole of the period'.

To recall these old merits would be merely a rhetorical exercise were it not for the fact that an ideal thread links that tradition with the world of contemporary research in astrophysics and cosmology. (In this respect, it is perhaps interesting to recall that the bee-hive reflector of Horn d'Arturo, preserved in the Astronomy Museum of the University, provided a model for this type of instrument.)

The University of Bologna has had its astronomical observatory ('La Specola') since 1727; the observatory of Loiano, a few kilometres from Bologna on the Apennines, had its first 60 cm telescope in 1936, sided by a 152 cm telescope in 1976 named after Cassini.

It is partly due to this tradition and the people who grew up within it, that Bologna had its first radiotelescope ('The Northern Cross') in Medicina, near Bologna, in 1965 and the more recent 32 m diameter antenna, part of the Italian VLBI project, in 1983 (Fanti et al. 1985). These instruments have made contributions to the development of modern cosmology, in particular in the field of radio source countings.

The history of science with the help of those who made it

Coming back to the meeting, it should be stressed that it was a very

[2] Cassini was already in his forties when he left Italy for France. His name has been recently revived with the approval of the CASSINI space mission to Saturn and its satellite Titan.

special one: it aimed at producing a reconstruction of the history of modern cosmology based also on accounts given by some of its main contributors. One may ask whether the participation of scientists – particularly the renowned and senior ones – is necessary in order to reconstruct the history of their discipline.

This is generally understood as the provision of unpublished documents and oral testimonies, solicited, recorded and edited by professional historians. In recent years, historians of science, and of physics in particular, have indeed engaged in wide-ranging projects, aimed at collecting documentary material for reconstructing periods in and fields of research from the scientists involved[3]

On the other hand, renowned scientists have often carried out historical work of their own, including scientific biographies of eminent figures, autobiographies, and surveys of events and characters.

There are of course exceptions, but it seems fair to conclude that professional historians have often relegated to scientists the role of being mere producers of raw and uncorrelated information[4], while, on the other hand, scientists acting as historians, conditioned by the necessity of reaching a wider audience, have often refrained from dealing with the more technical respects of their discipline.

The main point we want to make is that, for the sake of the history of science, a different, and perhaps altogether better use can be made of scientists, especially of those directly involved in the landmarks of development in their field: they can, in a word, be used to produce first-hand accounts of the events, and the history of science would benefit from their technical knowledge of the subject matter.

There are at least two good reasons why this objective should be pursued.

Firstly, professional historians are often too slow in tackling subjects in contemporary science, and physics in particular, that are in the process of becoming history: the beginnings of this process are often felt only by the leading figures in a field of research. As an example, let us consider the progress of General Relativity after the 1920s. This subject has received little attention from the historians (see, however, Eisenstaedt 1986), while aspects of it have been addressed repeatedly by active physicists. It was Chandrasekhar (1979) who analysed the reasons for 'the benign neglect to which this theory has been subjected over the years', and it is in the preface

[3] See, for instance: Kuhn et al. (1967); the activities of the 'Office for History of Science and Technology' led by J. Heilbron at Berkeley; and the setting up of contemporary scientific archives at Oxford (J. J. Roche, Round Table on the Evolution of Symmetries, in Doncel et al. 1987).

[4] It seems recently to have become a common attitude among historians to consider the historical accounts given by scientists as less reliable than is desirable. It is for this reason that they are asked to provide data only in the form of reminiscences. These will be looked upon 'as a testimony to be evaluated with respect to all the available knowledge', and possibly 'suggest to the historian paths of research' or give him 'hints for otherwise painstaking discoveries' (Giuliani 1988).

to their treatise 'Gravitation' that Misner, Thorne and Wheeler (1973) analysed the reason for its revival in the sixties. The epistemological and sociological aspects of these events were also first analysed by physicists: Bertotti (1979) has pointed out the change of paradigm (in Kuhn's sense, Kuhn 1962) of gravitation physics and the subsequent establishment of general relativity as 'normal science'; Hawking and Israel[5] and Wald (1984), while making observations similar to those of Chandrasekhar and Misner, Thorne and Wheeler, have stressed, respectively, the importance of the first Texas Symposium on relativistic astrophysics in 1963, which took place after the discovery of quasars, as a 'significant landmark', and the revival of a strong interest in general relativity in the late 1950s 'particularly by the Princeton group led by John Wheeler and the London group led by Herman Bondi'.

In general, the slowness of professional historians in tackling aspects of the development of contemporary science can become a serious drawback in an era in which twenty years are often sufficient to produce the replacement of real history with the rigid and sometime arbitrary reconstruction presented in textbooks, a phenomenon which obliges each generation of historians of science to re-write the more recent history of their discipline.

The second reason is that contemporary science – in particular, contemporary physics – is highly technical, often too technical, not only for professional historians with a general original scientific training, but also for active scientists who were not specifically trained in the subject (for a discussion of this point, see Carazza 1985). As an example, let us consider one taken from our own subject, the history of cosmology: de Sitter's model, which, for a while, was considered the only alternative to Einstein's static universe (on the Einstein–de Sitter controversy, see Kerszberg 1989). De Sitter's universe appeared for quite a long time to be like a mathematical unicorn, deceiving even the most experienced readers with riddles and mirages (Merleau-Ponty 1965). This is due essentially to the fact that, since it is based on a spacetime of constant curvature, there is no single choice of fundamental world lines for the average motion of cosmic matter (Ellis 1990). As a consequence, it can be written as a static universe, or a Friedmann–Lemaître–Robertson–Walker (FLRW) universe with $k = +1$, 0, -1, the second case corresponding to the spacetime of the steady-state model. This basic ambiguity affected early discussions of observed redshifts in terms of the model (Ellis 1988, 1990). Now, it can hardly be imagined that a thorough analysis of the technical

[5] See Hawking and Israel (1979); Israel (1987). Israel reports a significant passage of an after-dinner speech delivered by T. Gold during the Texas Symposium: 'This is of course a historic meeting. It will be remembered as the meeting where these great new astronomical discoveries were first discussed . . . We have a case that allowed one to suggest that the relativists with their sophisticated work were not only magnificent cultural ornaments but might actually be useful to science!.'

aspects of the model necessary to decipher the meaning of those discussions could be carried out by anyone but a real expert in the field.

Urgency and technical difficulties are not the only reasons why accounts given by active scientists can be relevant to the history of science: the selection of subjects of interest is another. The attention of professional historians often concentrates on as yet unpublished documentary material, currently considered to be a fundamental source of historical information. To illuminate phases of historical development through the use of unpublished or forgotten documents is one of the important tasks of the historian. But we also need to find someone who is able to tell us about the terms of the scientific problem in hand, and the way the problem was faced[6] – which, by the way, is what mainly interests a relevant portion of the public, namely, the scientists themselves – terms and ways which are seldom to be found in unpublished documents, and are more likely to be clearly stated in published papers. To read this literature, order its content chronologically and logically, and clearly expound it to readers who have become unaware of it does not necessarily require a professional historian, while scientists active in the field are often qualified to do the job (see Pais 1982, for an outstanding example). Personal recollections are also important[7] and can be illuminating. That this is particularly true of cosmology, due to the peculiar aspects of this science, is illustrated by Bertotti in his contribution to this book in which he casts doubts on the idea that reflection about the universe could dispense with metatheoretical elements, such as cultural influences, ideological motives, or even the transposition of framing ideas by way of mere analogy (think of Gold's question, quoted by Hoyle (1990): 'What if the universe is like that?').

If this is how and why active scientists can help in writing the history of science, the risks of such writing are also evident: personal accounts can be biased due to prejudices, unconscious removals or simple forgetfulness.[8] Furthermore, to a much larger extent than professional historians, who

[6] An interest for this type of approach was manifested by Einstein in an interview Shankland had with him in February 1950 (Shankland 1963): 'A means of writing must be found [said Einstein] which conveys the thought processes that led to discoveries'. An excerpt of the interview has been significantly taken by Miller as an epigraph for his book on the emergence and early interpretation of the special theory of relativity (Miller 1981). It should be added that in the interview Einstein charged professional historians with philologism and scientists of acting as historians with a lack of historical sense.

[7] As has been recently put by G. Tagliaferri (conclusions stated in Giuliani 1988), 'recording views and details available only to the protagonists of past enterprises seems all the more necessary today that the habit of entertaining significant scientific correspondence between research workers is being supplanted by ephemeral telephonic communications. Historians of the future will miss the cherished tools of perusing collections of letters!'

[8] See, for instance, Kragh (1987), p. 150: 'The scientist is not always a witness to the truth when it comes to his own actions. General forgetfulness and a tendency to rationalize after the event in the light of later developments will naturally play a part in retrospective accounts of the events that took place many years ago. In connection with conflicts of priority, for example, he might consciously or unconsciously overestimate his own contribution, change dates or in some other way suppress a reality that he might have wished were otherwise.'

are protected against this flaw by their methodological and philosophical training, scientists acting as historians are likely to indulge a natural tendency to replace history with a rational reconstruction based on the present-day understanding of the subject.

As a first step, one may simply use current views and models as tools to read previous attempts and results. This is more likely to happen if the scientist-historian presents somebody else's achievements, and the effect tends to be greater the further into the past he is looking. We dare say that this is to some extent unavoidable if what is at stake is not only a description of 'what actually happened' but also a real understanding of it: as philosophers of various tendencies (see also Bondi 1990) have often suggested, the present can provide clues for the past.

As a perhaps subsequent step, and using Kuhnian categories, one may thus be led to read what was elaborated within one paradigm with lenses provided by another. (On this point, see Ellis 1990). This leads one already very close to the worst of sins in the eyes of professional historians: the Whiggish attitude. Whiggish history, according to Schweber (1984), 'is the writing of history with the final, culminating event or set of events in focus, with all prior events selected and polarized so as to lead to that climax'. (Schweber recalls that the term was coined by H. Butterfield in his *The Whig Interpretation of History*, London, 1931). The Whig pitfall, continues Schweber, 'is particularly attractive in the history of physics, because the traditional view of the history of science, and the prevalent one among scientists, is to consider the history of science as the progressive record of the successful unravelling of the secrets of nature, each advance bringing with it a closer approximation to the TRUTH'. However censurable, Whiggish history is met with favour by scientists, because it tells them how their present paradigms have arisen in the past. Professional historians could be persuaded that it can have a cultural value and it is, to some extent, unavoidable,[9] provided, of course, it is not read as the real thing. Such a reading is not warranted and the risks inherent in Whiggish accounts are evident, particularly so in the case of cosmology in an epoch which has witnessed the rise of the paradigm of an expanding universe which had its origins in a hot big bang.

We believe that the authors of this book were fully conscious of what was at stake and presented a first-rate reconstruction of the development of modern cosmology.

The scope of this book

The structure of the book is the following. In part I, Bruno Bertotti discusses the peculiarity of cosmology among the other sciences and the

[9] This has been particularly emphasised by Bachelard (see, for instance, Bachelard 1965, pp. 24–6, where the cultural value of a history 'qu'on éclaire par la finalité du present' is stressed). We thank R. Dionigi for useful discussions on the matter. See also Barrow and Tipler 1986, pp. 9–11.

consequences of this for its history, while J. D. North analyses the epistemological status and the early history of modern cosmology. In part II, Edward Harrison, J. B. Barbour and John D. Barrow present stimulating discussions of the riddles of and clues to cosmology provided by Olbers' paradox, Mach's principle and the coincidences of large numbers. In part III, G. F. R. Ellis and Andrzej Krasiński introduce the geometrical cosmology, while Ralph A. Alpher, Robert Herman and Robert V. Wagoner contribute on the more physical aspects of the big-bang cosmology. The flavour of the debate between supporters of different cosmological models and of personal recollections can be found in the contributions by Hermann Bondi, William McCrea and F. Hoyle in part IV, while José M. Sánchez-Ron discusses in particular the Hoyle and Narlikar theory. As far as Professor Hoyle's contribution is concerned, his ideas about a cosmology without big bang are, of course, highly interesting and deserve careful consideration. However, it should be pointed out to readers who are not specialists that they have given rise to a controversy and have not obtained the general consensus of cosmologists. Part V is mainly devoted to the history of cosmological observations; the contributions by Donald E. Osterbrock, Robert M. Wilson, Woodruff T. Sullivan, III, Peter Scheuer and Maarten Schmidt on observatories, instruments and basic discoveries are sided by Virginia Trimble's chapter on the dark matter problem. Finally, part VI presents chapters by Waltraut C. Seitler and Hilmar W. Duerbeck, M. De Maria and A. Russo, Virginia Trimble, and Alessandro Braccesi on some of the actors of our story only marginally mentioned in the previous contributions.

Acknowledgements

We wish to thank G. Dragoni for comments on the historiography of science, J. D. North, who first suggested that the Bologna element should enter in the story, J. B. Barbour, who quoted in his lecture from the essay by J. L. Russell to the effect that there had been a Keplerian tradition in Bologna spanning the period between Kepler and Newton, and made Russell's paper available to us, and A. Braccesi, for several comments on the development of astronomy in Bologna.

R. Balbinot, S. Bergia, B. Bertotti, A. Messina

References

Armitage, A. (1972). *The World of Copernicus*, Open University Set Book. East Ardsly: EP Publishing Ltd.

Bachelard, G. (1965). *L'activité rationaliste de la physique contemporaine*. Paris: Presses Universitaires de France.

Barrow, J. D. and Tipler, F. J. (1986). *The Anthropic Cosmological Principle*. Oxford: Oxford University Press.
Bertotti, B. (1979). I fondamenti sperimentali della gravitazione, in *Astrofisica e Cosmologia, Gravitazione Quanti e Relatività negli sviluppi del pensiero scientifico di Albert Einstein*. Firenze: Giunti Barbera.
(1990). This volume.
Bondi, H. (1990). This volume.
Braccesi, A. (1987). L'astronomia a Bologna dall'XI al XVIII secolo. Bologna Astrophysics Preprints 6–1987–1 DDA. Quadrivium, forthcoming.
Carazza, B. (1985). Che fare con la storia della fisica? *Il Nuovo Saggiatore*, **4**, 8.
Chandrasekhar, S. (1979). Einstein and General Relativity: Historical Perspectives. *American Journal of Physics*, **47**, 212.
Copernico, N. (1984). *Commentariolus*. Roma: Theoria. The manuscript of *Commentariolus* was circulated by the author a few years before the publication – in 1543 – of 'De revolutionibus'. A copy was found by M. Curtze in 1877 and published by him the next year.
Doncel, M. G., Hermann, A., Michel, L. and Pais, A., eds. (1987). *Symmetries in Physics 1600–1980*. Barcelona: Servei de Publicacions UAB.
Eisenstaedt, J. (1986). La relativité générale a l'étiage: 1925–1955. *Archive for History of Exact Sciences*, **35**, 115.
Ellis, G. F. R. (1988). The History of Cosmology: 1917–1955, pre-print.
(1990). This volume.
Fanti, R., Grueff, G. and Setti, G., eds. (1985). *Radio-astronomy and the Physics of the Universe* (dedicated to M. Ceccarelli). Bologna: CLUEB.
Giuliani, G. (1988). Foreword to *The Origins of Solid-State Physics in Italy: 1945–1960*, ed. G. Giuliani. Bologna: Ed. Compositori.
Hawking, S. W. and Israel, W. (1979). An introductory survey in *General Relativity: An Einstein Centenary Survey*, ed. S. W. Hawking and W. Israel. Cambridge: Cambridge University Press.
Hoyle, F. (1990). This volume.
Israel, W. (1987). Dark Stars: the Evolution of an Idea. In *300 Years of Gravitation*, ed. S. W. Hawking and W. Israel. Cambridge: Cambridge University Press.
Kerszberg, P. (1989). *The Invented Universe*. Oxford: Clarendon Press.
Kragh, H. (1987). *An Introduction to the Historiography of Science*. Cambridge: Cambridge University Press.
Kuhn, T. S. (1962). *The Structure of Scientific Revolutions*. Chicago: University of Chicago Press.
Kuhn, T. S., Heilbron, J. L., Forman, P. L. and Allen, L. (1967). *Sources for History of Quantum Physics: An Inventory and Report*. Philadelphia: The American Philosophical Society.
Merleau-Ponty, J. (1965). *Cosmologie du XX siècle*. Paris: Gallimard.
Miller, A. I. (1981). *Albert Einstein's Special Theory of Relativity – Emergence (1905) and Early Interpretation (1905–1911)*. New York: Addison-Wesley.
Misner, C. W., Thorne, K. S. and Wheeler, J. A. (1973). *Gravitation*. San Francisco: Freeman.
Pais, A. (1982). *Subtle is the Lord: The Science and the Life of Albert Einstein*. Oxford: Oxford University Press.

Russell, J. L. (1964). Kepler's Law of Planetary Motions: 1609–1666. *The British Journal for the History of Science*, **5**, 2.
Schweber, S. S. (1984). In *Some Chapters for a History of a Quantum Field Theory: 1938–1952*, ed. B. S. de Witt and R. Stora. Amsterdam: North Holland.
Shankland, R. S. (1963). Conversations with Albert Einstein. *American Journal of Physics*, **31**, 47.
Wald, R. M. (1984). *General Relativity*. Chicago: University of Chicago Press.

I

The general framework

1

Cosmology, a peculiar science

Bruno Bertotti

1.1 An apologue

Once upon a time a very advanced civilization, during its interstellar explorations, discovered a single biological individual, made of carbon compounds, who was sadly living all by himself on an otherwise mineral planet. The scientific interest of the Explorers ran very high, not only for the study of that particular individual, whom they called Adam, but also for the general problem of how to set up a biological science based upon a single individual, without the benefit of the multiplication in time and space offered by a species. Adam was for the Explorers – whose bodies were made of stuff entirely different from our own – an utterly new being. Contrary to what happens in biology and, indeed, in any experimental science, in 'Adamology' the fundamental procedure of construction and verification of an assumption through the repetition of an experiment on different individuals or systems was precluded. The distinction, seemingly essential in science, between the Law and the accidental individual variations was in jeopardy; the august and cherished notion of Natural Law was in doubt. Moreover, any progress was crucially dependent on the ken of the observations, both in time and depth of detail, and the problem of selection effects was both essential and dangerous.

At first the Explorers carefully collected and classified Adam's responses to different stimuli and arrived at certain 'laws' with a reasonable predicting power. Adamology soon emerged as a 'normal' science and slowly grew by methodical application of results and the construction of more refined theories. The crucial assumption that Adam was in a static state, with no development, was tacitly taken for granted. After some time, however, the community was thrown in a great confusion by a well-known adamologist called Hibble, who pointed out that some discrepancies in Adam's responses and behaviour could easily be explained if one recognised that he *was getting older*. Hibble was also able to determine the time scale 1/H of his ageing. This meant not only that any effect occurring in a time scale longer than, or of the order of, 1/H was utterly unamenable to theoretical understanding; even for those phenomena occurring over a time scale $\tau < 1/H$ a precise analysis was precluded. Indeed, a fraction $H\tau$ of any change could be ascribed to the unknown evolution and escape a rational explanation. Every conclusion of the time-honoured 'Static Adamology' was affected by a small, but essential uncertainty. New paradigms and, indeed, a new epistemology, were needed for further progress.

The body of Adam was made up of cells, whose biochemistry was generally well understood; but it was impossible, even in principle, to predict the behaviour of Adam on the basis of biochemistry alone, because only a small and superficial part of his body, as long as he stayed alive, was accessible to direct analysis and observation. On the other hand, he appeared to be a complex, but well-integrated organism, in which each part had its proper function and purpose. Some people even believed that he had a 'mind', a concept difficult to understand on the basis of a mere collection of cells. He was indeed a small 'cosmos', in the Greek meaning of a well-ordered system. Other traditional adamologists of a reductionist mind, in particular a Frenchman called Cumte, refused to enter into wild and idle speculations and declared that any theory about Adam as a whole was not experimentally verifiable and hence could not be regarded as a scientific endeavour at all; only the cells of his body accessible to observation could be the subject of investigation, under the framework of well-known biological laws. Cumte even went as far as to say, 'nous avons seulement besoin de connaître ce qui peut agir sur nous, d'une manière plus ou moins directe', which certainly does not include a unique and lonely biological individual. As a consequence of this opposition, several research grants were refused and Adamology came to a standstill.

This state of affairs, however, did not prevent more independent and imaginative minds from addressing the problem of Adam in a much more aggressive way, following a paradigm of *a priori* rational simplicity pushed to the extreme. *Since he is unique*, they argued, it should be possible to single out his actual detailed structure and his evolution among all the infinite possibilities allowed by biochemical processes; in other words, they would have liked to believe that the individual they stumbled upon by chance was unique in theory as well as in fact, in the sense that any other realisation was precluded by some law of nature based on simplicity, but yet to be discovered.

Adamology has never since been a fully 'normal' science and has gone through several drastic and even painful changes. Its development has been characterised much more by a reliance upon *a priori* theory construction than by the collection and elaboration of experimental evidence. Lacking clear-cut experimental verification, its progress has been difficult to define and it has not been easy to falsify or to refute 'wild' speculations. The history of Adamology has been characterised by long-standing controversies about fundamental issues and by its susceptibility to the influence of ideas and themes outside the domain of science.

1.2 Cosmological controversies

The apologue of 'Adamology' (Bertotti 1982; Partridge 1969) vividly brings home the peculiar character of cosmology among all the other sciences and alerts the historian to take in his approach a wider and less conventional point of view (see also Munitz 1963). In the context of Kuhn's view of the development of a science as a succession of 'normal' phases interrupted by controversial periods of crisis and anomalies, cosmology is never in a fully normal state; its paradigms are always more or less unstable and strongly depend on external cultural factors. This is well illustrated by two great cosmological controversies which have shaken and, indeed, continue to shake, its foundations: the debate about the object of cosmology and Mach's Principle.

1.2.1 *The object of cosmology*

The immense broadening of the horizon of astronomy introduced in the century by the idea of the 'island universes' (according to which the nebulae really are distant galaxies; Kant 1755) provided for an entirely new idea of the cosmos. This idea was very familiar to W. Herschel, the great British astronomer who tried in his outstanding papers of 1784 and 1785 (see Hoskin 1963) to guess the distance of the nebulae and the scale of the universe. His work was basically similar to the modern research on source counts; if a system of sources has a finite extent, a cut-off will appear in the observed statistical distribution according to the apparent magnitude. At the end of the nineteenth century several thousand nebulae were known, providing enough material for statistics; in 1885 the first nova was identified in one of them, making it possible to draw a rough estimate of the distance. But the progress was frustratingly slow. The accuracy, the sensitivity and the reliability of the observing instruments were poor; the nature of the sources was unknown and their assignment to the same class of intrinsic magnitudes was uncertain. The conceptual framework and the mathematical formal tools for a proper statistical analysis were not available. (For a good history of premodern cosmology, see Jackisch 1979.)

In the prevalent positivism of the nineteenth century, the progress of cosmology was also hindered by philosophical objections (Merleau-Ponty 1983). Astronomers and 'natural philosophers' were repelled by the impossibility of drawing firm conclusions about the nature of single objects. The paradigm of astronomy was celestial mechanics, by which the motion of planets and their satellites was beautifully framed and well understood within the highly mathematical scheme of Newtonian and analytical dynamics. As Comte said explicitly (1835): 'Il faut concevoir l'astronomie positive comme consistent essentiellement dans l'étude géométrique et mécanique du petit nombre de corps célestes qui composent le monde dont nous faisons partie'. Astronomy was confined to the solar system; not only the emerging cosmology of nebulae, but also the study of stars was excluded by Comte from 'positive astronomy'. One wonders what the effect of this influential view was on the slow development of nebulae observation in Europe in the nineteenth century. The impasse was broken later by independent and imaginative individuals, like Einstein and de Sitter, who were not influenced by positivistic scepticism and developed in mathematical terms the concept of the universe. It is also worthy of note that the construction of better observing instruments and their use for extragalactic astronomy took place, to a large extent, in the United States.

On 26 April 1920 a 'great debate' took place between Shapley and Curtis at the National Academy of Sciences in Washington about the 'island universes'. The issues were: what is the order of magnitude of the distance to the stars, and whether the nebulae are part of the Galaxy and made up of stars themselves. Shapley maintained that the system of the globular clusters was about 300,000 light years across, about one order of magnitude larger than it was usually assumed and Curtis believed. Their estimate of the absolute luminosities of stars could therefore differ by a factor 100; if one accepted Shapley's view the nebulae would have to be much further away in order to be unresolved in single stars. Accordingly, Shapley believed that the nebulae are extragalactic, but gaseous masses at not too great a distance. By contrast, the unifying Kantian view of the universe as a set of similar galaxies underlined Curtis' position, who pointed out several similarities between the nebulae and our own Galaxy. This debate had the great merit of bringing out clearly the modern idea of the Cosmos; it is quite remarkable that this occurred as late as 1920, 135 years after Herschel's papers (Hoskin 1982).

The Washington debate ended up with Curtis' view prevailing; but, of course, the debate did not end there. A recent sequel to the receding horizon of cosmology was a similar debate on the nature of quasars. The enormous amounts of energy required if the redshift of quasars is to follow Hubble's law have encouraged researchers (such as Terrell 1967) to explore the possibility of quasars being 'local' objects, not far from our own Galaxy. Were this the case, the focus of modern cosmology would be drastically different.

1.2.2 *Mach's Principle*

As in Adamology, the uniqueness of the universe raises at once the problem of the meaning of 'Natural Law' for the cosmos and the issue of whether it is at all conceivable to ask any question about the differential equation governing its dynamics and evolution. Indeed, as Mach has pointed out, since local dynamical laws seem to depend on the distribution and the motion of matter in the universe, it appears difficult to reconcile the supposed universality of physical laws – as embodied, for example, in the theory of Special Relativity – with an evolving, and possibly inhomogeneous, universe. In this view it seems inappropriate to use for the dynamics of the universe those local laws which ultimately depend on its structure; a different, axiomatic approach to cosmology is called for. This consideration was at the basis of the 1948 paper by Bondi and Gold, whose steady-state theory was proposed on the basis of simplicity alone. An inherent instability is thereby built in to any dynamical theory of cosmology. The controversy about Mach's Principle is a recurrent, and probably unavoidable one. Since Barbour deals at length with this topic in the present volume, I shall not pursue it any further here.

1.3 The cosmology of Edgar Allan Poe

Imagination plays a paramount role in the development of cosmology. Attempts to exceed the familiar domain of experience and to invent new concepts and new modes of thinking are pervasive. Conventional logical tools, for example, are inadequate for dealing with problems of infinity, both in space and time – problems which, of course, are crucial in cosmology. Before the theory of general relativity showed that it is possible to have an unbounded universe of finite volume, the problem of spatial infinity was a great stumbling block. The pioneering work by Charlier (1912) showed, however, that it is possible to construct a hierarchical and infinite universe, with structures at progressively larger scales. For some antinomies arising in the steady-state cosmology (infinite both in time and space), see Schlegel (1965) and Rucker (1982: 241).

The striking paucity of outstanding cosmological work in the nineteenth century has a solitary, and even more striking, exception. The American novelist and poet Edgar Allan Poe wrote in 1848 a long essay 'Eureka', which is a cosmology entirely built from first principles, on the basis of an immense power of imagination. Poe had no professional knowledge of astronomy, but understood Newtonian gravity, the elliptical Keplerian orbits, the geometric propagation of light and Laplace's nebular hypothesis for the origin of the solar system. His method of approach required very little technical knowledge: 'We require something like a mental gyration on the heel. We need so rapid a revolution of all things that, while the minutiae vanish altogether, even the more conspicuous objects become blended into one' (p. 212). Poe was possessed by a visionary tension, an undaunted faith in his own reason ('A perfect consistency can be nothing but an absolute truth', p. 300), so radically different from the positivism of his times.

Poe's cosmology is essentially that of the big bang: the universe arose from a Single Particle made of pure radiation (p. 254) which exploded and fragmented in innumerable pieces. The ongoing explosion is generally uniform, but the 'atoms' differ by shape, size and spacing; it is these inequalities, and not differences in nature, which give rise to the variety we see in the universe. ('It would be supererogatory to predicate of the atoms all the other differences: we then establish the Universe on a purely *geometrical basis*', p. 229. This, of course, reminds one of the programme of reducing all physics to geometry which is at the basis of general relativity.) Gravity produces out of the expanding atoms agglomerations of all sizes – the worlds – and eventually will overcome the expansion and bring the universe back to a final collapse. He also dimly understood that, if ordered motion is the basic structure of the cosmos, distances are measured by times and 'Space and Duration are one' (p. 291). Poe's universe is a finite, roughly spherical cluster of clusters, irregularly dispersed; its finiteness is demanded for dynamical and optical reasons ('Were the succession of stars endless, then the background of the sky would present us an uniform luminosity', p. 278), although no specific mention is made of Olbers' paradox.

It is ironical that Comte, a high priest of nineteenth-century European science, was a positive and unimaginative obstacle to the development of our science, while a literary man of the New World was able to prophesise and prefigure modern cosmology. Later, when the cosmological point of view became fully accepted and extragalactic astronomy was the object of intense research, the static scenario was taken for granted, causing a great resistance to the idea of an ordered universal expansion, as indicated by Hubble's work. Even Einstein felt it necessary to corrupt the simplicity of his field equations, introducing the obstreperous cosmological term, in order to save the static universe. One wonders if Poe's work was known to him; perhaps it would have saved many useless detours and false steps.

1.4 Conclusion

The peculiarity of cosmology makes its history more interesting and also more difficult to understand. Subtle but important influences from other fields of human endeavour are at work and a satisfactory picture of its development can hardly include only scientific elements and discoveries. Of course, when we come to the standard modern cosmology, as taught in the textbooks and classes and the basis of large research projects, we are presented with an apparently normal science, with well-structured and practically undisputed paradigms. I believe that this is not the whole story: cosmology is peculiar. For this reason, the reconstruction of the history of modern cosmology will, as explained in the preface, greatly benefit from the active participation of those who have actually promoted its development and gone through its phases of epistemological instability. Some of them have contributed to this book; their personal witness will enlighten those important aspects of cosmology which are outside its 'normal' paradigm.

References

Bertotti, B. (1982). In *Il Problema del Cosmo*, ed. G. Toraldo di Francia, p. 81. Rome: Istituto dell'Enciclopedia Treccani.

Bondi, H. and Gold, T. (1948). *Monthly Notices of the Royal Astronomical Society*, **108**, 252.

Charlier, C. V. L. (1921). How an Infinite World may be Built Up. *Arkiv för matematik, astronomi och fysik*, **16, 1**.

Comte, A. (1835). *Cours de philosophie positive. Leçons sur l'astronomie.* Paris: Schlercher Frères.

Hoskin, M. (1963). *William Herschel and the Construction of the Heavens.* London: Olbourne.

(1982). *Stellar Astronomy. Historical Studies.* Chalfont St. Giles: Science History Publications. See, in particular, the articles 'Island Universes: An Overview' (p. 154) and 'The Great Debate: What Really Happened' (p. 175).

Jackisch, G. (1979). *Johann Heinrich Lamberts 'Cosmologische Brief' mit Beiträgen zur Frühgeschichte der Kosmologie.* Berlin: Akademie-Verlag.

Kant, I. (1755). *Allgemeine Naturgeschichte und Theorie des Himmels.* Königsberg.

Merleau-Ponty, J. (1982). In *Il Problema del Cosmo*, ed. G. Toraldo di Francia, p. 39. Rome: Istituto dell'Enciclopedia Treccani.

Munitz, M. K. (1963). The Logic of Cosmology. *British Journal for the Philosophy of Science*, **13**, 34.

Partridge, R. B. (1969). *American Scientist*, **57**, 1.

Poe, E. A. (1848). *Eureka.* I quote from the 1987 Penguin edition.

Rucker, R. (1982). *Infinity and the Mind.* Brighton: Harvester.

Schlegel, R. (1965). The Problem of Infinite Matter in Steady-State Cosmology. *Philosophy of Science*, **32**, 21.

2

The early years

J. D. North

The first decades of the history of modern scientific cosmology are as controversial in their way as the first minutes of the history of the universe, and there is a potentially infinite controversy as to when exactly the subject was born. Less than thirty years ago, no less a pundit than the late Herbert Dingle was often to be heard insisting that cosmology still did not exist, and this was not a matter of indifference to someone wishing to write its history. There is an element of convention in all birth certificates, but Dingle would certainly have dismissed as a forgery one bearing the date 1915, especially if it had Einstein's name on it. If you stipulate that legitimate cosmology must be born of observation out of theory, then a date of 1915 means birth outside wedlock – not that this is likely to scandalize a modern audience, or indeed affect the vitality of the offspring. As I shall show, relations between the parents were very casual at first, and years passed before they really settled down together respectably, but the question of paternity is not in doubt: you only have to consider the relativistic character of the mature offspring to see that.

I shall take as an arbitrary date of conception 25 November 1915, when Albert Einstein submitted for publication the paper containing his general theory of relativity in its generally covariant and more or less final form (Einstein 1915).[1] Some of the best astronomers had by this time come round to accepting, at least as the most reliable working hypothesis, that the Sun is a constituent of one of a multitude of 'island universes' (Eddington 1914). The signs were auspicious, and the first relativistic model for the universe was announced by Einstein on 8 February 1917 – the so-called 'cylindrical' model (cylindrical if time is taken into account).

The western world was of course then divided by war, and was again so within scarcely more than two decades. There was no absolute hiatus, but there is a strong sense in which the period I am considering is a historical island, with a character of its own. This was the heyday of what one might

[1] This was the last of a series of five papers. Two earlier studies (1913) were written jointly with Marcel Grossmann and published elsewhere.

call 'geometrical cosmology', when positional, kinematic, and gravitational considerations ruled supreme. There are to be found in this period some of the seeds of a cosmology better integrated with the rest of physics, but they are often to be properly appreciated only with the advantage of hindsight.

Whether or not we take modern cosmology to have been born in 1915, we cannot ignore its ancestry. It is clear that observational evidence was a *sine qua non*, but so, in a very different way, was the discovery that Newton's theory of gravitation was difficult to reconcile with certain astronomical evidence (in particular the anomalous perihelion advance of Mercury), and that, when applied to cosmological problems with an infinite Euclidean universe, it seemed to lead to inconsistencies.[2] These theoretical difficulties might seem remote from cosmological schemes to come, but they were at least as important as the realization that the 'white nebulae' – as the non-gaseous nebulae were called – were comparable in status to our own Galaxy, and that they exhibited large radial velocities. After all, with new observations or old, there is always the problem of making sense of them, and this Carl Neumann and Hugo von Seeliger – to name only two – found it singularly difficult to do, granted an infinite distribution of matter of finite mean density behaving according to Newtonian laws. At the end of the nineteenth century, we find them modifying the Newtonian potential function in such a way as to introduce what was effectively a cosmical repulsion comparable with that later introduced by Einstein.[3]

At first sight, it seems that this they did simply to ensure a well-tempered theory, a consistent theory: and yet at the root of their concern were presuppositions about the actual distribution of matter that were not in any way guaranteed by the astronomical observations then available. They found paradoxes in a finite mean density within an infinite universe, and tried – unsuccessfully – to eliminate them by changing the laws of nature. Others, such as C. V. I. Charlier, had more confidence in the laws, and accordingly modified the assumptions made as to the distribution of matter, replacing a hypothesis of homogeneity with the assumption that the universe is hierarchical. Again, the empirical evidence was virtually non-existent.

Successful or not, in all these cases the approach was truly cosmological, in the sense that it concerned the totality of matter in the cosmos. Scientifically, however, all had something of the instability of a three-legged stool balanced on one leg, or at most two. The tripod I have in mind rests ideally on the body of natural law independently established (to a

[2] On the anomalous perihelion advance see N. T. Roseveare (1982).

[3] For this and other objections to traditional assumptions concerning the overall distribution of matter in the universe, and its behaviour, see North (1965), chapter 2, and for attempts to modify Newton's law of gravitation, see chapter 3.

reasonable degree) on a less than cosmic scale; on presuppositions as to the contingent distribution of matter on the large scale, and evidence for those presuppositions; and finally, on relevant observations of matter on all scales. I do not, of course, suggest that any of these supports can be taken independently of the others, but looking at the problem in this rough and ready way, you might be inclined to ask yourself how one should ideally respond, when faced with a choice between three alternatives: tampering with the assumption of large-scale homogeneity in the distribution of matter, tampering with a law of gravitation that could always be modified, up to a point at least, with impunity,[4] or introducing other physical laws into your explanations.

What is obvious for one generation need not be so for the next. The very existence of the Milky Way had long been thought to challenge homogeneity. Hierarchical systems of sorts had first been seriously proposed in the eighteenth century, by such writers as Thomas Wright, Immanuel Kant, and Johannes Lambert. In the mid-1920s, after Edwin Hubble's claim to have detected a 'Local Group' of galaxies to which our Galaxy belongs, talk of a hierarchical universe was revived, as on more recent occasions. One standard response by the homogeneity club has been to enlarge the unit of volume used in averaging the distribution of matter. This draws our attention to the nature of all the best presuppositions: they are often made invincible by fiat. If you are excessively fond of Euclidean geometry, and if the empirical evidence seems to be placing it under threat, then – as Henri Poincaré realized – you may retain the old geometry if you are prepared to modify the laws of optics. To take another example of a widespread axiom from the first three decades of our century: almost everyone then made the assumption that the universe is on average *static*. Pertinent observations were in relatively short supply, but no one had given much thought to the alternatives, since the generally unchanging pattern of stars and galaxies seemed to be guaranteed by centuries of observations – it seemed only marginally less obvious than that classic of the obvious, the darkness of the night sky. In the absence of large proper motions, who would have expected large radial velocities? So strong was this conviction that we inhabit a static universe that long after the discovery of the galactic redshifts there was a veritable industry in finding alternatives to interpreting them as Doppler shifts indicative of a real velocity. As we shall see, even Hubble had his doubts.

I have collected together here a somewhat arbitrary set of cosmological hypotheses in order to draw attention to the uncertainties and generally

[4] As everyone knew, there are limits to the liberties one may take with Newton's laws if one is not to jeopardize an extremely well-elaborated theory of planetary perturbations. For the sake of brevity I have greatly simplified the choices, and especially Seeliger's complex standpoint. He investigated many hypotheses, for example, that of gravitational absorption and that of interstellar cosmic dust. By the latter he tackled the problem of the paradox of the dark night sky.

under-determined state of cosmological thinking at the beginning of our century. It was not that astronomers and others were lacking in cosmological ambitions, but rather that there were too many unknowns in the case for them to be able to make a start. I should like to be able to say that Einstein's general theory narrowed down the alternatives, but in a sense it did the opposite. Only hardened rationalists had ever expected the distribution of *matter* in the universe to be anything but contingent, but it took a long time for the lesson of the general theory of relativity to sink in, namely that this contingency should be extended to *space and time*, and that the infinity of possible distributions of matter corresponded to an infinity of possible developmental models for the universe in a deeper, spatio-temporal, sense. This made life more difficult for the astronomers, who in due course had to learn a new language and interpret much more carefully their observational findings. This they did more readily in planetary than in galactic astronomy, but the former was actually highly relevant to the latter, for it concerned a crisis of confidence in the theory of gravitation itself.

Explanations of the anomalous advance of Mercury's perihelion were, roughly speaking, of three sorts. Some postulated the existence of invisible or barely visible matter, such as asteroids round the Sun (Le Verrier 1859) or zodiacal light. Others tried modifying Newton's law of gravitation. Asaph Hall was probably the first to do so in this context, in 1894. Ironically, in view of his cosmological objections to Newton's law, it was von Seeliger who first proposed the zodiacal light hypothesis, in 1906. A third group attempted to bring into the reckoning physical forces other than gravitation – electrical forces, for example (Roseveare 1982: passim). These various hypotheses were subject to intense discussion, especially between the years 1906 and 1920, and many new ideas were tried out. Gravitational absorption, for instance, was not only discussed by theoretical astronomers, but tested for, in elaborate experiments; indeed, it provided a cosmological apprenticeship for Willem de Sitter, who in 1909 and 1913 made a critical study of the principle and of its application by K. F. Bottlinger, a pupil of von Seeliger's. What characterized most of this work, however, was its *ad hoc* nature: von Seeliger even proposed a light-absorbing ether to explain the darkness of the night sky, but as for the Mercury perihelion, a rather arbitrary patching up of belief was the order of the day, whether it was of belief as to fundamental law or belief as to the distribution of matter. There is nothing particularly surprising or 'unscientific' about this, but it was destined to take on a rather sad appearance after the superb architectonic of Einstein's general theory, which more or less put an end to it – for a time, at least.

The impression is often given that during these early years of the century there was little professional contact between the applied mathematicians and the astronomers, but this was far from being the case. Einstein was much helped on astronomical questions by Erwin Freudlich. De Sitter and Arthur Eddington in particular deserve mention for their role in integrating the general theory of relativity into astronomy. Both were eminently qualified for the task. De Sitter had worked under David Gill at the Royal Observatory in Cape Town, and then as assistant to Jacobus Kapteyn in Groningen – a university at which he had earlier studied. He moved to Leiden in 1908. Eddington had studied mathematics at Manchester and Cambridge with great distinction before moving to the Royal Observatory at Greenwich, where he remained from 1906 to 1913. He had led an eclipse expedition to Brazil in 1912, and it is not surprising that he was the man who organized the 1919 expeditions that offered the first empirical support for Einstein's predicted deflection of light grazing the Sun's surface.

Both de Sitter and Eddington were experienced in the statistical analysis of proper-motions and star counts in the Kapteyn tradition, and in galactic modelling on that basis, and both were *au courant* with the best of current opinion on the constitution of the visible universe.[5] It was, however, within the context of Einstein's general theory that each made his first important contribution to cosmology proper.

From about 1911, de Sitter occupied himself with the potential repercussions of relativity theory on practical astronomy (Kerszberg 1987b). He interested himself in a variety of fundamental problems, such as the interrelationship of Mach's Principle and general covariance, the Ritz theory of light emission, and the astronomical relevance of the relativity of time. He was regularly in correspondence with Eddington in these years, and met and discussed problems of common interest with Einstein, P. Ehrenfest, and H. A. Lorentz in Leiden. Einstein valued this contact, because it allowed him – through two papers by de Sitter in 1916 – to make his ideas known in Britain, and when anti-German feeling there was running high, Eddington was able to excuse his acceptance of the two communications with the comment that 'he was interested to hear [from de Sitter] that so fine a thinker as Einstein is anti-Prussian' (Kerszberg 1987b: 55).[6] It is interesting to contrast the situation with that during the Second World War, when communication between astronomers on opposing sides was virtually impossible: the first (1942) edition of Otto Heckmann's *Theorien der Kosmologie*, for example, is to be found in virtually no British library.

Eddington appreciated the revolutionary nature of Einstein's new work as soon as he learned of it from de Sitter. He threw himself wholeheartedly into a study of the absolute differential calculus of C. G. Ricci and T. Levi-Civita, and wrote a masterly *Report on the Relativity Theory of Gravitation* in 1918, the proof-sheets of which were read, incidentally, by de Sitter. We

[5] For more details of this, see p. 000 below.

[6] On Einstein's time in Berlin as a 'prize hen', as he said, and his wartime relations with Ehrenfest and others in Holland, see Klein (1970), especially chapter 12.

may see this *Report* as a test-run for Eddington's *Mathematical Theory of Relativity* of 1923, a work described by Einstein in 1954 as the finest presentation of the theory in any language. Towards the end of it, in the context of de Sitter's model – of which more, shortly – he was able to refer to 'the very large observed velocities of spiral nebulae, which are believed to be sidereal systems', and to add that it 'is not possible to say as yet whether the spiral nebulae show a systematic recession, but so far as determined up to the present receding nebulae seem to preponderate' (Eddington 1918: 89).[7] This, as I said, was in 1918. In 1914 he had been limited to confessing that direct evidence on the nature of the spirals was entirely lacking, that is, as to whether they were within or without our own stellar system, but he was of the opinion that the 'island universe' theory was a good working hypothesis.[8] It is significant that he was then still speaking of the structure of the *sidereal* universe: the significant unit member of the universe was still the *star*, and not until he was within twenty pages of the end of *Stellar Movements and the Structure of the Universe* did he touch briefly on the nature of the spirals.

Einstein was by no means the first to use non-Euclidean geometries in physics. To name but three earlier instances: Nicolai Lobachevskii had proposed an astronomical test for the curvature of space – unrealistic since so few parallax measurements were available; von Seeliger's pupil Karl Schwarzschild, around about the end of the century, was able to use similar arguments with later parallax measurements to set lower limits to space curvature, for both elliptic and hyperbolic space; and A. Calinon had as early as 1889 suggested that the discrepancy between our space and Euclidean space might be a function of time. Einstein, however, was the first to make use of the idea that *gravitation* is explicitly related to the geometrical structure of Riemannian space-time, the metric properties of space-time now being seen as affected by the presence of matter and energy, with free test particles following geodesic paths.[9] As soon as Einstein began to think about the universe as a whole, it became clear that he could not think of gravitational masses as merely providing small local deformations of an otherwise Minkowskian space-time. Since he wanted gravity and inertia to be linked, as in Newtonian theory, we find him from around 1912 trying out appropriate hypotheses to effect the link, hypotheses such as that of the induction of mass by other masses, inspired by

[7] Those with an interest in Eddington's future concern with fundamental constants should note his derivation of a fundamental unit of length (4×10^{-33} cm) on the last page of the *Report*. In the 1920 edition (pp. ix–xi), he was able to refer to the 1919 eclipse material and to the notable extensions of Einstein's theory by Hermann Weyl, who included electromagnetic and gravitational forces in a single scheme.

[8] See n. 2 above.

[9] See North (1965) for material on many early sources. From the time of a paper written with M. Grossmann in 1913, Einstein took as his fundamental invariant the line element ds (first introduced by Minkowski for the flat space-time of the special theory) with coefficients $g_{\mu\nu}$ regarded as gravitational potentials.

what he took to be a principle due to Mach. He believed very firmly that there should be no solution of the field equations applied to the whole universe and describing a space-time empty of matter, since, empty or not, it would have geodesics and hence allow inertial motion.

In 1916 Paul Ehrenfest had suggested to de Sitter that difficult problems of boundary conditions in an open, infinite, universe might be avoided if one took instead a closed model, with spherical space-time. In 1917 Einstein tried to find a static solution with just such a spatially finite character, but could not do so without introducing his notorious 'cosmical term', a universal constant of unknown but necessarily very small value.[10] Within a couple of years he gave an alternative interpretation, seeing his cosmical term as a constant of integration arising in the solution of certain equations (for zero tensor divergence) that the energy tensor has to satisfy. For a decade and more, there were many surprisingly strong views aired as to the virtues and vices of the new constant. De Sitter retained it, but referred to it as 'a term which detracts from the symmetry and elegance of Einstein's original theory, one of whose chief attractions was that it explained so much without introducing any new hypothesis or empirical constant' (quoted in North 1965: 86).

Einstein supposed that his field equations had no solution for empty space, but he was very soon shown that he was mistaken: de Sitter, using the conditions of isotropy, a static state, and the demand that spatial sections be of constant curvature, found a trio of solutions. One of these was Einstein's own, with finite density (related to the cosmical term) and zero pressure; one was Minkowskian, with zero density, pressure, and cosmical term; and one was the solution we now know by de Sitter's name, with zero density and pressure. This model had the interesting property that there should be a lowering of the frequency of distant sources of light within it (overlooking their etherial zero-mass character). This seemed highly relevant to the redshifts of the spirals, then being found in increasing numbers. De Sitter's model, however, alarmed some of his readers, including Einstein, since it revealed a 'horizon' for every observer in it, a distance at which any finite value of ds would correspond to an infinite value of dt. There, nature would appear to be at rest, so to speak. In Eddington's words, 'the region beyond . . . is altogether shut off from us by this barrier of time' (Eddington 1918: 89). Only gradually were the distinctions between proper time and coordinate time carefully drawn. That they were sorted out at all owed much to this important paper by de Sitter, to the great clarity of Eddington's perceptions, and to an interesting paper by Cornelius Lanczos (1918), who in 1922 transformed the de Sitter line-element to a non-static form.[11]

The 'de Sitter effect' was not strictly interpretable as a Doppler effect

[10] See North (1965), chapter 5, esp. pp. 80ff, for these developments.
[11] Cf. North (1965), pp. 111–13 for various interpretations of the words 'stationary' and 'static'.

indicating a general recession of the spirals, but both de Sitter and Eddington did argue, along different routes, for recessional movements. It was calculated that a number of particles initially at rest in the de Sitter world will tend to scatter, up to a certain limit, at which their velocities will be comparable with that of light. The conclusion was much criticized, for instance by L. Silberstein (1924, 1930) and later by E. T. Whittaker (1931), and there was in the same decade much more polemic over the interpretation of other points in the same theory, although interest in de Sitter's solution waned as it was supplemented by others.

If Eddington and others seemed to be holding on tenaciously to talk of a universal recession, it was not for theoretical reasons alone. The evidence being gathered with the help of the great telescopes made these more than academic exercises. There was a consensus among the best astronomers, around the time that Einstein's general theory was being evolved, that Jacobus Kapteyn's picture of the stellar universe was more or less acceptable, or if not, then Karl Schwarzschild's rewriting of it would do. Before Kapteyn's time, attempts to derive parallaxes on the basis of proper motions were chaotic, resting on a hypothesis that he himself at first accepted, namely that real stellar motions are equally frequent in all directions. He succeeded in framing statistical relationships between parallaxes and proper motions and magnitudes, having been led to a hypothesis of star-streaming: stars move in one of two distinct and diametrically opposed directions. This was 1902. In 1904, the year in which he made public his discovery, he proposed his crucial 'plan of selected areas', and asked for the assistance of observatories in all parts of the world in photographing the stars in 206 limited areas of the sky.[12] This project, I suspect, has much to do with the fact that astronomers the world over became exceedingly 'universe-minded' in the 1910s and thereafter.

An analysis of the data on proper-motions challenging the star-streams hypothesis was offered soon afterwards by Karl Schwarzschild, who argued for a single population with an ellipsoidal velocity distribution. Eddington, in his 1914 book, tended to support this second scheme, and Kapteyn's star-streams passed out of favour after about 1918, when Shapley's work became known. The velocity ellipsoid (stream I stars) became the basis of most later work – that of B. Lindblad and J. H. Oort in particular – and yet still more recent work rather seems to support Kapteyn's distinction, with older and more widely dispersed stars being now recognized as representative of his stream II.[13]

After much refinement of his methods, Kapteyn's life-work was com-

[12] On Kapteyn's life and statistical studies generally, see Hertzsprung-Kapteyn (1928); Eddington (1914); the chapter by E. R. Paul in van Woerden et al. (1985).

[13] For a short but important note reappraising the value of Kapteyn's analysis of star stream II, see S. V. M Clube's comment in van Woerden et al. (1985), p. 37.

pleted with two classic papers, one published in 1920 (with P. J. van Rhijn) and the other in 1922, the year of his death. His model of the universe was, broadly speaking, an ellipsoid, roughly 16 kpc in length, in which density decreases rapidly with distance from a centre relatively near to the Sun (650 pc). In the 1922 paper he presented a dynamical theory, explaining stellar distributions and motions in gravitational terms, and making much use of the knowledge he had derived earlier in his studies of star-streaming.

Kapteyn had long recognized that one of the great unknowns in his theory was the absorption of light by clouds of obscuring matter – matter of a sort in evidence, it was rightly believed, in the dark regions of the Milky Way. In 1915, Harlow Shapley failed to detect any reddening in light from the Hercules globular cluster, which he estimated to be at a far greater distance than the major axis of the Galaxy, and this encouraged Kapteyn to persist with his general picture, and to persist especially in the view that the Sun was in a near-central position. But then in 1918 Shapley put the cat among the pigeons by arguing for a vast increase in the diameter to be assigned to the Galaxy.

Working at Mount Wilson, and applying the discoveries of Solon Bailey and Henrietta Leavitt on Cepheid variables, Shapley used Hertzsprung's statistical method to calibrate the period-luminosity relation, and was thus able to make estimates of the distances of clusters containing them. Since those clusters appeared to share a plane of symmetry with the Milky Way, he took them to outline our Galaxy. This implied a vast increase, not only in the size of our Galaxy, but in the distance to its centre. As Eddington said in a letter to him, 'this work marks an epoch in the history of astronomy, when the boundary of our knowledge of the universe is rolled back to a hundred times its former limit'.[14] It is noteworthy, all the same, that Kapteyn and a few others were at first able to more or less sidestep Shapley's work.[15] In 1920 Heber D. Curtis of Lick Observatory attempted a public debate with Shapley on the whole question, but what could have been a firecracker went off like a damp squib. As R. W. Smith has pointed out, Shapley made a very slow and over-cautious start, taking seven out of nineteen pages of his script before reaching the definition of a light-year, a fact conceivably not unrelated to the presence at the lecture of those who

[14] Quoted by R. W. Smith, in van Woerden et al. (1985: 48–9). See also Smith (1982). Eddington was not here thinking primarily of Kapteyn, whose distances were smaller than Shapley's but much larger than the standard estimates then current. Shapley found clusters at more than 50 kpcs from the Sun, and made out the whole distribution to be about 70 kcps across, with the Sun at about 17 kpcs from its centre (cf. Kapteyn's 16 kpc for the stellar universe). He found no globular cluster within 1 kpc of the galactic plane. Many of his distances were overestimated, thanks to his failure to allow for interstellar absorption.

[15] Kapteyn and van Rhijn made the basis of their attack Shapley's requirement that the RR Lyrae variable stars had intrinsic brightnesses a hundred times that of our Sun, and yet had large proper motions, which they thought pointed to a more or less comparable brightness. Shapley proved to be right; the stars do have large linear velocities.

might appoint him Director of the Harvard College Observatory, for which he was then applying.[16]

Shapley had in some people's eyes unwittingly effected a neo-Copernican revolution: the Sun was no longer at or near the centre of the Milky Way system. In another respect, however, he was mistakenly conservative, for whereas Curtis, with his extensive experience in photographing them, instinctively appreciated the 'island universe' character of the spirals, Shapley at that time – misled by Adriaan van Maanen's spurious measurements of the rotations of M 101, M 33, and M 81 – saw them as truly nebulous, and as only marginally beyond the main Milky Way system. The Gordian knot was cut by Edwin Hubble, who began a study of the constitutions and structures of the spiral and other 'nebulae', and of their distributions and motions. A key date is early 1924, when he was able to write to Shapley to tell of his discovery of two Cepheid variables in M 31, the Andromeda nebula, from which it was evident that this – and by implication other spirals – must be far beyond the confines of our own galactic system.[17] As explained in the chapter by Professor Osterbrock,[18] Hubble proceeded to use the 100-inch telescope, first with the Cepheid criterion, then with the brightest star criterion, then with the brightnesses of galaxies themselves, to estimate the distance of galaxies beyond the local group. He knew the risk that the 'brightest stars' might not be of comparable intensity from galaxy to galaxy; as Allan Sandage showed in 1958, they were not stars at all, but clouds of ionized hydrogen. To his method of finding distances through galactic brightness Hubble added some simple statistical safeguards, such as taking the fifth brightest galaxy in a cluster.

By 1929, Hubble had distances enough to allow him to announce his law of the linear relationship between the distances and radial velocities of the galaxies. For such a law, not only distances were needed, but also velocities, and these at first he had from Vesto Slipher of the Lowell Observatory.[19] Slipher had been the first to measure the radial velocity of a 'nebula' (M 31), in 1912. I have already alluded to the use Eddington made

[16] The post was offered to H. N. Russell who eventually declined. Shapley – who had written his doctoral thesis under Russell at Princeton – was somewhat reluctantly given a trial year in the office, but became full director in October 1921.

[17] He first recognized a Cepheid variable in the outer regions of M 31 the year before. His 1924 figure for the distance was 280 kpcs. Kurt Lundmark and Heber Curtis had in 1917 both considered novae in spirals as a means to determining their distances, and Hubble's Cepheid was a star he first thought to be a nova.

[18] See pp. 247–289 below.

[19] In his extended study of 1928–1936 he was greatly assisted by Milton L. Humason as spectroscopist, using a new type of lens. Hubble gave an account of his discovery in his Silliman Lectues at Yale (1935), published in Hubble (1936). It is worth giving prominence here to a significant technique often entirely overlooked, namely Baade's (1926) method for the distance of a pulsating star (RR Lyrae or Cepheid types): temperature having been found from its spectrum, assuming a black body, and radiant intensity from Planck's law, the observed intensity gives angular radius (a function of time) and distance. This important method was modified by A. J. Wesselink in 1947.

of Slipher's data, and others were to pay them much attention, with results I shall shortly explain.

As we all know, the law of velocity and distance rightly or wrongly became associated with Hubble's name. If I draw attention to the irony of this, it is not in any way to diminish the importance of his observational work, but merely to point out that his role was to put the capstone on an edifice that had long been ready for it. By 1914, Slipher had thirteen velocities – or redshifts, if you want to play safe – and by the end of 1925 forty-five. Apart from a few spectral shifts to the violet, indicating approach, all were to the red, and some implied velocities so great that it seemed evident to those who pondered the *dynamics* of the case that the nebulae concerned must be outside the gravitational influence of our own Milky Way system.

Earlier I quoted Eddington from 1918, when he asked whether there was *systematic* recession of the nebulae. He seems to have had in mind a double effect: what he called a 'spurious' systematic radial velocity, presumably in the sense of a misreading of the de Sitter effect, and a superimposed systematic scattering of nebulae, with velocities growing from the time when they separate, until they are comparable with the velocity of light. Quite apart from such cosmological calculations, however, there were astronomers who simply hoped to derive the *solar* velocity from Slipher's results: following Carl Wirtz (1918), they spoke of a certain K-term that had to be eliminated from the nebular velocities before they could arrive at the solar velocity. Unfortunately Wirtz and Lundmark soon realized that they were dealing with a term much greater in value than the solar velocity, and Wirtz suggested that it might point to a systematic recession of nebulae from the Sun – even that it may be a function of distance.[20] In 1925 Lundmark used data from forty-four nebulae to express the K-term velocity as

$$(513 + 10.365\,r - 0.047\,r^2)\ \text{km/sec.}$$

To these early velocity-distance laws we must now add others, but in doing so let me make one point quite explicit; we are now considering laws with an underlying theoretical foundation. Hubble gave relatively little attention to the latest theoretical developments. At first he thought that with his systematic recessional velocities he had found the de Sitter universe. As he wrote later (Hubble 1936: 109), when commenting on the vital need to obtain reliable distances: 'This fact, together with perhaps a natural inertia in the face of revolutionary ideas couched in the unfamiliar language of general relativity, discouraged immediate investigation'. The theoreticians, for all their gravity, had no such inertia. Already, in a paper published by Howard Robertson in 1928, we find a claim that there is a

[20] For further details of this discussion of the K-term, see North (1965: chapter 7). Hubble (1936: 108–13) gives full recognition to this work.

linear relationship between assigned velocities and distances of the extragalactic nebulae (see North 1965: 117–19). Even earlier, Lemaître had put forward a similar idea. We must return, now, to the theoretical context of such claims. What justification they had in observation was provided entirely by Slipher, Hubble, and their cohorts, of course.

After the first trio of solutions to the relativistic field equations, found by Einstein and de Sitter in 1917, we find that while Arthur Eddington, Ludwik Silberstein, Hermann Weyl, Richard Tolman, and others were probing the kinematics, not to say the physics, of the de Sitter model (North 1965: 92–104), the young Aleksandr Friedmann, newly appointed head of the mathematics department in Petrograd, was making considerable progress with Einstein's equations. In 1922 he drew attention to the possibility of a model with space-curvature dependent only on time, while in 1924, the year before his premature death, he investigated the cases of stationary and non-stationary worlds with negative curvature (North 1965: 113–22). He accepted most of the assumptions of Einstein and de Sitter – in particular the cosmical term – but made the explicit assumption that at any instant the model should have a space of constant curvature. He tacitly assumed that there is a cosmic time with space orthogonal to it.[21] On this new basis, Friedmann derived the set of models for which he is now best remembered – with the radius of curvature increasing monotonically or changing as a periodic function of the time, and classified according to the values of the space-curvature constant and the cosmical constant. In his second paper he showed – and he was the first to do so – that a non-stationary world of constant *negative* curvature with positive material density was possible.

Friedmann's work received surprisingly little attention from the scientific community. Einstein criticized it in a short note, and then retracted his criticism, which had been based on an arithmetical error of his own. Tragically, Friedmann's fame was posthumous, and was a consequence of a renewal of interest in these matters as a result of the work of Georges Lemaître (1925, 1927) and H. P. Robertson (1928).

Lemaître was not quite thirty when in 1923–4 he went to study with Eddington in Cambridge. When he left, it was for a stay of nine months at the Harvard College Observatory, and from there he wrote his first paper on cosmology (Lemaître 1923). In this he raised objections to de Sitter's 1917 model. It lacked matter, of course, but also curvature, and was thus unable to yield a finite space. This is a remarkable illustration of the way in which a new generation, trained in the methods of general relativity could be dismissive of Euclidean geometry. (But Euclid still had his fans.)

[21] H. P. Robertson later made the same assumptions quite explicitly, while R. C. Tolman preferred another approach, showing that the orthogonality condition could be derived from another set of assumptions, including that of isotropy.

Lemaître objected further against de Sitter's model that coordinates were used for it that seemed to imply that the universe has a centre. Insisting on the equivalence of all (four-dimensional) points, he derived a model with the radius of curvature constant over space but everywhere a function of time. Unlike Friedmann, but just as Silberstein had done the year before, he derived a formula for the redshift of spectra proportional to distance. If this did not claim much attention, perhaps it was because in the 1920s the mathematical translation of de Sitter's solution to produce an 'apparently' non-static model was becoming almost a pastime. Weyl, Lanczos, Silberstein, and Robertson were often supposed to have done just this, when they produced what, with the advantage of hindsight, we can take to be non-static models. There was a similar feeling of unreality about the mathematical singularities in Friedmann's models and Lanczos' transformed de Sitter model: was expansion from an initial singularity not just an illusion created by the mathematics?

Hermann Weyl's contribution to this subject, especially as it appears in the fifth edition of his *Space – Time – Matter* of 1923, is highly significant, for as J. Ehlers has emphasized, he had a clear conception of causal structure in cosmology. Weyl pointed out that to have a cosmological model one must specify, besides a space-time, a congruence of time-like curves to represent the mean motion of matter. He constructed such a congruence on the de Sitter hyperboloid, and required not only that all these curves emerge from a common origin in the infinite past, but that the causal futures of all curves be identical.

Lemaître had been ordained abbé in 1923, and it is not difficult to understand why his science had strong theological implications for him. An initial singularity was not something to be avoided, but a positive merit, a token of God's creation of the world. In 1927 we find him solving the relativistic field equations in a form containing – as Friedmann's had not – a term for the radiation pressure.[22] Einstein's 1917 solution, and Lanczos' version of de Sitter's, were seen to be only special cases. He now introduced as a more general solution an expanding universe with radius increasing asymptotically from a constant value in the infinite past (Lemaître 1927).[23]

From a formal point of view Lemaître's paper adds little to what we can learn from Friedmann's, of which he was ignorant, but the two men were plainly of different mentalities: Friedmann was at heart a mathematician, and Lemaître a physicist, as he showed even more plainly by his subsequent writings. It seems that in 1927 Lemaître met Einstein, who told him that his paper was mathematically sound. Einstein made it clear,

[22] But not the material pressure, which he thought could be ignored.

[23] For more details of the background to Lemaître's work, see Kragh (1987) and North (1965), esp. pp. 119–21. Lemaître (1950) collects together (in English translation) various works of his dated between 1929 and 1945.

though, that he did not believe in the expanding universe that it entailed (see Kragh 1987: 125). As Lemaître wrote afterwards, he had the feeling that Einstein was not *au courant* with the latest astronomical facts. It was from him, however, that Lemaître learned of Friedmann's earlier work. The information must have disappointed him, but no doubt it explains his excessive reticence about his own achievements. With due humility, when in 1929 he made a survey of recent cosmological progress, he said very little about his own original ideas.[24] He cautiously referred to an 'apparent Doppler effect' where 'the receding velocities of extragalactic nebulae are a cosmical effect of the expansion of the universe' (North 1965: 119–21). He did not at the time fit either his exact or his approximate formula ($c = 1.732\,cr/R_0$) to empirical data, but that is not to say that he could be accused of disregard for the empirical side of his work. He would no doubt have thought any such claim to be premature.

Not until Lemaître read a report of a meeting at the Royal Astronomical Society in London, at which Eddington had advocated paying more attention to non-static models, was he prompted to write to remind his former mentor of the 1927 paper. Eddington was at the time working with his research student G. C. McVittie on the problem of the instability of the Einstein spherical world, and he saw at once the implications of Lemaître's work: the Einstein world is *intrinsically* unstable (North 1965: 122–5). He sent a copy of the paper to de Sitter, who was in turn clearly more or less convinced by its truth, as we know from a letter to Shapley.

Through Eddington and de Sitter the world now learned of theoretical developments that had lain dormant for three years and more. Einstein gave the expanding universe his blessing; popular works by James Jeans made it newspaper-worthy; and Eddington polished Lemaître's rough diamond, to produce what became known as the Lemaître–Eddington model – a universe evolving from a stagnating Einstein world of indeterminate age. Accepting Hubble's parameter for the expansion, Eddington supposed that this expansion began about 2×10^9 years ago. When, in later years, ages became an embarrassment to cosmologists – with the Earth seemingly as old as the 'Hubble time' – there was a certain propaganda value in the vague 'eternity' of the Einstein state from which Eddington took everything to evolve. This situation evaporated only when in 1952 Walter Baade, then at Palomar, discovered Hubble's mistake in regard to galactic distances: the 'Hubble-time' expanded overnight by a factor of 5 or 10, and the Lemaître–Eddington world was no longer at a premium. It experienced something of a revival in the 1960s, when it was used to explain the concentration of the redshifts of quasi-stellar objects in the neighbourhood of 2.

Long before any of these things happened, Lemaître had lost faith in the

[24] In the lecture held in Brussels in January 1929 ('The Size of Space', Lemaître 1950) he gave a few empirical data and mentioned Friedmann's work but made no reference whatsoever to his own.

model, and was working on another tack entirely. The 'discovery' of his paper had resulted in a certain interest in the physical problem of how the expansion from the Einstein state was initiated. Eddington, McVittie, W. H. McCrea, Tolman, N. R. Sen, and R. O. Lifshitz, for instance, all investigated the problem, but it would not be true to say this cosmic physics set the tone for the cosmology of the 1930s.[25] The approach generally adopted by the more mathematical cosmologists was then strongly geometrical – in the relativistic sense – beginning from general principles of isotropy and homogeneity. H. P. Robertson, E. A. Milne, A. G. Walker, and many others after them, took this way forward, and of course the style has been canonized with the so-called 'Robertson–Walker metric' of the text-books.

One might loosely distinguish between three different styles during this period – and I do not mean to suggest that an Eddington or a Tolman could not practise more than one. There were the 'physicists', and the 'geometers', in the distinctive styles to which I have just alluded, and then of course there were the observing astronomers. They, it must be said, tended to take a relatively cavalier view of theory. Hubble, for instance, was no Philistine, and he obtained much theoretical help from Tolman, but his book of 1936 has just three or four pages out of 200 on theories of cosmology, and they come at the very end. The irony is that for a decade everyone had been waiting for reliable *distances*, and whilst they waited, intense discussions of the significance of the redshifts had been going on. As soon as Hubble provided the distances, discretion tended to be thrown to the winds. The redshifts suddenly became Doppler shifts for most ordinary astronomers, that is, they were taken to be indicative of velocities pure and simple.

At the other extreme of theoretical sophistication comes Eddington, whose way of uniting seemingly disparate branches of knowledge was extraordinary – and far outside the compass of this short survey of mine. One of his abiding interests was in the fundamental contants of nature – the gravitational constant, the velocity of light, Planck's and Rydberg's constants, and so on – and it was his ambition to unite general relativity with quantum theory. The ambition, it seems, was infectious, for Lemaître, on reading an address given by Eddington to the Mathematical Association on the origin and end of the world, printed in *Nature* in 1931, was prompted to object in terms that are now of great interest (Lemaître 1931).

Eddington had asked what the world would have been like when it was in a state of zero entropy. Although this was outside the realm of scientific reasoning, he said, 'philosophically the notion of a beginning of Nature is repugnant to me'. Lemaître's intuitions were quite otherwise:

[25] I shall have little opportunity to mention R. C. Tolman here, but his work was of great value. He was the great pioneer of relativistic thermodynamics, and introduced great sophistication into the basic materials, the 'perfect fluids', of the models discussed.

> I would rather be inclined to think that the present state of quantum theory suggests a beginning of the world very different from the present order of Nature. Thermodynamic principles from the point of view of quantum theory may be stated as follows: (1) Energy of constant total amount is distributed in distinct quanta. (2) The number of distinct quanta is ever increasing. If we go back in the course of time we must find fewer and fewer quanta, until we find all the energy of the universe packed in a few or even in a unique quantum (Lemaître 1931).

So was born the idea of the Primeval Atom, an idea that undoubtedly made many of his readers feel distinctly uncomfortable. He went on, however, to talk in a language that now looks somewhat prophetic. The unique atom might be seen as one with an atomic weight equal to the entire mass of the universe. Highly unstable, it 'would divide in smaller and smaller atoms by a kind of super-radioactive process', and 'Some remnant of this process might, according to Sir James Jeans' idea, foster the heat of the stars until our low atomic number atoms allow life to be possible'. In November of the same year, 1931, he returned to the theme, explaining in broad outline how cosmic radiation, with its enormous energy content, might be understood as 'glimpses of the primeval fireworks of the formation of a star from an atom of atomic weight somewhat greater than that of the star itself'.[26]

These ideas were speculative, but not irresponsible. He had high hopes of supporting them, and for this he needed two things: a theory of nuclear structure applicable to atoms of extreme weights and better information about cosmic radiation. He thought his hypothesis of super-radioactive cosmic radiation was supported by A. H. Compton's claim that the radiation consists of charged particles, and his faith in it was increased when it was gradually revealed that the energies involved were much greater than had at first been suspected.

Lemaître was a man of great instinct and few words. He was to modern physical cosmology what the Primeval Atom was to the modern universe. This is not to say that his ideas have survived these fifty or sixty years – he died, after all, in 1966, only a year after the discovery of the cosmic microwave background radiation by Arno Penzias and Robert Wilson – but he perceived that there was much more to cosmology than gravitational geometry. Star formation, galaxy formation, and the relative abundance of the various chemical elements, all figured in his scheme.[27] He was too far ahead of his time: no real progress on this front was possible before the late 1930s, however, with Carl von Weizsäcker's and Hans Bethe's independent solutions to the problem of generation of stellar

[26] A note reprinted in Lemaître (1950: 80–6).

[27] Lemaître's account in the early 1930s was necessarily in one respect on the wrong lines, since he accepted Eddington's mechanism (1917) for the conversion of matter into radiation, namely, by electron-proton annihilation. In 1934, after the discovery of the positron, Eddington put forward an electron-positron annihilation process, and this was integrated into later thermonuclear theories.

energy. Only with the subsequent studies by George Gamow, assisted by Ralph Alpher, of nucleo-synthesis in a hot big bang, were plausible estimates of abundances possible. Plausible, but not accurate: as is explained in a later chapter in this volume, they were able to use big-bang cosmology to explain abundances of nuclei of only the lightest elements. Alpher, with Robert Herman, did nevertheless throw up a prediction of the existence of radiation surviving from the early history of the universe with a present temperature of about 5° K. Lemaître would have been delighted by that, and delighted too to read what has by now become almost a platitude: data on the abundances of the elements are the most reliable data we have about the history of the universe.

This is hardly relevant, though, to the situation in the early 1930s, and we must not obscure from view the excellent and extensive work done at that time on relations between other 'observed quantities'. I have already touched on this problem obliquely, in referring to the naive assumption that redshifts must be Doppler shifts indicative of velocities. Of course there were always the cautious few who reserved judgment – Jeans held out for some time, for instance – as well as the resolutely sceptical, who rejected a Doppler interpretation almost as a point of honour, and sought alternative explanations. (One favourite was the 'ageing of light'.)[28] There was much ambiguity in the way many spoke about the expansion of the universe: for some it meant only that the distances separating the galaxies were increasing with time, for others it meant that the relativistic metric was time-dependent. It was widely assumed – especially by observing astronomers – that measurements based on observation could be passed from theory to theory.

For all this, I think it is true to say that in no other branch of science was so much care given to the analysis of the concepts employed in it. This deep analytical tendency owes more to Eddington, I believe, than to any other man, and it had many beneficial consequences. One of these concerns observation: by a painstaking theoretical interpretation of what was actually involved in observation, such theoreticians as E. T. Whittaker, R. C. Tolman, and G. C. McVittie were able to establish a whole range of relations, such as those of number-counts of galaxies against apparent magnitude or redshift, of redshift or apparent magnitude against apparent size of galaxies, and so on. 'Distance' had become a subsidiary concept, or rather had divided into many concepts. This work was not spectacular, but it was of the very essence of scientific cosmology. It should in principle have allowed a decision to be made between the various theoretical models then available. That the time was not ripe for any remotely decisive elimination of alternatives was not the fault of the theoreticians.

[28] For several hypotheses – for instance F. W. Zwicky's notion of a gravitational analogue of the Compton effect – see North (1965: 229–34). Even Hubble lost his nerve as to the reality of the velocities: he was later criticized by McVittie and Otto Heckmann.

Alternatives there were in plenty, in these early years, especially after the work of Friedmann and Lemaître was systematized, and models were classified by the cosmical constant, the curvature constant, and assumptions about density and pressure. Einstein and de Sitter jointly produced another model (with all of these zero, apart from the pressure). Oscillatory models, introduced by Friedmann, later had a certain vogue – and Tolman took a strong interest in its thermodynamic properties.[29] Then came a revival of so-called 'Newtonian cosmology', by E. A. Milne and W. H. McCrea, and later G. J. Whitrow. Their 'kinematic relativity' is one example of an alternative to Einstein's. Others were the theories of gravitation developed by G. D. Birkhoff, A. N. Whitehead, and J. L. Synge. All of these had potential cosmological ramifications.[30] They were all symptomatic of an exceedingly vital scientific movement, where thought tended to precede the amassing of data. These theories came, though, in the wake of the general theory of relativity, and it was with the Einsteinian version that the initiative remained – even though the boat was not infrequently rocked by pirates from without and disgruntled passengers from within. The circumstances of some of these perturbations are the subjects of the chapters that follow.

References

Eddington, A. S. (1914). *Stellar Movements and the Structure of the Universe.* London: Macmillan.

(1918). *Report on the Relativity Theory of Gravitation.* London: The Physical Society of London. (2nd edn. 1920).

(1931). The End of the World from the Standpoint of Mathematical Physics. *Nature*, **127**, 447–53.

Einstein, A. (1915). Die Feldgleichungen der Gravitation. *Sitzungsberichte der Preussischen Akademie der Wissenschaften*, 844–7.

Hertzsprung-Kapteyn, H. (1928). *J. C. Kapteyn, zijn Leven en Werken.* Groningen: P. Noordhoff.

Hubble, E. (1936). *The Realm of the Nebulae.* London: Oxford University Press.

Kapteyn, J. C. (1922). First Attempt at a Theory of the Arrangement and Motion of the Sidereal System. *Astrophysical Journal*, **55**, 65–91.

Kerszberg, P. (1987a). On the Alleged Equivalence between Newtonian and Relativistic Cosmology. *British Journal for the Philosophy of Science*, **38**, 347–80.

(1987b). The Relativity of Rotation in the Early Foundations of General

[29] Tolman showed that each cycle in a series of oscillating phases of the model may be larger (in radius of curvature) than the preceding cycle, by virtue of the increase in entropy. It is worth remembering that when news of Penzias' and Wilson's discovery of the microwave background radiation came through to R. H. Dicke, in 1965, he had been independently preparing to search for a residuum of black-body radiation which he thought might come from an earlier *compression* phase of the universe.

[30] These theories are discussed at length in North (1965: chapters 8 and 9). For a valuable discussion of the alleged equivalence of Milne's and Einstein's theories, see also Kerszberg (1987).

Relativity. *Studies in the History and Philosophy of Science*, **18**, 53–79.
Klein, M. J. (1970). *Paul Ehrenfest. Vol. 1: The Making of a Theoretical Physicist.* Amsterdam: North-Holland.
Kox, A. J. (1988). Hendrik Antoon Lorentz, the Ether, and the General Theory of Relativity. *Archive for History of Exact Sciences*, **38**, 67–78.
Kragh, H. (1987). The Beginning of the World: Georges Lemaître and the Expanding Universe. *Centaurus*, **32**, 114–39.
Lemaître, G. (1925). Note on De Sitter's Universe. *Journal of Mathematics and Physics*, **4**, 188–92.
(1927). Un univers homogène de masse constante et de rayon croissant. *Annales de la Société Scientifique de Bruxelles*, **47**, 49–56. Translated in *Monthly Notices of the Royal Astronomical Society*, **91**, 483–90.
(1931). The Beginning of the World from the Point of View of Quantum Theory. *Nature*, **127**, 706. Quoted *in extenso* in the introduction to the work below, pp. 17–19.
(1950). *The Primeval Atom. An Essay on Cosmogony.* New York: Van Nostrand.
North, J. D. (1965). *The Measure of the Universe. A History of Modern Cosmology.* Oxford: Clarendon Press; reprinting, New York: Dover Books.
Roseveare, N. T. (1982). *Mercury's Perihelion, from Le Verrier to Einstein.* Oxford: Clarendon Press.
Shapley, H. and Curtis, H. D. (1921). The Scale of the Universe. *Bulletin of the National Research Council*, **11**.
de Sitter, W. (1911). On the Bearing of the Principle of Relativity on Gravitational Astronomy. *Monthly Notices of the Royal Astronomical Society*, **71**, 388–415.
(1916). On Einstein's Theory of Gravitation and its Astronomical Consequences. *Monthly Notices of the Royal Astronomical Society*, **76**, 699–724, and **77**, 155–84.
(1932). *Kosmos.* Cambridge: Harvard University Press. A revised edition in Dutch (translated from the English of the original) was published in The Hague in 1934.
Smith, R. W. (1982). *The Expanding Universe: Astronomy's 'Great Debate', 1900–1931.* Cambridge: Cambridge University Press.
van Woerden, H., Allen, R. J., and Burton, W. B., eds. (1985). *The Milky Way Galaxy* (I.A.U. Symposium No. 106, Groningen, 1983). Dordrecht: Reidel.

Discussion

Ehlers:

Concerning the history of the theory underlying Hubble's law I would mention that H. Weyl in 1923 (*Space – Time – Matter*, 5th edn.) points out that to have a cosmological model one has to specify, besides a space-time (M, g), a congruence of timelike curves to represent the mean motion of matter. He then goes on to construct such a congruence on the de Sitter hyberboloid and requires, not only that all these curves emerge from a common origin in the infinite past, but that the causal features of all these curves should be identical. This appears to be the first use of causal structure in relativistic cosmology.

II

Riddles of and clues to cosmology

3

Olbers' Paradox in recent times

Edward Harrison

Since the sixteenth century astronomers have conjectured that the universe consists of unbounded space populated with either a uniform distribution of stars or a uniform distribution of clusters of stars. This assumption that seemed eminently plausible led directly to the celebrated night-sky riddle now popularly known as Olbers' paradox.[1] The riddle is as follows: every line of sight from the eye that extends into the depths of space must, in an unbounded and starfilled universe, ultimately intercept the surface of a star. Hence, the night sky should be bright with starlight at every point and no dark gaps should separate the stars. If most stars resemble the Sun, as astronomers generally supposed, the sky at every point should shine as bright as any point on the Sun's disk.[2] Jean-Philippe Loys de Chéseaux, a Swiss astronomer, in 1744 showed that the sky is 180,000 times larger than the Sun's disk and therefore the starlight incident on Earth should be 180,000 times more intense than sunlight.[3] Hence the riddle: why is the sky dark at night?

Heinrich Wilhelm Olbers of Bremen, although not the first astronomer to discuss this startling contradiction between cosmological theory and observation, originated in 1823 (Olbers 1823) the line-of-sight argument that I have used above.[4] The riddle, for historical reasons, is now popularly known as Olbers' paradox.

[1] Extensive references may be found in Jaki (1969) and Harrison (1987). The discussion in the present paper parallels the treatment in Harrison (1987).

[2] The minimum angle resolvable by the eye is about 1 minute of arc. Each visible star forms a diffraction disk subtending 1 minute of arc, and hence 1.5×10^8 stars can cover the whole sky. Telescopes with their larger apertures resolve smaller angles, say 1 second of arc, and in this case 5.3×10^{11} stars can telescopically cover the whole sky. In either case, of course, the sky would not blaze with the brightness of the Sun's disk. Olbers' paradox neglects diffraction and considers only the geometrical size of a star. If the density of stellar matter is the canonical 10^{-30} g cm^{-3} (the density is probably smaller), and if all stars are sunlike, the number geometrically covering the sky is 5×10^{60}, extending to a distance of 10^{23} light-years. The stars farther away than 10^{23} light-years are occulted by nearer stars. When sunlike stars cover the sky geometrically and not diffractionally, the sky at every point then attains the brightness of the Sun's disk. I am indebted to George Ellis for raising this question.

[3] J. P. Loys de Cheśeaux (1744), appendix 2.

[4] I have followed the custom of referring to Olbers' rather than Olbers's paradox.

Absorption of light by interstellar gas and dust complicates but does not invalidate the line-of-sight argument. Both Chéseaux and Olbers attributed the darkness of the night sky to interstellar absorption. John Herschel in 1848 showed that absorption fails to solve the riddle because an absorbing medium in a universe filled with intense starlight rapidly heats up and emits as much radiation as it absorbs.

The riddle of cosmic darkness – the problem of why the sky is dark at night – emerged during the sixteenth century in the aftermath of the Copernican revolution, and has since occupied a prominent position in the history of Western cosmology. Many different solutions have been proposed, each related to a particular set of cosmological assumptions. The general solution, presented in a quantitative form by Lord Kelvin in 1901, applies to most cosmological models; it is rather simple, does not require an advanced understanding of science, and was anticipated in a rudimentary form by the poet and essayist Edgar Allan Poe. The physical concepts essential for the general solution became known after Ole Roemer had established in 1676 that light travels at finite speed; the general solution could therefore have been discovered even in the days of Isaac Newton and Edmund Halley, and by astronomers such as Chéseaux, Olbers, and Herschel. The general solution recognizes that:

(1) light propagates at finite speed;
(2) either the universe is of finite age or stars have a finite luminous lifetime;
(3) stars have separating distances of a few light-years.

Statement (1) means that when we look out in space we look back in time. No astronomer after the discovery of aberration by James Bradley in 1729 doubted Roemer's result that light travels at finite speed (Cohen 1940).

Statement (2) means that an unimpeded line of sight extends back to either the birth of luminous stars or the origin of the universe; if the universe is infinite in space, each observer sees only a finite number of stars extending to a horizon at a distance determined by the lifetime of luminous stars or the age of the universe; this horizon is the limit of the visible universe.[5] According to the Book of Genesis and Mosaic chronology the universe has an age of only a few thousand years; though this cosmic time-scale was extended in the eighteenth century by Georges Buffon to 100,000 years, and the age of luminous stars was found in the nineteenth century by Kelvin to be tens of millions of years, almost all estimates of the age of the universe and of luminous stars were small by modern standards.

Statement (3) leads by straightforward calculation to the conclusion that if the age of the universe or the luminous lifetime of stars is less than $10^{14}L^3$

[5] The horizon of the visible universe is known as the particle horizon. See Rindler (1956).

years, where L is the mean separating distance between stars measured in light-years, the number of luminous stars within the visible universe is insufficient to cover the sky.[6] Newton knew from James Gregory's photometric comparisons with the outer planets that the nearest stars (assumed to be the brightest) are at a distance of about 300,000 astronomical units or 5 light-years (Hoskin 1977). After the death of Newton, this knowledge became generally known and Gregory's photometric method was used by both Chéseaux and Olbers to estimate the distances of nearby stars.

The information needed to solve the riddle has been available since the early eighteenth century. It seems surprising that scientists in the eighteenth and nineteenth centuries, believing that the universe had an age measured in thousands or at most millions of years, failed to publish calculations showing that the number of visible stars is insufficient to cover the sky.

In this brief historical survey of Olbers' paradox I dwell mainly on the developments in recent times. First, I make comments on the state of cosmology in the nineteenth century and early years of the twentieth century until the time of Einstein's theory of general relativity. Then I consider the recent revival of interest in Olbers' paradox and comment on how twentieth-century developments in cosmology have changed our view of the darkness-at-night riddle.

Agnes Clerke (1890) in her classic *The System of the Stars* surveyed the astronomical evidence and concluded that most stars in the universe were gathered into one vast cluster: the Galaxy. This one-island picture of the universe, in which the Sun occupied a position near the center, consisted of a finite system of stars surrounded by boundless and empty space. The

[6] The age of the universe or of luminous stars, whichever is the less, must exceed λ/c, where

$$\lambda = \frac{L^3}{\pi R^2},$$

in order to cover the sky with stars, where L is the mean separating distance between stars, R the radius of a typical star, and c the speed of light. When L is measured in light-years, then $\lambda = 10^{14}L^3$ light-years for sunlike stars. Hence, when the age of the universe or the age of luminous stars, whichever is less, exceeds $10^{14}L^3$ years, the sky is geometrically covered with stars. The number of stars needed to cover the sky is

$$N = \frac{4\pi\lambda^3}{3L^3},$$

or $10^{42}L^6$ when L is measured in light-years. The number of visible stars in a static universe of age t is

$$N_{\text{visible}} = \frac{4\pi(ct)^3}{3L^3}$$

equal to $4\pi t^3/3L^3$ when t is measured in years and L in light-years. Later in the text λ is referred to as the mean free path of light rays between emission and absorption by stars, and is the background distance of the stars that supposedly cover the sky. The travel time λ/c is the mean free time of light rays between emission and absorption by stars, and is referred to as the fill-up time (the time to fill the universe with radiation in equilibrium with the stars).

one-island or single-galaxy universe accounted for the known facts of astronomy and had served as the principal cosmological model of the nineteenth century. Quite incidentally it provided a straightforward solution to the riddle of cosmic darkness: we look out between the stars and clusters of stars of our Galaxy into starless extragalactic space.

The rival many-island or multiple-galaxy picture of the universe consisted of countless other milky ways similar to our Milky Way. This picture, much less popular among astronomers, was little more than an echo of Thomas Wright's eighteenth-century novel speculations. William Herschel in his telescopic surveys of the heavens employed the concept of a many-island universe, then later, early in the nineteenth century, after realizing that his assumptions had become untenable, he argued in favor of a one-island universe. John Herschel, who succeeded his father as the most eminent astronomer in Europe, vigorously promoted the one-island universe, which then became the standard cosmological model of the century. All the astronomical evidence, wrote Clerke, consistently supports this picture of the universe; moreover, were this picture incorrect and an infinity of stars occupied boundless space, constituting an endless Milky Way or innumerable separate milky ways, the "intense inane" of a continuous background of starlight would bewilder our feeble senses (Clerke 1890: 368). The controversy between advocates of a one-island universe and a many-island universe flared up in the twentieth century and developed into the "great debate" (Hoskin 1976; Osterbrock 1990; Smith 1982). Harlow Shapley remained until late in the 1920s a staunch advocate of the one-island universe.

Lord Kelvin, in a little-known paper,[7] solved the riddle using the conditions in statements (1), (2), and (3). His calculations gave the first (to my knowledge) correct quantitative solution within the frame of a static universe. When we make allowance for expansion, his solution holds true for an expanding universe of finite age.

In these introductory remarks, I cannot resist commenting on what seems to me an inhibition in the minds of astronomers that must surely have affected the development of cosmology from the seventeenth to the nineteenth century. This inhibition was a marked reluctance to recognize or acknowledge the principle that when we look out far in space we also look back far in time.[8] This inhibition has received scant recognition in the

[7] Thomson (Lord Kelvin) (1901). For a discussion on this paper see Harrison (1986).

[8] We take for granted that we see things in the external world as they are, that we see them instantaneously, and this belief dates back to prehistory. René Descartes was fully convinced that light travels at infinite speed. How confused our visual reconstruction of the world would be, he said, if the rays of light entering the eye and coming from different distances were emitted at different times. His fears were groundless, at least in everyday life, for the speed of light is much greater than he imagined possible. Not until the end of the eighteenth century, with few exceptions, do we find acknowledgment of the fact that we see the heavens as they were long ago. Francis Roberts (1694) wrote "Light takes up more time in Travelling from the Stars to us, than we in making a *West-India* Voyage (which is ordinarily performed in six Weeks)."

history of science and may, among its many effects, have delayed the general solution to the riddle of cosmic darkness.

The heavens gave visual evidence that stars observed at distances of thousands of light years have existed for thousands and even tens of thousands of years.[9] Other worlds, possibly still existing, are seen as they were long ago. This evidence showing that stars originated at least many thousand of years ago controverted scriptural testimony on the age of the heavens. Debate raged in the nineteenth century on the significance of fossils and the age of the Earth (Haber 1959), and in his work on natural selection Charles Darwin followed the lead of the geologist Charles Lyell and assumed that the Earth was of indefinite age. Although astronomers had evidence demonstrating the great age of the stars that contradicted a too literal interpretation of the Book of Genesis, they remained mostly silent and contributed little or nothing to the debate.

We can easily imagine the astonishment of the public if astronomers had openly said: when we look out in space we look back thousands and perhaps millions of years to a time when the heavens were created! This helps us to understand why they stood aside and did not participate in the debate on the age of the cosmos. Many astronomers, perhaps most in Britain, were ordained ministers and few had the temerity of Lyell and Darwin. In publications and from pulpits they lavished praise on the glories of the heavens that revealed the handiwork of the Creator, and had no desire to disturb their admirable security by casting doubt in public on the details of popular belief. The cosmological significance of looking out in space and back in time, though occasionally addressed in guarded terms in the nineteenth century, was not openly discussed in Victorian drawing rooms and did not enter astronomical literature until the twentieth century.

The theory of relativity established the twentieth-century habit of measuring astronomical distances in terms of light-travel time. George Gamow's prediction in the late 1940s that the universe is flooded with the afterglow of the big bang and its confirmation by the discovery of the microwave background radiation in 1965 made secure the picture of an evolving universe of finite age. The principle that when we look out in space we look back in time – and therefore an unimpeded line of sight extends back to the birth of stars, galaxies, and even the universe – no longer seriously threatens prevailing religious belief.

Relativistic cosmology started with Einstein's general theory of relativity and blossomed in the early 1930s with the recognition of the expansion of

[9] William Herschel (1802) reported that light rays from a certain nebula had been "almost two millions of years on their way." In the nineteenth century astronomers avoided such provocative statements. John Herschel, a pillar of the scientific establishment, in his *Treatise of Astronomy* (Herschel 1830: 354) cautiously wrote that among the countless multitudes of stars "must be many whose light has taken at least a thousand years to reach us."

the universe. The old dark night-sky riddle lay forgotten amidst the confusing complexities of the new cosmology. Then, in 1952, Hermann Bondi reawakened interest in the cosmological consequence of a dark night sky.[10] Unaware of the history of the riddle, he referred to it as "Olbers' paradox."

Bondi's discussions tended to be nonmathematical, and yet, despite the intricacy of radiative transfer with luminous sources and sinks in expanding universes of non-Euclidean metric, he showed that a uniform spatial curvature cannot solve the riddle, and that the universe may be open or closed without affecting the problem. Bondi also showed that the solution proposed by Olbers – the interstellar absorption of starlight – fails to explain darkness at night.

Bondi identified and examined the following assumptions in Olbers' argument:

(i) the universe is unbounded in space and uniformly populated with stars or clusters of stars;
(ii) the universe is unlimited in age and the luminosity of the stars remains unchanged in time;
(iii) the universe is static, neither expanding nor contracting.

These assumptions, when accepted, lead inescapably to the conclusion that the sky blazes with starlight. A line of sight on the observer's backward light cone stretches out until it eventually intercepts the surface of a luminous star, therefore the sky is covered with stellar disks, and the universe contains radiation everywhere in equilibrium with the stars.[11]

Olbers explicity acknowledged assumption (i). He argued that stars may be uniformly distributed or clustered to form uniformly distributed systems (perhaps he had in mind external milky ways), and the clustering of stars would not affect the problem of a dark night sky. Twenty-five years later, John Herschel countered Olbers' statement with the argument that if stars extend endlessly, darkness could still prevail provided that stars form clusters that are members of larger clusters that are members of even larger clusters, and so on, in an unlimited hierarchical arrangement. Such a system, first suggested by Immanuel Kant a hundred years previously, had decreasing density when averaged on increasing scales; a line of sight could extend unimpeded to infinity when certain conditions were satisfied, and the night sky would therefore be dark.[12]

[10] Bondi (1952). Further references can be found in the works cited in n. 1.

[11] The bright-sky conditions envisaged in Olbers' paradox correspond to thermal equilibrium: hence, the temperature of thermal radiation in space equals the temperature at a stellar surface. Main sequence stars have large internal temperature gradients and must radiate energy from their surfaces; therefore stars cannot exist in radiative equilibrium with the external world. One possible answer to the dark night-sky riddle is that stars could not exist otherwise. Stars did not form in the radiation era of the early universe, but later when the universe had become dark.

[12] In the early decades of the twentieth century, the Swedish astronomer Charles Charlier developed a theory of astronomical hierarchy. A more recent treatment can be found in Harrison (1987).

Olbers assumed (ii) implicitly and not explicitly. Assumption (ii) contradicts statement (2). I am inclined to think that Olbers would have rejected assumption (ii) in the precise form stated by Bondi. Rather than assuming that stars shine with undiminished luminosity for eternity, it is more likely that he would have assumed without critical examination that we see the heavens as they are in the present and not as they were in the past. If this is true, he overlooked the significance of a finite speed of light when observing the heavens. This mental habit of primitive origin persists even in the twentieth century. When I observe the sky at night, I realize only by conscious effort that I am looking back into the past and seeing other worlds as they were long ago. The automatic supposition that light travels instantaneously reinforces the religious inhibition I referred to earlier. Not inconceivably, Olbers may have unconsciously assumed that light travels at infinite speed and thereby avoided conflict with prevailing religious belief.

Bondi emphasized that the cosmological principle (the universe on the average is the same everywhere in space) consists of assumption (i) and the perfect cosmological principle (the universe on the average is the same everywhere in space and time) consists of assumptions (i) and (ii) together. He showed that Olbers' paradox can be solved by rejecting (ii) and assuming that stars have a finite luminous lifetime. But this resolution of the riddle was not explored in much detail. As one of the originators of the perfect cosmological principle, Bondi accepted the truth of (i) and (ii) within the context of the steady-state theory.

In the twentieth century, assumption (iii) is no longer correct. The principal omission in Olbers' treatment, said Bondi, unavoidable in the nineteenth century, was the redshift of distant sources in an expanding universe. Bondi proposed that the cosmological redshift solves Olbers' paradox: stars cover the sky, but most are invisible by virtue of their extreme redshift.

Many readers with memories of cosmology in the 1950s and 1960s, when asked why the sky is dark at night, will recall Bondi's solution of the riddle and answer, "Because the universe is expanding and distant stars therefore have large redshifts and cannot be seen. The darkness of the night sky proves that the universe is expanding."

One of the most puzzling problems besetting cosmology in the 1930s and 1940s was the brevity in the age of the universe derived from the recession of the galaxies. An expanding universe originating at high density has an age of the same order as the expansion time H^{-1}, where H is the Hubble term in the velocity–distance law (recession velocity = $H \times$ distance). The Hubble term, which had changed little in value since Edwin Hubble's original determination, indicated that the universe had an age of order 10^9 years. This short time span conflicted with the estimated ages of the

Earth and main sequence stars. The Lemaître and Eddington–Lemaître universes delayed expansion and achieved greater age by employing Einstein's cosmological constant, but most investigators viewed the mysterious cosmological constant with great misgiving. The expanding steady-state universe proposed by Bondi, Thomas Gold, and Fred Hoyle in 1948 ingeniously avoided the age problem without recourse to the cosmological constant and became in the 1950s a popular product in supermarket cosmology. The steady-state universe expanded exponentially ($H = \dot{R}/R$); it had infinite age and consisted of infinite flat space uniformly populated with galaxies. The continuous creation of nascent matter maintained a constant average density and luminosity.

Bondi's redshift solution is valid for the steady-state universe. An observer's line of sight (on the backward light cone) in the steady-state universe never extends beyond the Hubble sphere (in this case the event horizon) of radius c/H. The line of sight stretches out in space, reaches back in time, and asymptotically approaches the Hubble sphere in the infinite past. Thus, the number of stars on the observer's backward light cone is infinitely great. The estimated value of c/H is now of order 10^{10} light-years, and the number of visible stars (nearest and uneclipsed) covering the sky is of order 10^{35}. Calculation shows that the average redshift of starlight is of order 10^{13}, and the average energy density of radiation at the surface of a star is roughly 10^{13} times the average density of starlight in space.

Two comments must be made concerning Bondi's redshift solution of Olbers' paradox.

First, we note that expansion is necessary but not sufficient for a dark night sky in the Bondi–Gold–Hoyle universe. Apparently, advocates of the redshift solution never entertained the possibility of a bright night sky in an expanding universe. For the steady-state universe, we find by calculation that when the mean free path between emission and absorption of light rays by stars exceeds the Hubble distance c/H, or the mean free time between emission and absorption of light by stars exceeds the expansion time H^{-1}, the sky at night is dark because of the redshift effect; when, however, the mean free path is less than a Hubble distance, or the mean free time between emission and absorption is less the expansion time, visible stars cover the sky and redshift fails to darken the sky. Thus, to secure a dark night sky in the steady-state universe, a necessary condition is expansion and a sufficient condition is that the mean free time of a light ray exceeds the expansion time.[13] The mean free path is 10^{23} light-years and the mean free time is 10^{23} years. Hence, an increase in the average number of stars per unit volume by a factor 10^{13} creates a bright

[13] More precisely, $4H\tau > 1$, where $\tau = \lambda/c$ is the mean free time between emission and absorption by stars and is also the fill-up time, and $\lambda = L^3/\pi R^2$ is the mean free path between emission and absorption by stars and is also the background distance of stars covering the sky.

night sky in a steady-state universe. A reduction in the value of the Hubble term from 100 km s^{-1} Mpc^{-1} to 10^{-6} cm s^{-1} Mpc^{-1} increases the expansion time by 10^{13}; however, the sky remains dark because for every luminous star there are 10^{13} nonluminous stars covering the sky (assuming a stellar luminous lifetime of 10^{10} years).

Second, following in Bondi's footsteps, the majority of astronomers assumed that the cosmological redshift of distant sources automatically solves the riddle of cosmic darkness for all expanding universes. Many writers and lecturers took for granted that Olbers' paradox has a unique solution and accepted without question that the solution for the steady-state universe of infinite age must be the same for a big-bang universe of finite age. Almost everybody, including myself, accepted the premise that the sky is covered with stars, and most stars are invisible because of their extreme redshift. "Why is the sky dark at night?" Here was a riddle that enthralled a wide audience and contributed greatly to the popularity of modern cosmology.

Alexander Friedmann in 1922 used the theory of general relativity and proposed that the universe has expanded from an early state of high density. By the early 1930s the extragalactic redshift and distance determinations by Vesto Slipher, Milton Humason, and Edwin Hubble, and the theoretical studies by Georges Lemaître, Arthur Eddington, Howard Robertson, and Richard Tolman had securely established the expanding universe. An evolving and expanding universe originating in a state of high density has since become the dominant cosmological theme of the twentieth century. Hoyle, in a radio talk in the late 1940s, dubbed Lemaître's primordial atom and George Gamow's big crunch the *big bang* (a better term is *early universe*[14]). Gamow proposed that the early universe was not only dense but also hot, and Gamow's able colleagues Ralph Alpher and Robert Herman investigated the physics of the early universe.

Gamow at first hoped that nucleosynthesis in a hot and dense early universe would explain the cosmic abundance of the elements. This hope was dashed by the gap 5 in atomic weight, and by the realization that carbon and heavier elements can be synthesized in an evolving stellar medium of increasing density and temperature, but not in an expanding cosmic medium of decreasing density and temperature. When the universe is only seconds old, weak and strong interactions transform a quarter by mass of the hydrogen into helium; further substantial nucleosynthesis demands higher not lower temperatures and densities. The discovery by the pioneers in stellar nucleosynthesis (Alistair Cameron, Geoffrey Burbidge, Margaret Burbidge, William Fowler, and Hoyle) that evolved stars

[14] Harrison (1968). The term *early universe* properly denotes a period in the history of the universe, whereas the term *big bang* improperly denotes a discrete object or event and confuses many people.

manufacture the heavy elements added to the appeal of steady-state cosmology.

But clouds already were gathering on the horizon of steady-state cosmology. Extensive revisions in the extragalactic distance scale by Walter Baade and Allan Sandage in the 1950s greatly reduced the value of the Hubble term. The expansion time H^{-1} now exceeded 10^{10} years and big-bang cosmology at last seemed capable of accommodating astronomical chronology. The death-blow to steady-state cosmology came in 1965 when Arno Penzias and Robert Wilson detected the thermal afterglow of the early universe that had been predicted by Gamow. The low-temperature radiation flooding the universe provided clear and convincing evidence of a dense and hot early universe. Hoyle and Roger Tayler (1964), James Peebles (1966), Robert Wagoner, Fowler, and Hoyle (1967) showed that nucleosynthesis in the early universe accounted for the cosmic abundance of helium.

Popular accounts and even learned treatments continued to attribute the darkness of the night sky to the effect of redshift in an expanding universe. Bondi's redshift solution, however, which remains valid in the steady-state universe, fails in a big-bang universe. Kelvin's finite-age solution for a static universe applies more generally and is valid for most big-bang universes. In the steady-state universe the luminosity per unit volume stays constant for unlimited time owing to the continual creation of matter, and in the big-bang universe the luminosity per unit volume evolves and endures for only a limited time owing to a finite cosmic age and the conservation of energy. For dark night skies, we have Bondi's redshift explanation in a steady-state universe, and Kelvin's finite-age explanation in a static universe, and Kelvin's explanation, slightly modified, applies also in a big-bang universe. In fact, with appropriate modifications, Kelvin's explanation applies generally to expanding, static, and contracting dark-sky universes of finite age.

Energy considerations reveal a fundamental flaw in Olbers' paradox. The paradox claims that bright stars should cover the sky. This means the universe should brim with starlight in thermodynamic equilibrium: the radiation density at the surface of stars should equal the radiation density in space, and thermal radiation everywhere in space should have a temperature of order 10^4 degrees Kelvin. This would be possible only if stars had access to sources of energy sufficiently great to create a state of thermodynamic equilibrium. Let us imagine that all stars can be abruptly and totally annihilated and their energy $E = mc^2$ converted into thermal radiation evenly spread throughout space. We find the temperature of this radiation equals only a few degrees Kelvin and not the several thousand at stellar surfaces. Our universe contains insufficient energy resources to create the bright sky claimed in Olbers' paradox.

Cumbersome time-retarded integrations in an expanding metric tend to cover up the energy discrepancy and obscure the essential physics of the problem. As shown in 1964,[15] the thermodynamic differential equations analogous to the equations for radiation in an expanding cavity are much simpler to understand; these equations immediately reveal, for example, the possible contribution of radiation before the birth of stars, the importance of the luminous lifetime of stars and the age of the universe, and the irrelevance of spatial curvature. We consider a typical star in an expanding cavity having perfectly reflecting walls. The volume of the cavity equals the mass of a typical star divided by the average cosmic density of stellar matter. The differential equations of thermodynamics show that a dark night sky exists when the star fails to fill the cavity with radiation in a time less than the luminous lifetime of stars, or the age of order H^{-1} of the universe, whichever is the smaller. The *fill-up time* of the cavity is identical with the fill-up time of the universe; it equals the mean free time between emission and absorption of light rays by stars and is the characteristic time for radiation in space to come into equilibrium with the surfaces of stars.

In general, the night sky is dark – that is, uncovered by visible stellar disks – when the mean free path of starlight (between emission and absorption by stars) exceeds the size of the visible universe. Alternatively, the night sky is dark when the *background distance* (the average distance of the stars needed to cover the sky) exceeds the size of the visible universe. Here is the most general solution of Olbers' paradox. The cause of darkness, however, is not the same in different cosmolgical models. In the steady-state model of indefinite age the visible universe is the Hubble sphere, and though stars cover the sky, most of them have very large redshifts. This is Bondi's solution. In a static model of finite age the visible universe spans the maximum distance traveled by light during the lifetime of luminous stars, and through the gaps between the stars the observer sees the darkness that existed before their birth. This is Kelvin's solution. In an expanding big-bang model of finite age the visible universe has a size roughly the same as the Hubble sphere, and between the stars the observer sees the early universe. This is a modification of Kelvin's solution. The redshift of stars in an expanding universe of finite age merely achieves a state of greater darkness in a universe already dark. Through the empty gaps between the stars the observer sees the big bang darkened by redshift.

The solution in the previous paragraph supersedes most solutions proposed in the past. For example, in the hierarchical model, the mean free path of light rays is infinitely large, but the proposed solution to Olbers' paradox assumes that stars at distances of at least 10^{23} light-years

[15] Harrison (1964). This paper draws attention to the energy argument, the utility of differential equations, the failure of the redshift solution in big-bang universes, and the generality of the finite-age solution.

are visible, and hence assumes that stars have luminous lifetimes of at least 10^{23} years. Hierarchy merely achieves greater darkness in a universe already dark.

The riddle of cosmic darkness has challenged astronomers since the Copernican revolution and many solutions have been proposed. In the twentieth century, the redshift solution for an expanding universe has enjoyed wide popularity. Like many other persons, I was fascinated by the idea that invisible stars covered the sky and their invisibility proved that the universe was expanding. Later, I realized that darkness failed to prove expansion and felt, as a result, that cosmology had lost some of its dramatic appeal. But, more recently, I have realized that the darkness of the night sky has acquired more dramatic meaning than ever before in the history of astronomy. The sky is covered not by stars but by what existed before the birth of stars. We look out in space and back in time and see in every direction the big bang covering the entire sky – an incandescent big bang, enfeebled and redshifted into an infrared gloom by the expansion of the universe. Thus the twentieth century has brought Olbers' paradox to a startling conclusion: at night, covering the sky, we see the dark loom of the big bang.

References

Bondi, H. (1952). *Cosmology*. Cambridge: Cambridge University Press. (2nd edition 1960).

Clerke, A. M. (1890). *The System of the Stars*. London: Longmans.

Cohen, I. B. (1940). Roemer and the First Determination of the Velocity of Light (1676). *Isis*, **31**, 326.

Haber, F. C. (1959). *The Age of the World: Moses to Darwin*. Baltimore: Johns Hopkins Press.

Harrison, E. R. (1964). Olbers' Paradox. *Nature*, **204**, 271.

(1968). The Early Universe. *Physics Today*, **21**, 31.

(1986). Kelvin on an Old and Celebrated Hypothesis. *Nature*, **322**, 417.

(1987). *Darkness at Night: A Riddle of the Universe*. Cambridge: Harvard University Press.

Herschel, J. (1830). *Treatise of Astronomy*. London: Longmans.

Herschel, W. (1802). Catalogue of 500 New Nebulae, Nebulous Stars, Planetary Nebulae, and Clusters of Stars; With Remarks on Construction of the Heavens. *Philosophical Transactions*, **92**, 477.

Hoskin, M. A. (1976). The "Great Debate": What Really Happened. *Journal of History of Astronomy*, **7**, 169.

(1977). Newton, Providence and the Universe of Stars. *Journal of History of Astronomy*, **8**, 77.

Jaki, S. L. (1969). *The Paradox of Olbers' Paradox*. New York: Herder and Herder.

Loys de Cheśeaux, J. P. (1744). *Traité des comètes*. Lausanne: Bousequet.

Olbers, H. W. M. (1823). Ueber die Duerchsichtigkeit des Weltraumes. In *Astronomisches Jahrbuch für das Jahr 1823*, ed. J. E. Bode. Berlin: Späten.
Osterbrock, D. E. (1990). This volume.
Peebles, P. J. E. (1966). Primeval Helium Abundance and the Primeval Fireball. *Physical Review Letters*, **16**, 410.
Rindler, W. (1956). Visual horizons in World Models. *Monthly Notices of the Royal Astronomical Society*, **116**, 662.
Roberts, F. (1694). Concerning the Distance of the Fixed Stars. *Philosophical Transactions*, **18**, 101.
Smith, R. W. (1982). *The Expanding Universe: Astronomy's "Great Debate" 1900–31*. Cambridge: Cambridge University Press.
Thomson, W. (Lord Kelvin) (1901). On Ether and Gravitational Matter through Infinite Space. *Philosophical Magazine*, **2**, 161.
Wagoner, R. V., Fowler, W. A. and Hoyle, F. (1967). On the Synthesis of Elements at Very High Temperatures. *Astrophysical Journal*, **148**, 3.

4

The part played by Mach's Principle in the genesis of relativistic cosmology

J. B. Barbour

The expression "Mach's Principle" was coined in 1918 by Einstein (Einstein 1918) to denote a principle which he claimed was a "generalization of Mach's requirement that inertia be traced back to an interaction between bodies." Any talk about Mach's Principle is fraught with problems, since it is notoriously difficult to get any two physicists – let alone philosophers – to agree on the precise content of the principle or on the extent of its validity. Since we are concerned here with the birth of modern cosmology, it would be quite inappropriate for me to attempt either an extended critique or a detailed analysis of the extent to which Einstein succeeded in his own Machian aims. Instead, I shall merely allow myself a few comments on why, in my view, discussion about Mach's Principle is so confused and then concentrate on what Einstein took the Machian requirement to be and how the attempt to fulfill this requirement helped him, first, to find his general theory of relativity, and then, in the framework of that theory, to construct in 1917 the first scientifically based model of the universe, creating simultaneously in a brief paper of just eleven pages the strikingly bold framework of relativistic cosmology.

Let me first make a general remark. Cosmology is a unique subject and, perhaps more than any other, forces us to consider the question of foundations. The viewpoint that Mach expressed so cogently and influentially was that there are in fact no external foundations. There can be no foundation on which the world rests. The observed world must itself supply the terms of its own description. It must be conceived to reside and unfold *self-referentially* in nothing. The Machian requirement that Einstein so intensely felt he had to meet sprang from such a conviction. Through it he wanted, in his own expression (Einstein 1922: 62), to close "the series of causes of mechanical phenomena." He wanted to construct a causally self-contained world with observable causes of all observable effects.

In this connection, before turning to details, I should like to draw attention to a much earlier occasion in the history of physics when distinctly Machian ideas played a role that was especially relevant for the eventual emergence of relativistic cosmology. I am referring to the discovery by Kepler in the period 1600 to 1605 of his first two laws of planetary motion. These were, of course, crucial for the discovery by Newton of his law of universal gravitation, without which there could clearly never have been either the general theory of relativity or relativistic cosmology. Careful reading of Kepler's *Astronomia Nova* (1609), in which he recounted how he made his great discoveries, reveals many remarkable parallels with Machian attitudes to the basic problems of motion. The similarity is not at all fortuitous and is closely related to the point made in the last paragraph, namely, that the world and its parts do not rest on any "external" foundations. Mach rejected the notion of an *invisible* background space as the agent responsible for the phenomenon of inertial motion and was therefore forced to seek some other real (and *visible*) agent to fulfill this crucial role. In his time, the distant masses of the universe were the only natural candidates, and this led to his conjecture that they, and not absolute space, were what guided local bodies in their inertial motions. But, centuries earlier, Kepler was forced to embark on a remarkably similar quest for the ultimate determinants of observed motions by Brahe's assertion, deduced from the apparent motions of the comets, that the crystal spheres hitherto supposed to carry the planets simply could not exist. Kepler completely accepted Brahe's argument and it profoundly influenced his attitude to the problem of the planetary motions. For it appeared that the planets must somehow find their way through the completely featureless ether of interplanetary space along quite definite paths but without anything to guide them or move them. He was forced to develop a conceptual scheme in which the planets's motions were directly determined (both generated and guided) by the bodies known to exist in the universe, above all the Sun and the stars. This makes the parallel with the Machian program evident. In its time, Kepler's basic approach was every bit as radical and revolutionary as Mach's much later proposal. For in the framework of the ancient astronomical techniques which Kepler inherited from Copernicus the vital role of the Sun in governing the planetary motions was completely hidden. (For very understandable reasons, which have to do with ancient astronomical traditions and the specific eccentricities of the various planetary orbits, Copernicus actually centered the entire planetary system on the void second focus of the Earth's orbit. The Sun is not even shown in any of his diagrams explaining the motions of the planets!) Moreover, Kepler's "Mach's Principle" was carried through to successful implementation in the form of his first two laws of planetary motion, whereas Mach only made his proposal but took it no further himself. It would be very easy to say a great deal more about this intriguing parallel, including also some very striking similarities between Kepler's long Odyssey in search of the laws of planetary motion and Einstein's equally long one in search of his general theory of relativity, but that would leave no time at all for my main topic and I have written about the subject elsewhere (Barbour 1989: chapter 6). Here, I shall say only that Kepler's achievements, which were not properly appreciated for over half a century, were recognized very early in the University of Bologna, at which there seems to have been "a more or less continuous Keplerian tradition" (Russell 1964) spanning most of the period between the discovery by Kepler of his laws and their dynamical interpretation by Newton three-quarters of a century later.

Coming now to the main topic, let me first mention and, I hope, dispose of the prime source of confusion on the subject of Mach's Principle. It is Einstein's curious assertion, first made in 1912 (Einstein 1912) and then repeated or implied on numerous occasions in his subsequent writings (for example, Einstein and Grossmann 1913; Einstein 1917) that Mach wished to establish the *relativity of inertia*. By *inertia* in this context Einstein quite definitely meant the inertial mass that appears in Newton's second law, i.e., the coefficient that multiplies the acceleration. It is not too difficult to see how Einstein came to believe he could explain inertial resistance to acceleration (his argument will be sketched near the end of this paper), but I am still at a loss to understand why Einstein thought that Mach sought to establish relativity of inertia in such a sense. There is nothing in Mach's writings to suggest it; quite the opposite. I almost think that on this point Einstein was the victim of a semantic confusion. For Mach, who, after all, created a beautiful operational definition of inertial mass (Mach 1960: 264ff), saw no problem at all in that use of the word inertia. What always concerned him were concepts at a much more primitive level, those of position and velocity. His concern was not with *inertial resistance* but with the *law of inertia* (Mach 1960: 271ff). He was concerned solely with what is often called *kinematic relativity*. Bodies can be observed only relative to other bodies. In a universe in which all bodies are in a state of relative motion, how can objective meaning be given to the idea of any definite motion, let alone a uniform one in a straight line? It was this problem that made Newton introduce the notions of absolute space and time.

If we examine Einstein's work closely, we see that in reality his notion of relativity of inertia did not significantly influence the basic structure of the general theory of relativity, though, as we shall see, it did decisively influence the construction of his cosmological model (and, perhaps even more strongly, de Sitter's rival model). The central problem in the creation of the general theory of relativity did not revolve around relativity of inertial mass but around *relativity of frames of reference*. There is no doubt that in addressing the question of frames of reference and the fact that some appeared to be distinguished compared with others Einstein was addressing the same problem as Mach. However, even here there was a very important difference of approach that was brought about by two factors. The first was the evolution of ideas during the more than thirty years which elapsed between Mach's original questioning of the law of inertia and Einstein's attempt to do something about it. The second was the dramatic effect of what Einstein was later to call "the happiest thought of my life" – the discovery of the equivalence principle.[1] This had a most profound influence on the way in which Einstein attacked the problem of the frames of reference.

Einstein defined an inertial frame of reference, which was absolutely central to his special theory of relativity, either as a frame in which Newton's laws were found to hold or, more vaguely and more generally, as one in which the laws of nature were found to take their simplest form. For quite some time their existence had been felt to be paradoxical in view of the manifest relativity of motion that Mach, above all, stressed so strongly. Although Mach himself produced no concrete theory of the origin of these mysterious frames of reference, two very characteristic remarks of his give a pretty clear hint of the direction in which he was thinking. One was made in connection with his criticism of Newton's bucket experiment (Mach 1960: 284): "The universe is not *twice* given with an earth at rest and an earth in motion; but only *once*, with its *relative* motions, alone determinable." The second remark (p. 286) was: "When we say that a body preserves unchanged its direction and velocity *in space*, our assertion is nothing more or less than an abbreviated reference to *the entire universe*." I would like to draw special attention to the great prominence given in these remarks to the universe as a whole, which Mach evidently regarded as a single dynamical entity. Mach reformulated the aim of dynamics: it was not to provide laws of motion of individual bodies in space and time but rather to study the evolution in time of the relative separations of the bodies of the universe. In modern terms, the fundamental dynamical variables are not position vectors of bodies in space but all the relative distances between the various bodies in the universe.

The logic of Mach's approach was to dispense altogether with coordinate systems and frames of reference and concentrate instead directly on the *universe as a whole*, describing it by a relational law containing only relative distances and relative velocities. Distinguished frames of reference would then arise only if we fix our attention on local bodies and attempt to describe them in coordinate systems chosen to make their motion appear particularly simple, as in Newton's first law. Since the motions are in reality taking place with respect to the universe at large, our distinguished coordinate systems will actually be tied to and determined by the same universe. This, in essence, was the Machian explanation for the

[1] For the charming account of how it happened, see Pais (1982: 177ff).

observed coincidence of the family of inertial frames with frames nonrotating with respect to the distant stars. In such an approach one must clearly distinguish between a basic, let us call it primordial, relational law of the universe as a whole and effective local laws that are recovered from the primordial law by referring local motions to special frames. Clearly, the first step, which Mach never took unfortunately, was to find the primordial law of the universe as a whole.[2]

I have sketched this Machian approach to highlight the fact that Einstein adopted a totally different one. This I believe is a second major reason why discussion of Mach's Principle is so tangled. Although there are several quite clear hints in Einstein's papers that local physics is crucially influenced by the matter in the universe at large (especially in Einstein 1912, 1916) they remained only implicit (in the background inspiration), and the *explicit* consideration of the universe as a single dynamical entity occurs remarkably late in Einstein's papers and only *after* his theory was essentially complete. As we shall see, this had an important consequence: whereas in the period up to 1916 Einstein believed that he was implementing Mach's Principle automatically and dynamically, by the very structure of his equations, in the later period, when the dynamical structure of his theory was already complete, he was forced to attempt to implement his Machian idea by means of boundary conditions.

To highlight the lateness at which Einstein turned his attention seriously, that is, as a matter of first priority, to the relationship between the local dynamics and the universe at large, let me remind you of Aristotle's spherical cosmos, created well over 2,000 years before Einstein's. The similarities between the two are really rather striking: both are spatially spherical and self-contained and both extend in time infinitely far into the past and the future. Both exist in nothing. Aristotle is careful to point out that outside his cosmos there exists nothing at all – not even space or time. What is especially interesting is that both cosmological models – Aristotle's and Einstein's – were created in response to what may be called Machian considerations. There was, in fact, a most interesting and illuminating pre-run of the absolute/relative debate in antiquity. It was stimulated very largely by the atomists. Aristotle was unhappy about their idea of atoms in a void and often criticized the void in distinctly Machian terms (especially in his *On the Heavens* and *Physics*; for a fuller discussion of these questions and detailed references, see Barbour 1989). His cosmology, with its well-defined outer shell and equally well-defined center was conceived explicitly to provide a framework for his laws of motion, just as Newton invoked absolute space. Thus, in Aristotle's case the form of his cosmos and the specific laws of motion crystallized together. The relation between the two entered explicitly from the

[2] Bertotti and I have shown how a Machian program of the kind just outlined can be carried out in detail (Barbour 1974; Barbour and Bertotti 1977, 1982).

beginning. In Einstein's case it was quite different – the laws of motion came first, the explicit cosmological considerations very late.

It is not hard to find reasons for this difference. Jacques Merleau-Ponty (Merleau-Ponty 1982) has noted that throughout the nineteenth century cosmology was almost totally ignored by the physics and astronomy community. A rather positivistic approach to the natural sciences was adopted – it was all to do with measurement of the properties of matter as observed around us and description of the results obtained by mathematics by a process of induction. I believe a most important factor in this connection was the practical success of Newton's concepts of absolute space and time, particularly after they had been made epistemologically more respectable by the concepts of inertial frames of reference. In fact, absolute space and time provided a kind of surrogate cosmology and this is what made it possible for serious concern with cosmological questions to be deferred for such a remarkably long time.

A final point to be made about the difference of approach between Einstein and Mach is that one can find remarkably little explicit concern anywhere in Einstein's writings for the issues relating to the *relativity of motion* that loom so large in Mach's writings. The problem I have referred to as kinematic relativity is seldom, if ever, directly addressed or even mentioned. Einstein simply does not consider the question of the practical determination of position of a given body by means of the other bodies in the universe. In contrast to his contemporary Weyl, for example, he hardly uses the expression "relativity of motion" (he almost always refers to "relativity of inertia"). In fact, Einstein circumvented the basic problems of position determination by his use, from the very beginning, of the concept of *frames of reference*, and he attacked the problem of their distinguished nature, not along the lines just outlined, i.e., from a primordial relational law, but by questioning whether they were distinguished at all. Moreover, for a very long time his outlook remained essentially local – he attempted to abolish the distinction at a local level. There are no doubt several good historical reasons that can explain Einstein's preference for a local rather than a global approach. First among these must obviously have been the rejection of instantaneous action at a distance and the associated rise of field theory, with which Einstein himself was so closely associated. In addition, there was the general absence of direct concern with cosmology just mentioned. However, equally if not more important was the totally new principle that Einstein found – the equivalence principle. I have said enough about what Mach would have done and Einstein might have done. It is time to consider what Einstein actually did.

His overall strategy, from which he never really wavered until after the theoretical structure of general relativity was complete, is clearly revealed in his first comments on the subject of gravitation in 1907. At the end of a review of the special theory of relativity (Einstein 1907), he commented that hitherto one had required the laws of nature to be independent of the state of motion only for *unaccelerated* frames of reference. He then asked: "Could one suppose that the principle of relativity is also satisfied for systems moving relatively to each other with acceleration?" He added that "This question must occur to anyone who has followed the applications of the principle of relativity up to the present time." It was, of course, the equivalence principle, only given that name a few years later, that enabled Einstein to pose such a seemingly absurd question. From the fact that "in a gravitational field all bodies are accelerated equally" Einstein argued that "at the present state of our knowledge" there are no grounds for believing that there are any respects in which a frame of reference accelerated uniformly in a region free of a gravitational field differs from one at rest in a homogeneous gravitational field.

Discussion of Einstein's work has tended to concentrate on the way in which he used the equivalence principle to draw some first conclusions about the laws of physical processes in homogeneous gravitational fields. But the fact that he was from the start simultaneously following a further aim becomes clear from his second paper on the subject, published in 1911. There he said (Einstein 1911): "In such an approach one cannot speak of the absolute acceleration of the coordinate system any more than in the special theory of relativity one can speak of the *absolute velocity* of the system" (original italics).

The drift of Einstein's thought is now clear. Whereas the logic of Mach's comments called for explicit derivation of the distinguished local frames of reference from a relational law of the cosmos as a whole, Einstein is working towards elimination of the problem of the distinguished frames by asserting that they are not really distinguished at all. The first step, more or less successfully accomplished in the earliest papers, was to dispose at least of the frames in a state of uniform acceleration.

But the success was in fact paid for at a price and this price casts a lot of light on the problems Einstein faced in his approach to the Machian problem. For the original relativity principle purports to be a universal principle, that is, it asserts that all physical processes unfold in exactly the same way in all the allowed equivalent frames of reference. However, in the case of Einstein's extension to uniformly accelerated frames of reference or homogeneous gravitational fields the laws of the gravitational field itself are simply not covered. The principle merely gives one information about the way in which nongravitational processes unfold in a given and, in fact, homogeneous gravitational field. Whereas in the case of the original principle of relativity external factors were of no concern, the success of Einstein's extension depends crucially on things outside the system.

There is an essential incompleteness in all of Einstein's attempts to explain the effects of so-called inertial forces in terms of a gravitational field. Indeed, I do not think Einstein ever addressed this question thoroughly and directly. He never attempted to spell out in explicit detail precisely how the universe at large produced the particular gravitational field that would permit him to say that all apparently inertial effects are really gravitational and that therefore distinguished frames do not occur at all. This was a problem that, probably wisely, Einstein kept on deferring.

The equivalence principle and the scope he saw in it for solving the Machian problem were above all important for Einstein because they gave him the confidence to think the unthinkable.

One of the clearest examples of this can be seen in what Stachel (1980) has called the "missing link," namely, the precise insight that guided Einstein to the notion of a four-dimensional Riemannian manifold with a genuinely non-Euclidean geometry. A passage in a letter to Sommerfeld written in September 1909 shows which consideration above all it was that kept moving him forward into new territory. He wrote (Stachel 1980): "The treatment of the uniformly rotating rigid body seems to me to be of great importance on account of an extension of the relativity principle to uniformly rotating systems along analogous lines of thought to those that I tried to carry out for uniformly accelerated translation in the last section of my paper published in the *Zeitschrift für Radioaktivität.*" Einstein was probably prompted to this remark by a "paradox" that had just been discovered by Ehrenfest when he attempted to consider the geometry of a rotating disk in the framework of special relativity. For the length of a rod placed radially on a rotating disk should not undergo any Lorentz–Fitzgerald contraction relative to a nonrotating frame of reference because it would always be moving transversely. However, rods placed around the rim should be subject to a contraction, and therefore such rods would reveal a circumference of the circle that was more than π times the diameter. Stachel has marshalled evidence which shows that Einstein took this result, with its implications of the need to consider non-Euclidean geometry, very seriously. He had to, since it was his serious intention to show that all frames of reference should be equally valid for the description of nature.

Einstein's conclusion from the example of a rotating disk of the need to consider non-Euclidean geometry is rather ironic, since in the light of mature general relativity he had actually drawn an invalid conclusion. This came about because Einstein at that time used the concepts of frames of reference and transformations between them in a manner that does not correspond to the correct transformation laws of tensor calculus in general relativity.[3] Specifically, he was attempting to measure the geometry in a

[3] Norton (1984) makes some interesting comments on Einstein's use of frames of reference in his detailed study of Einstein's discovery of the equations of general relativity. His paper contains references to all of Einstein's important papers on general relativity in the period 1912 to 1915 together with unpublished material and letters.

spacelike hypersurface with measuring rods that do not move orthogonally to it in four-dimensional space-time. As a result, he generated a spurious non-Euclidean geometry of the instantaneous spacelike hypersurfaces, which must, of course, remain flat in both the original frame of reference and the rotating frame. However, this error, if we can call it such, was extremely helpful.

The definitive breakthrough on Einstein's part in the summer of 1912 to a four-dimensional space-time manifold with non-Euclidean geometry undoubtedly owed much to non-Machian factors but here again the possibility of extending the relativity principle to all frames of reference seems to have been the most important reason for Einstein's belief that he had at last found the correct framework. The key idea was his ansatz for the law of motion of a test particle in the form of the geodesic principle

$$\delta \int ds = 0, \; ds = (g_{\mu\nu} dx^{\mu} dx^{\nu})^{\frac{1}{2}},$$

where $g_{\mu\nu}$ is the metric tensor of the manifold, and his realization that such a law would take the identical form in absolutely any system of coordinates (Einstein and Grossmann 1913). In several places Einstein expressed the opinion that by itself this result already eliminated the problem of the distinguished frames of reference (see, for example, Einstein 1914). Einstein's concept of the Machian requirement had now become much more precise but it also, I believe, underwent a certain modification of content which prepared the way for Einstein's acceptance in 1918 (Einstein 1918) of Kretschmann's argument (Kretschmann 1917) that general covariance had no physical content but was merely a formal requirement of mutually consistent description of a unique object from different points of view.

Despite his assertion that his general framework by itself solved the Machian problem, Einstein was very well aware that he had solved only half of his problem. It was also necessary to find the equations of the gravitational field itself. We have here the remarkable story of Einstein's initial instinctive belief that they too must be generally covariant, the amazingly near miss on the part of Grossmann and himself when they attempted in their first joint paper in 1913 to find satisfactory field equations of such form (Einstein and Grossmann 1913), and then the invention by Einstein of his notorious argument by which he attempted to prove that the field equations of the general theory of relativity, the name he already gave to his incipient theory, could not themselves be generally covariant. This is a most tangled story into which it would be impossible to delve deeply here.[4] All that I will say is that despite all his difficulties and mistakes Einstein could never get the ideal of complete general covariance out of his head. He was quite capable of giving a proof that his current equations with only restricted covariance must be correct but then almost

[4] For a detailed discussion, see Norton (1984) and also Earman and Glymour (1978).

with the same breath saying the theory should be completely covariant. And the reason for the attachment to general covariance was clear – it alone would guarantee realization of his Machian ideal.

To cut a long story short, Einstein did finally return to general covariance and by the end of November 1915 had at last completed his great work, the general theory of relativity. In the famous summary of his theory published in 1916 in the *Annalen der Physik* (Einstein 1916) the very greatest emphasis is placed on the need for complete general relativity in order to resolve the problem of distinguished frames of reference. At the conclusion of his discussion of his famous example of two liquid spheres in a state of relative rotation, the one flattened but the other not, Einstein says:

> Of all imaginable spaces ... in any kind of motion relatively to one another, there is none which we may look upon as privileged *a priori* without reviving the above-mentioned epistemological objection. *The laws of physics must be of such a nature that they apply to systems of reference in any kind of motion.*

This statement marks the apotheosis of the relativity principle in its role as implementation of the Machian requirement directly and dynamically through the basic structure of the theory. It is the *point omega* towards which Einstein had been working methodically for nearly nine years – yet two years later he was to concede that he had not distinguished sufficiently clearly between the relativity principle and the Machian requirement and he was led to introduce a quite separate Mach's Principle (Einstein 1918). Before we consider this surprising turn and its consequences, a brief review is appropriate. Precisely how important was the Machian factor to Einstein as compared with the numerous other strands that he wove together in his theory? One must here distinguish between tools used to do a job and the inspiration to undertake it. Dealing with the tools first, I was very struck when going through all the Einstein papers just how many solid and well-established results from modern physics and mathematics Einstein did use – and, moreover, how effectively he used them. What is almost breathtaking and lends his theory such grandeur is the way in which he consistently applied the lesson he had learnt from special relativity – namely, to achieve the result you want, do not be afraid to tamper with space and time. In fact, it seems to me that Einstein exhibited an almost ruthless willingness to do just whatever he pleased to space and time provided only he could then show that the laws of nature would take exactly the same form at every point of space-time and in any frame of reference in which he might care to examine them. In terms of the concrete steps taken and the actual tools used in the process, special relativity and a whole slew of results that came with it – above all ones relating to Maxwell's theory and Minkowski's space-time formulation – were vastly more effective in establishing the final shape of the creation than was the Machian ideal. But if we ask how it was possible that Einstein ever came to create such an incredible theory, so utterly unlike anything even his most brilliant contemporaries were prepared to consider, the answer is clear. It was the Machian inspiration, never really precisely grasped, that provided the touch of magic and constantly drew him on. In the words of Keats, it was a case of "Heard melodies are sweet, but those unheard are sweeter."

Now we come to the cosmology. The manner in which Einstein wrote the introduction to the review of the *Annalen der Physik* indicates that at that time he still believed the Machian requirement was automatically satisfied by the very structure of the theory, i.e., by its general covariance. The first seeds of doubt arose from reflection on the solutions to his equations by means of which the planetary motions were described. For in these solutions the $g_{\mu\nu}$ were assumed to tend at infinity to the Galilean values

$$\begin{pmatrix} -1 & 0 & 0 & 0 \\ 0 & -1 & 0 & 0 \\ 0 & 0 & -1 & 0 \\ 0 & 0 & 0 & 1 \end{pmatrix},$$

the local concentration of matter in the sun merely giving rise to rather insignificant distortions in its vicinity. It seems to have dawned on Einstein rather slowly that the Machian ideal which had sustained him through such travail appeared to be almost as remote as ever.

This brings us to the final part of the story and Einstein's last attempt to grasp his goal. This attempt too was just as ironic as the first failed attempt. Yet again there was no implementation of Mach's Principle but once again there was a momentous consequence. This time it was the science of relativistic cosmology. It was born because, for the very first time in his work, Einstein addressed himself directly to the undertaking that had from the beginning been implicit in the Machian enterprise, namely, the derivation of local physics from the dynamics of the universe as a whole.

At this stage of the story Einstein's notion of the relativity of inertial mass became really decisive and truly dictated the steps he took. He expressed his article of faith in these words in his 1917 paper on cosmology (Einstein 1917): "In a consistent theory of relativity there can be no inertia *relatively to 'space,'* but only an inertia of masses *relatively to one another.*" From this, Einstein drew an important conclusion: "If, therefore, I have a mass at a sufficient distance from all other masses in the universe, its inertia must fall to zero." This single remark was to give rise to not only Einstein's cosmological model but also de Sitter's, as we shall shortly see.

Einstein's first idea, which he was developing in September 1916 at the time of a visit to Leiden in Holland, where he had several very important discussions with de Sitter, was to find a solution for his equations such that sufficiently far from all matter, i.e., at infinity, the criterion just formulated should hold.

He noted that in the general theory of relativity the equation of motion of a particle indicated that for a particle of rest mass m and 4-velocity dx^α/ds the quantity

$$m\sqrt{-g}\, g_{\mu\nu}(dx^\nu/ds)$$

can be identified as the negative momentum of the particle. Einstein then supposed the case of a spatially isotropic metric and wrote the line element in the form

$$ds^2 = -A(dx_1^2 + dx_2^2 + dx_3^2) + Bdx_4^2.$$

From this he concluded that the momentum of the particle would, for small velocities, be proportional to

$$mA/\sqrt{B},$$

so that $mA/\sqrt{B}$ was a measure of the inertia of the particle. He now wanted to achieve that this inertia would tend to zero at spatial infinity. He imposed the coordinate condition $\sqrt{-g} = 1$ and concluded then that A must diminish to zero while B should tend to infinity. He also considered more general situations without assumption of spatial isotropy of the metric, and, as de Sitter reports, contemplated general boundary conditions at infinity in space and time of the symbolic form

$$\begin{pmatrix} 0 & 0 & 0 & \infty \\ 0 & 0 & 0 & \infty \\ 0 & 0 & 0 & \infty \\ \infty & \infty & \infty & \infty^2 \end{pmatrix}.$$

For Einstein, the virtue of such values of the metric was that they were invariant for all transformations $x_\mu \rightarrow x'_\mu$ for which at infinity x_4 is a pure function of x'_4. Such boundary conditions would thus come very close to preserving the general covariance of the field equations. According to de Sitter,[5] Einstein took such boundary conditions very seriously at the time of his visit to Holland. He called them *natural values*.

Back in Germany, Einstein attempted with the help of the mathematician Grommer to see if any realistic matter distribution could lead to the

[5] De Sitter published several important papers in 1916/1917 (de Sitter 1916, 1917). Besides their intrinsic value in giving the de Sitter solution (1917b, 1917c) they were instrumental in acquainting English scientists with Einstein's theory and also cast a very interesting light on Einstein's approach to Mach's Principle at the time.

metric he required. He did this by the simple device of assuming the metric he required and then calculating from it the energy–momentum tensor that must correspond to it in accordance with the field equations. However, the matter distribution which Einstein and Grommer obtained seemed to be in crass contradiction with the astronomical facts as then known. The status of the nebulae later clearly recognized to be island universes like our own Galaxy and endowed with remarkable velocities of recession was still very obscure (this topic is amply covered by other contributors to this volume), and the galaxies played no part at all in Einstein's thoughts at that time. In fact, his "universe" was effectively not much larger than our Galaxy and its dominant feature, to which Einstein attached much importance in his 1917 paper, appeared to be the remarkably small (compared with the velocity of light) velocities of the stars relative to each other. Einstein concluded from this that, in a suitable frame of reference, the energy–momentum tensor of the matter in the universe must have all components very small compared with the energy density (the 44 component). This did not at all agree with the theoretical requirements that followed from his and Grommer's calculations, and Einstein, seriously misled by the then readily available astronomical knowledge, abandoned such an approach altogether. However, the boundary condition idea did have an important influence on de Sitter, as we shall see.

Einstein then hit upon the idea of doing without boundary conditions altogether. "For," he said, "if it were possible to regard the universe as a continuum which is *finite* (*closed*) *with respect to its spatial dimensions*, we should have no need at all of any such boundary conditions." Thus, we have the explicit formulation of the idea that the universe is a completely self-contained entity and everything that happens within it exists merely in relation to other happenings. All exists truly within nothing.

I shall not attempt to describe in detail Einstein's cosmological model, which he published in February 1917 but mention merely one or two key aspects. First, to construct it at all, Einstein was forced to introduce his famous cosmological term by adding to the field equations

$$G_{\mu\nu} = \varkappa T_{\mu\nu}$$

the term $-\lambda g_{\mu\nu}$ on the left-hand side with an undetermined coefficient λ. The aim of this term, which introduced an effective repulsion of matter, was to ensure that the equations could not have a reasonable cosmological solution without the presence of matter. The necessary presence of matter would then ensure, in Einstein's opinion, that inertia was not merely influenced by matter but completely determined by it. Simultaneously, the presence of the cosmological term would counteract the gravitational attraction of the matter and permit a stable solution. The resulting solution, which only later was recognized to be unstable, represented a finite spherical world in its spatial dimensions but existed from the infinite past to the infinite future in time. It was therefore called the cylindrical world.

Although the model was explicitly constructed with the primary aim of realizing the Machian requirement, one cannot help marvelling when reading the rather brief paper how much of the basic structure of modern relativistic cosmology is contained in it either explicitly or implicitly. I suspect de Sitter deserves a fair measure of credit for this. One gets the impression that in his discussions with Einstein he shifted him towards a scheme that was simultaneously realistic (for its time) and extremely fruitful. Although Einstein apologized to the reader for taking a "rather rough and winding road," the paper is wonderfully lucid and elementary. One wonders if a complete new science was ever created with such effortless ease and, moreover, as a mere by-product. Of course, the work of Riemann and other geometers, coupled with the new dynamical theory, was what made it possible, and the consideration of cosmological models that exploit closed spherical spaces was nothing new. In fact, the youthful Schwarzschild wrote a beautiful little paper on the subject in 1900 (Schwarzschild 1900), which, since it is referred to by de Sitter in one of his papers, may well have had an important influence on de Sitter and, through him (or directly), on Einstein.

What the 1917 paper does show is how naturally the complete structure of relativistic cosmology appeared as soon as there was a genuine stimulus to consider a topic that, prior to the scientific revolution, had been as central to scientific thought as it is possible to imagine but had then been banished to the periphery for about two centuries while scientists busied themselves with seemingly more mundane matters. There is a rather pleasing pattern to the whole period from Aristotle to Einstein – from the one spherical cosmology to the other. Ancient science began with cosmology, which provided a conceptual framework in which solid results of genuine science were gradually accumulated. Eventually, these results blew the old cosmology to pieces, and for a long time scientists worked away in their laboratories without turning round to consider the wider world. No one concentrated more intently on the local laws of nature than Einstein himself. Finally, Mach's insistence on the ultimate question – whence comes the basis of all these local laws? – forced Einstein to look wider and, almost miraculously, all the pieces came together in a general cosmological framework that was every bit as pleasing as the old one but, in contrast to it, had been built from the ground up, not the heavens down. What is especially interesting in the present connection is that all three of the cosmological frameworks that have dominated scientific enquiry in the entire period of its existence – Aristotle's cosmos, Newton's absolute space and time, and relativistic cosmology – crystallized out of deep consideration of the nature of motion (for a detailed discussion of the arguments that guided Aristotle and Newton see Barbour 1989: chapters 3 and 11).

I must now come to the final part of the story. Whereas Einstein's model proved, because of its instability, to have only transitory interest as an actual model of the universe – which in no way diminished its value as a paradigm of model construction – the model which de Sitter was led to construct through his explicit distrust of Einstein's dream of the "relativity of inertia," has proved to be one of the most important in the history of modern cosmology, in particular, for example, in steady-state theory, and, much more recently and still very topically, in the inflation hypothesis.

De Sitter's point of departure was the boundary conditions (display above, p. 58) that Einstein was considering in September 1916. De Sitter had several reservations about Einstein's whole approach. He thought the Machian "relativity of inertia" was a will-o'-the-wisp with which one could dispense without any detriment to the general theory. He disliked the extraordinary masses that at the time Einstein was forced to invoke. He called them "supernatural masses" and "quite as objectionable as absolute space." He was also disturbed by the unsymmetrical treatment of space and time forced upon Einstein by his Machian requirement, according to which the inertia of a particle should become zero at spatial infinity, at which particular boundary conditions needed to be imposed, whereas there was no analogous requirement at temporal infinity. This seemed to him to go right against the spirit of relativity theory. He felt that the only natural and appropriate boundary conditions which one could impose were that the $g_{\mu\nu}$ should vanish at both spatial and temporal infinity. De Sitter was therefore thinking in terms of a cosmological model with a space-time symmetry of a higher order than the one towards which Einstein was working. Indeed, the fact that he found a solution at all, especially one having the very high degree of symmetry which makes it such a remarkable (if not to say mysterious) geometrical entity and world model (cf. Schrödinger 1956) is a direct result of de Sitter's reaction to the incomplete, Machian-dictated symmetry of Einstein's solution. De Sitter was fully conscious of this, and his papers not only gave the solution but also initiated the study of its properties.

Einstein and de Sitter remained in touch and almost as soon as Einstein had constructed his cosmological model de Sitter produced his rival model. It was presented in a paper (de Sitter 1917b, expanded account in de Sitter 1917c) only seven weeks later than Einstein's.

The key idea of the paper, which de Sitter credited to Ehrenfest, was that of making "the four-dimensional world spherical in order to avoid the necessity of assigning boundary conditions." De Sitter's world was actually hyperbolical but by employing an imaginary time coordinate he made it formally spherical to bring out a very close parallel between the cylindrical model of Einstein with its three-dimensional spherical space. Einstein had embedded his three-dimensional space as the surface of a sphere in a four-dimensional Euclidean space. De Sitter noted that if as coordinates to express Einstein's solution one employed the stereographic projections in this higher dimensional space then the metric at infinity tended to the values

$$\begin{pmatrix} 0 & 0 & 0 & 0 \\ 0 & 0 & 0 & 0 \\ 0 & 0 & 0 & 0 \\ 0 & 0 & 0 & 1 \end{pmatrix}.$$

In order to eliminate the asymmetrical and hence objectionable unity in the 44 position, de Sitter obtained his model by embedding a four-dimensional hypersphere in a five-dimensional space. He then obtained a solution in which all the components of the metric in the stereographic projections had the values zero at infinity. He further emphasized the analogy between the two models by giving the explicit parallel expressions for the metric. For Einstein's model the metric was

$$g_{ij} = -\delta_{ij} - \frac{x_i x_j}{R^2 - \Sigma x_i^2},\ g_{44} = 1,\ g_{0i} = 0,\ \Sigma x_i^2 = x_1^2 + x_2^2 + x_3^2,$$

while for his own it was

$$g_{\mu\nu} = -\delta_{\mu\nu} - \frac{x_\mu x_\nu}{R^2 - \Sigma x_\mu^2},\ \Sigma x_\mu^2 = x_1^2 + x_2^2 + {}_3^2 + x_4^2.$$

It should be noted that the extremely close parallel is a little spurious, having been achieved by the use of one imaginary component. However, it does emphasize the way in which the de Sitter solution arose as an explicit reaction against Einstein's Machian requirement.

For models with the very high degree of symmetry assumed by Einstein and de Sitter it is, of course, comparatively easy to solve the field equations. In fact, what the field equations did for both Einstein and de Sitter was to indicate what (averaged) matter was needed to permit the space-time geometry required by the respective models. The answer that de Sitter found in his case was truly the final death knell for Einstein's attempts to implement Mach's Principle. For he found "the remarkable result, that now no 'world-matter' is required." His four-dimensional analogue of the Einstein cosmos did not contain any matter at all. It was completely empty! But for Einstein the whole point of the introduction of the cosmological term had been to enforce the presence of such matter.

Before concluding I should mention the two remarkable papers published by the Russian Friedmann[6] in the *Zeitschrift für Physik* in 1922 and 1924 (Friedmann 1922, 1924). Together with the work we have just discussed, these two papers must surely rank as among the most important ever published in the field of cosmology. Friedmann's work, like that of Einstein and de Sitter, belongs to the more purely theoretical stage of relativistic cosmology that predates the central concern with galactic redshifts. (The redshifts started to become a major topic of discussion between the publication of Friedmann's two papers but are not discussed by him.) Although Friedmann refers to the Machian problem only to the extent of a passing remark that "the problem of centrifugal force" might cast light on the problem of choosing between the numerous different world models that his analysis had revealed as being possible, his work is to be seen as a very natural extension of the model building of Einstein and de Sitter. The overall pattern of both these papers clearly owes a very great deal to their work, and Friedmann says explicitly that his aim is to generalize their work and find solutions to Einstein's equations, (including Einstein's and de Sitter's as special cases), in which the curvature with respect to three coordinates, which serve as space coordinates, is constant with respect to these coordinates but is a function of the fourth, which serves as the time. I think it is therefore fair to say that the last positive service performed by Mach's Principle was that it helped to bring forth, yet again as a by-product, cosmological models that were explicitly evolutionary.[7] This is perhaps the most distinctive feature of modern cosmology. Incidentally, Friedmann's papers really emphasize to an extraordinary degree the point that I made at the start about the universe being conceived to unfold self-referentially in nothing. Rather over half way through Friedmann's first paper, which is written throughout in a very matter-of-fact way but with a lucidity worthy of Einstein, there comes an innocent sounding little definition which must, I think, bring any reader up with a jolt. The only concession to effect that Friedmann makes is to put the concept he is defining in the expanded type often used for a definiendum in German. He considers the time of increase of the radius R of the world from the value 0 to some given radius R_0 and calls it *die Zeit seit der Erschaffung der Welt* – the time since the creation of the world. The cosmos that Friedmann described is completely self-contained yet evolves entirely lawfully. It surely resides in nothing. It even springs out of nothing.

Let me go back to de Sitter's paper. It carried a title with possibly the

[6] The editors of the *Zeitschrift für Physik* generated some confusion about the spelling of the surname: Friedman in 1922, Friedmann in 1924. Today the transliteration from Russian into English would be Fridman. All three spellings can be found, but Friedmann is the most common.

[7] It seems to me quite likely that Friedmann was stimulated to his study by the almost casual sentence with which Einstein ended his 1917 paper: "It should however be emphasized that a positive curvature of space still results from the presence of matter in it even when the extra term [the cosmological term] is not added; we need that term only to permit a quasistatic distribution of matter, as corresponds to the fact of the small velocities of the stars."

very slightest hint that Einstein was too prone to speculation, namely: "On the relativity of inertia. Remarks concerning Einstein's latest hypothesis." It would in fact be an interesting exercise to count up precisely how many different fruitful hypotheses about gravitation and inertia Einstein did advance between 1907 and 1917. One thing at least is certain: de Sitter's paper on March 31 1917 brought the stream of hypotheses to a definite end. Einstein did not after that add anything of enduring significance to the fundamentals of his gravitational theory, and the long-running saga of Mach's Principle finally petered out into almost nothing, since Einstein was forced to admit that he had not achieved his aim of showing that a cosmological model without matter was impossible. Only gradually could he bring himself to abandon the dream. In fact, it was only a year after de Sitter gave his solution that Einstein, responding to criticism by Kretschmann (Kretschmann 1917), at last gave a formal definition of Mach's Principle (Einstein 1918):

> The G-field [the metric] is *completely* determined by the masses of the bodies. Since mass and energy are identical in accordance with the results of the special theory of relativity and the energy is described formally by the symmetric energy tensor ($T_{\mu\nu}$), this means that the G-field is conditioned and determined [*bedingt und bestimmt*] by the energy tensor of the matter.

However, despite asserting that he personally regarded its fulfillment as absolutely necessary, Einstein was finally forced to dissociate himself from Mach's Principle, though I do not think the yearning for it ever left him. In his "Autobiographical Notes" published in 1949 (Einstein 1949) he points out how mistaken he was on the subject, but the very next paragraph begins by saying that Mach's critique was in essence very sound, and Einstein illustrates his point by means of an analogy[8] that he originally was given by his close friend Michele Besso. It is interesting to note that Einstein first published the analogy in the dark days of 1914 when he felt obliged to work with noncovariant field equations but still had his irresistible hankering for general covariance. Incidentally, the journal in which it appeared was *Scientia*, published in Bologna (Einstein 1914).

[8] Einstein's analogy was as follows:

> How sound, however, Mach's critique is in essence can be seen particularly clearly from the following analogy. Let us imagine people construct a mechanics, who know only a very small part of the Earth's surface and who also can not see any stars. They will be inclined to ascribe special physical attributes to the vertical dimension of space (direction of the acceleration of falling bodies) and, on the ground of such a conceptual basis, will offer reasons that the Earth is in most places horizontal. They might not permit themselves to be influenced by the argument that as concerns the geometrical properties space is isotrope and that it is therefore supposed to be unsatisfactory to postulate basic physical laws, according to which there is supposed to be a preferential direction; they will probably be inclined (analogously to Newton) to assert the absoluteness of the vertical, as proved by experience as something with which one simply would have to come to terms. The preference given to the vertical over all other spatial directions is precisely analogous to the preference given to inertial systems over other rigid co-ordination systems.

I quoted earlier a line from Keats' "Ode on a Grecian Urn." A few lines after the one quoted Keats gives what can be seen as a poetic summary of the whole saga of Einstein and Mach's Principle:

> Bold Lover, never, never canst thou kiss,
> Though winning near the goal – yet, do not grieve;
> She cannot fade, though thou hast not thy bliss,
> For ever wilt thou love, and she be fair!

What is most singular about the whole affair is that although in Einstein's view the passion never came to consummation two extraordinarily robust and handsome children were born: the general theory of relativity and modern relativistic cosmology.[9]

References

Barbour, J. B. (1974). Relative-Distance Machian Theories. *Nature*, **249**, 328.

(1989). *Absolute or Relative Motion?* Vol. 1 *The Discovery of Dynamics*. Cambridge: Cambridge University Press.

(1990). Einstein and Mach's Principle. In *History of General Relativity* (Einstein Studies, Vol. 3, ed. Howard, D., and Stachel, J.), Proceedings of the Second International Conference on the History of General Relativity, Marseille-Luminy, September 1988, Boston: Birkhäuser.

Barbour, J. B., and Bertotti, B. (1977). Gravity and Inertia in a Machian Framework. *Nuovo Cimento*, **B38**, 1.

Barbour, J. B., and Bertotti, B. (1982). Mach's Principle and the Structure of Dynamical Theories. *Proceedings of the Royal Society of London*, **A382**, 295.

De Sitter, W. (1916a). On Einstein's Theory of Gravitation, and its Astronomical Consequences, Part 1. *Monthly Notices of the Royal Astronomical Society*, **76**, 699.

(1916b). On Einstein's Theory of Gravitation, and its Astronomical Consequences, Part 2. *Monthly Notices of the Royal Astronomical Society*, **77**, 155.

(1917a). On the Relativity of Rotation in Einstein's Theory. *Proceedings of the Koninklijke Nederlandse Akademie van Wetenschappen Amsterdam*, **19**, 527.

(1917b). On the Relativity of Inertia. Remarks Concerning Einstein's Latest Hypothesis. *Proceedings of the Koninklijke Nederlandse Akademie van Wetenschappen Amsterdam*, **19**, 1217.

(1917c). On Einstein's Theory of Gravitation, and its Astronomical Consequences, Part 3. *Monthly Notices of the Royal Astronomical Society*, **78**, 3.

Earman, J., and Glymour, C. (1978). Lost in the Tensors: Einstein's Struggles with Covariance Principles 1912–1916. *Studies in the History and Philosophy of Science*, **9**, 251.

Einstein, A. (1907). Über das Relativitätsprinzip und die aus demselben gezogenen Folgerungen. *Jahrbuch der Radioaktivität und Elektronik*, **4**, 411.

(1911). Über den Einfluss der Schwerkraft auf die Ausbreitung des Lichtes. *Annalen der Physik*, **35**, 898.

[9] The present paper was not the occasion to put forward my own view on the matter, which is that Einstein did in fact create in his general theory of relativity a dynamical structure that is far more Machian (though in a sense rather different from what Einstein intended) than is generally believed (see Barbour and Bertotti 1982; Barbour 1989; Introduction; and Barbour 1990).

(1912). Gibt es eine Gravitationswirkung, die der elektrodynamischen Induktionswirkung analog ist? *Vierteljahrsschrift fur gerichtliche Medizin,* **44**, 37.
(1914). Zum Relativitätsproblem. *Scientia*, **15**, 337.
(1916). Die Grundlage der allgemeinen Relativitätstheorie. *Annalen der Physik*, **49**, 769 (translation published in: *The Principle of Relativity*, collection of papers by A. Einstein et al., New York: Dover [1952]).
(1917). Kosmologische Betrachtungen zur allgemeinen Relativitätstheorie. *Sitzungsberichte der Preussischen Akademie der Wissenschaften*, February (1917) (English translation in the collection of Einstein, 1916).
(1918). Prinzipielles zur allgemeinen Relativitätstheorie. *Annalen der Physik*, **55**, 241.
(1922). *The Meaning of Relativity*. Methuen: London.
(1949). Autobiographical Notes. In *Albert Einstein – Philosopher – Scientist*, ed. P. A. Schilpp. Evanston, Illinois: The Library of Living Philosophers, Inc.
Einstein, A., and Grossmann, M. (1913). Entwurf einer veraligemeinerten Relativitätstheorie und einer Theorie der Gravitation. *Zeitschrift für Mathematik und Physik*, **62**, 225.
Friedmann, A. (1922). Über die Krümmung des Raumes. *Zeitschrift für Physik*, **10**, 377.
(1924). Über die Möglichkeit einer Welt mit konstanter negativer Krümmung des Raumes. *Zeitschrift für Physik*, **21**, 326.
Kretschmann, E. (1917). Über den physikalischen Sinn der Relativitätspostulate, A. Einsteins neue und seine ursprüngliche Relativitätstheorie. *Annalen der Physik*, **53**, 575.
Mach, E. (1960). *The Science of Mechanics*. Open Court: La Salle, Illinois.
Merleau-Ponty, J. (1982). Le Trasformazioni del Concetto di Cosmo nella Filosofia e nella Scienza. In *Il Problema del Cosmo*, ed. G. Toraldo di Francia, pp. 39–50. Rome: Istituto dell'Enciclopedia Treccani.
Norton, J. (1984). How Einstein found his Field Equations: 1912–1915. *Historical Studies in the Physical Sciences*, **14**, 253.
Pais, A. (1982). "*Subtle is the Lord . . .*" *The Science and Life of Albert Einstein*. Oxford: Oxford University Press.
Russell, J. L. (1964). Kepler's Laws of Planetary Motions: 1609–1666. *British Journal for the History of Science*, **2**, No. 5.
Schrödinger, E. (1956). *Expanding Universes*. Cambridge: Cambridge University Press.
Schwarzschild, K. (1900). Über das zulässige Krümmungsmass des Raumes. *Vierteljahrsschrift der Astronomischen Gesellschaft*, **35**, 337.
Stachel, J. (1980). Einstein and the Rigidly Rotating Disk. In *General Relativity and Gravitation. One Hundred Years after the Birth of Albert Einstein*, Vol. 1. New York: Plenum Press.

5

The mysterious lore of Large Numbers

John D. Barrow

Although we talk so much about coincidence we do not really believe in it. In our heart of hearts we think better of the universe, we are secretly convinced that it is not such a slipshod, haphazard affair, that everything in it has meaning.

J. B. Priestley

Any coincidence is always worth noticing. You can throw it away later if it is only a coincidence.

Miss J. Marple

5.1 Introduction

It has been said that though God cannot alter the past, historians can; it is perhaps because they can be useful to him in this respect that He tolerates their existence.

Samuel Butler

Modern physics means *mathematical* physics and mathematics means numbers. This preoccupation with the numerical as the route to understanding the physical seems to have begun with the Pythagoreans' conviction that the true meaning of Nature was to be found only in those harmonies and patterns which mathematics displays. We are the heirs to this prejudice but we have found it expedient to deviate from it in one crucial respect. Whereas the Pythagoreans were persuaded that numbers themselves were possessed of some especial significance we have found it most fruitful to place significance upon the mere fact that there exist numerical relationships *between* things. Our 'laws of Nature' are the lists of these habitual relationships. This approach matured after the Renaissance, in parallel with the mechanical world-view engendered by the work of Galileo, Newton and their like-minded followers. The establishment of a picture of the universe as a clockwork mechanism whose workings could be described by mathematical relations consigned to oblivion what was left of the mystical hermetic tradition which saw the universe as a code to be deciphered or a portentous sign to be read and interpreted. None the less, there has remained a small subculture within modern cosmology that has continually laid great stress upon the existence of particular numbers which characterise the large-scale structure of the universe and relate it to the rest of physics. A little of the story of how this culture has evolved and what it has taught us about the universe is the subject matter of this chapter. Whereas other chapters will tell a specific story circumscribed by particular events in time or space, this one follows diverse contributions which span nearly a century and introduces in one guise or another some of the leading figures of twentieth-century physics. This is not by any means a complete historical account of the cosmological treatment of the lore of Large Numbers, a fuller discussion of which can be found in Barrow (1981) and Barrow and Tipler (1986). Rather, it is intended to illustrate several historical trends which have an unexpected topicality in view of recent developments in fundamental physics and cosmology.

5.2 Constants in physics

> Men occasionally stumble over the truth, but most of them pick themselves up and hurry off as if nothing had happened.
>
> Winston Churchill

Although science is predicated upon discovering the numerical relationships *between* things, there are particular numbers which take on a special status in the labyrinth that results. These are termed variously 'constants of Nature' or 'constants of physics' (the latter no doubt if one is a physicist). The mathematical equations which describe the world most effectively possess three ingredients: some algorithmic structure which enables new states to be computed from old ones, proportionality constants and starting conditions (Barrow 1986b). The proportionality constants come in two forms: *pure numbers* such as $e = 2.718\ldots$ or $\pi = 3.14\ldots$, and *dimensional constants* such as the speed of light or Newton's constant of gravitation. It is generally believed, perhaps wrongly, that the latter are not simply artefacts of this particular way of describing the world but possess some fundamental significance. In practice, the pure numbers appear in physical equations in combinations that are close to unity in value. As a result the method of dimensional analysis devised first by Newton and Fourier is an accurate approximate method of analysis. In 1911 Einstein[1] recognised this to be an unexplained piece of good fortune and challenged physicists to explain it. Elsewhere (Barrow

[1] Einstein maintained a long-term interest in the question of constants of physics. His most complete statement on the matter occurs in private correspondence with Ilse Rosenthal-Schneider, a former undergraduate student, which is now published in Rosenthal-Schneider (1980). There, in a letter written in 1945, he distinguishes 'basic numbers' (*Rationelle Zahlen*) such as 'pi' or dimensional '"universal" constants' (*'universellen' Konstanten*) such as the velocity of light, from the 'genuine universal constants' (*eigentlichen universellen Konstanten*), like the fine structure constant, which are the dimensionless combinations that can always be formed from the dimensional constants.

and Tipler 1986) we have argued that it is a consequence of the fact that the observed universe possesses relatively few (i.e., three) spatial dimensions. The pure numbers appearing in physical equations are at root geometrical in origin. When one computes the volume of an N-dimensional sphere of radius R one obtains

$$V_N = 2\pi^{(N+1)/2} R^N / \Gamma[(N+1)/2], \tag{1}$$

where $\Gamma[\ldots]$ is the Euler gamma function. We see that the numerical factor premultipling the dimensionful factor R^N diverges from unity as N^N for large N. If we lived in a world of $N > 3$ dimensions we would not find dimensional analysis such a useful tool for estimating physical magnitudes.

The dimensional constants can always be combined so as to produce combinations that are dimensionless. The challenge of explaining the numerical values of these dimensionless constants of Nature combinatorically or geometrically is the greatest unsolved problem of physics. A belief in their ultimate explicability is rooted deep in many scientists' faith in the unity of Nature. Einstein believed that

> In any reasonable theory there are no [dimensionless constants of Nature whose values are only empirically determinable] . . . I cannot imagine a unified and reasonable theory which explicitly contains a number which the whim of the Creator might just as well have chosen differently, whereby a qualitatively different lawfulness of the world would have resulted . . . A theory which in its fundamental equations explicitly contains a non-basic constant would have to be somehow constructed from bits and pieces which are logically independent of each other; but I am confident that this world is not such that so ugly a construction is needed for its theoretical comprehension.[2]

The Victorian era saw a huge increase in the systematic study of mechanical devices and the classification and calibration of everything and anything. Units of measurement proliferated. The most significant contribution to this proliferation was made by the distinguished Irish physicist George Johnstone Stoney at the Belfast Meeting of the British Association for the Advancement of Science in August 1874.[3] This meeting was dedicated to the systematic organisation of the multitude of units of measurement current at the time but Stoney pointed to a preferred set of

[2] Rosenthal-Schneider (1980: 37–8). This belief was to manifest itself later in Einstein's search for a 'unified field theory'. This investigation of possible sets of field equations employed a function-counting procedure which gauged the 'strength' of the resulting theory by the number of additional pieces of data that were necessary to determine the solutions of the equations (Einstein 1956). The smaller this number the stronger the system of equations and the more in tune with Einstein's desire to minimise arbitrariness.

[3] These results were read again at the meeting of the Royal Dublin Society on 16 February 1881 and published later that year, Johnstone Stoney (1881); see also Johnstone Stoney (1894). Stoney wrote on this subject again in the foreword to the first edition of Fournier d'Albe's *The Electron Theory* (1906) although the discussion was removed from later editions. D'Albe's book contains a chapter on natural units.

units of mass, length and time that Nature picked out independent of any artefact or anthropocentric standard. He showed that the velocity of light (as an average of extant experimental data he takes $c = 3.10^8\,ms^{-1}$), Newton's gravitational constant (he uses John Herschel's value of $G = 2/3 \times 10^{-11}\,m^3\,Kg^{-1}\,s^{-2}$) and finally, the electric charge of the electron ($e = 10^{-20}$ Ampère, a constant proposed and derived by Stoney himself[4]), can be combined uniquely to give his 'natural' units of mass, length and time:

$$m_J = (e^2/G)^{1/2} = 10^{-7}\,gm, \tag{2}$$

$$l_J = (Ge^2/c^4)^{1/2} = 10^{-37}\,m, \tag{3}$$

$$t_J = (Ge^2/c^6)^{1/2} = \tfrac{1}{3} \times 10^{-45}\,s. \tag{4}$$

This idea was rediscovered by Max Planck in 1899. He proposed that natural units of mass, length and time be constructed from G, c and the new constant of action, h, which now bears his name. In addition, the incorporation of Boltmann's constant, k, allowed him to define a natural unit of temperature, T_{pl}. The so-called 'Planck units' are then

$$m_{pl} = (ch/G)^{1/2} = 5.37 \times 10^{-5}\,gm, \tag{5}$$

$$l_{pl} = (Gh/c^3)^{1/2} = 3.99 \times 10^{-33}\,cm, \tag{6}$$

$$t_{pl} = (Gh/c^5)^{1/2} = 1.33 \times 10^{-43}\,s, \tag{7}$$

$$T_{pl} = (hc^5/Gk^2)^{1/2} = 3.60 \times 10^{32}\,K. \tag{8}$$

These quantities are defined in an identical discussion which appears in two of his papers (Planck 1899, 1900) and were publicised in a series of lectures delivered in Berlin in 1906/7. These lectures were later published as *Theorie der Wärmestrahlung* in 1906 where the same discussion of natural units ('Natürliche Masseinheiten') again appears (Planck 1906).

The reason for the co-existence of these two different sets of natural units of mass, length and time, each differing in value by roughly a factor of a hundred, is that in our three-dimensional world the combination e^2/hc is a dimensionless constant of Nature roughly equal to $1/(2\pi \times 137) = 1/860$ using currently determined values of the constants. The Johnstone Stoney

[4] Johnstone Stoney invented the name 'electron' for the basic unit of electric charge in 1891 (*Scientific Transactions of the Royal Dublin Society*, **IV**, series 11) and it was eventually adopted despite J. J. Thomson's desire to call it the 'corpuscle'. In his talk to the British Association in 1874, and his paper of 1881, Johnstone Stoney labels it 'E1', refers to it as the 'electrine' and gives his first calculation of its expected value. He divides the quantity of electricity required for the electrolysis of 1 cc of hydrogen by the number of hydrogen atoms in 1 cc, which is given by the Loschmidt–Avogadro number. Johnstone Stoney was a distant older cousin of Alan Turing. Turing recalled that as young children they always rather irreverently called him 'electron Stoney' (Turing 1959); see Barrow (1983, 1988c) for further details and also *Encyclopaedia Britannica* entries (1924–36) under 'Electron' written by R. A. Millikan (1926) who credits Johnstone Stoney's 1881 paper with the first calculation of the charge on the electron.

and Planck units of mass, length and time thus all differ by the constant factor $\sqrt{860} \simeq 29$.

Through the existence of his natural units Stoney saw a way of cutting the Gordian knot of subjectivity in the choice of units whereas Planck used his natural units to underpin a non-anthropomorphic basis to physics, the progressive revelation of which was for him the hallmark of progress 'towards as far-reaching a separation as possible of the phenomena in the external world from those in human consciousness'. He saw (Planck 1899) his units as a

> means of establishing units of length, mass, time and temperature, which are independent of special bodies or substances, which necessarily retain their significance for all times and for all environments, terrestrial and human or otherwise, and which may, therefore, be described as 'natural units'.

Unlike Einstein, Planck did not believe in any attainable all-encompassing theory of physics which would explain all the constants of Nature for if such a theory were believed in 'then physics would cease to be an inductive science, and it will certainly always be that'. Others, like the instrumentalists Pierre Duhem and Percy Bridgman, regarded the promised Planckian separation of scientific description from human conventions as unattainable in principle, viewing the constants of Nature and the theoretical descriptions that they underpin entirely as artefacts of a particular human choice of representation to 'save the appearances' (Bridgman 1920; Duhem 1954).

It is also salutary to witness how few leading physicists appreciated the significance of these systems of natural units. An interesting case is that of Heisenberg who was greatly concerned over many years with the determination of a fundamental 'smallest length' in Nature in terms of known or unknown fundamental constants. Having selected c and h as ingredients of this length he then argues (Heisenberg 1958) that

> there must exist a third universal constant in nature ... for purely dimensional reasons ... The most appropriate way of introducing the third universal constant would be by the assumption of a universal length the value of which should be roughly 10^{-13} cm, that is, somewhat smaller than the radii of the light atomic nuclei.

The third constant was G and his 'smallest length' had in fact been found eighty years earlier by Stoney and Planck.[5]

[5] According to Focken (1953), Eddington claimed that the Planck length must be the key to some essential structure because it is so much smaller than the radii of the proton and electron. Focken does not reference Eddington's claim but he probably refers to the report on general relativity prepared for the Physical Society (Eddington 1918). On the last page of that report, which later matured into his text on relativity (Eddington 1923) Eddington derives Planck's natural unit of length, (6), as 4×10^{-33} cm and then finishes his report with the remarkable statement that

> There are other natural units of length – the radii of positive and negative unit electric charges – but these are of an altogether higher order of magnitude ... no theory has

5.3 Large numbers, Eddington and Milne

Little laws have bigger laws
From which they're forced to follow.
But a bigger law's a bigger guess
That's harder still to swallow.

Anon.

Psychologists tell us that the writing of large numbers by young children is a sign of intelligence, but of what is it a sign in adults? When the first twentieth-century cosmologists began to ponder the dimensionless constants of Nature that emerge when one considers at once the large- and small-scale structure of the universe they discovered a catalogue of unlikely coincidences between numbers of enormous magnitude. These coincidences provoked a series of *ad hoc* approaches to cosmology that in retrospect are interesting for a variety of reasons:

* They signal the beginning of a gradual recognition of the relationship between the global properties of the universe and the local environment necessary to evolve and sustain life, culminating in the development of the so-called 'Anthropic Principle'.
* They provoked the first investigations of how physical, geological and biological data could constrain the structure of cosmological theories.
* They led to the development of new theories of gravitation and, thereby, to a deeper understanding of the physical content of Einstein's theory of general relativity, and renewed the experimental investigation of gravity with unprecedented accuracy.
* The early attempts to explain Large Numbers in cosmology display a premature example of a naive theory whose goal was to unify all the forces of Nature.
* They gave birth to serious consideration of how traditional 'constants' of Nature might in fact be changing slowly over cosmic timespans.

The most striking thing about these five historical lessons is that they have matured into contemporary ideas of great cosmological significance: the Anthropic Principle, the symbiotic relationship between terrestrial physics and cosmology, the behaviour of gravity and constants in Kaluza–Klein and superstring theories, and the perennial quest for a 'theory of everything'.

> attempted to reach such fine-grainedness. But it is evident that this length must be the key to some essential structure. It may not be an unattainable hope that some day a clearer knowledge of the processes of gravitation may be reached; and the extreme generality and detachment of the relativity theory may be illuminated.

Bridgman (1920) also pointed out that the huge value of the Planck temperature, even by astrophysical standards, and indicated that it might be associated with some new and fundamental level of cosmic structure.

Although 'Large Numbers' have come to be associated with the names of Dirac and Eddington, it was Herman Weyl who first noticed the presence of large dimensionless numbers in physics. In 1919, at the end of an early study of general relativity Weyl (1919) noticed that the ratio of the classical electron radius (e^2/m_ec^2) to the gravitational radius of the electron ($2Gm_e/c^2$) was of order 10^{40} and he speculated (correctly) that the ratio of the gravitational radius of the visible universe to that of the electron might be of similar magnitude:

> It is a fact that pure numbers of magnitude totally different from 1 appear with the electron; for example, the ratio of the electron radius to the gravitational radius of its mass, which is of order 10^{40}; the ratio of the electron radius to the world radius may be of similar proportions.

During the next twenty years Weyl returned to these coincidences, and others, on several occasions (Weyl 1934, 1949). Even two years after the introduction of Dirac's 'Large Numbers Hypothesis' in 1937, Fritz Zwicky (1939) would still refer to the idea that large dimensionless numbers of similar magnitude are causally related as 'Weyl's hypothesis'.

Many others followed this early example and searched for coincidences between any and every combination of fundamental and cosmological constants.[6] But only two, Arthur Eddington and Paul Dirac, sought to derive from these coincidences a deeper and more unified understanding.

Arthur Stanley Eddington was the greatest astrophysicist of his day, a Quaker of deep conviction, and a brilliant writer who used his literary talents to popularise science with unparalleled eloquence and to articulate his own idealistic neo-Kantian philosophy of science. His particular brand of 'selective subjectivism' enabled him to believe that the world can be understood by pure deductive thought because it cannot be the ultimate and unattainable 'thing-in-itself'. His famous illustration of the fish and the fishnet (Eddington 1939) is strikingly similar to Wittgenstein's picture of the spots and the mesh in *Tractatus Logico-Philosophicus*.[7]

[6] The earliest was the American astronomer John Q. Stewart (1931) who noticed a numerical coincidence between the current Hubble radius of the universe cH_o^{-1} and the product of the classical electron radius (e^2/m_ec^2), the fine structure constant ($2\pi e^2/hc$), and the ratio of the electromagnetic and gravitational forces between two electrons (e^2/Gm_e^2). This connection between Hubble's newly measured constant H_o and microscopic constants suggested to him the likelihood of a tired-light explanation for the redshift. Stewart's coincidence is essentially the same as that noticed by Weinberg (1972: 619).

[7] *Tractatus* paragraph 6.341, see the translation by D. F. Pears and B. F. McGuinness, Routledge: London (1961): 'Let us imagine a white surface with irregular black spots on it. We then say that whatever kind of picture these make, I can always approximate as closely as I wish to the description of it by covering the surface with a sufficiently fine square mesh, and then saying of every square whether it is black or white. In this way I shall then have imposed a unified form on the description of the surface.' Eddington's analogous parable of the ichthyologist and the fishnet occurs on pp. 16–20 of Eddington (1939). Eddington and Wittgenstein, like Russell, were contemporaries at Trinity College, Cambridge. However, Wittgenstein evidently had a very low view of Eddington's philosophical excursions ('vague philosophical ramblings . . . typical of middle-aged physicists bored with their subject'). Varied assessments of Eddington's philosphy of Nature can be found in Joad (1932), Stebbings (1944), Whittaker (1949), Dingle (1954), Witt-Hansen (1958) and Yolton (1960).

Eddington's precise viewpoint is hard to pin down because it appears to develop with time.[8] In *Space, Time and Gravitation* (Eddington 1920), he asks whether it is not 'possible that laws which have not their origin in the mind may be irrational' to the extent that 'we can never succeed in formulating them'. Elsewhere he claims that 'Those laws of nature which have been woven into a unified scheme – mechanics, gravitation, electro-dynamics and optics – have their origin, not in any special mechanism of nature, but in the workings of the mind.' Eddington's position is that much of what passes for objective knowledge about the world is really partially epistemological in character. Some of his writings and certainly those of commentators give the impression that he believed that we could have no objective knowledge at all, but this is probably too extreme a claim and he firmly denied being a Kantian. His subjectivism was selective: although the mesh of the fishnet determines the smallest size of fish that you can catch it does not bias the properties of the larger ones that are caught. At root Eddington sought to understand why theoretical schemes prove so successful in describing Nature and to reduce physics to its lowest terms as Russell and Whitehead had done with mathematics. Is it because they are creations of our own minds that our theories work so well or is the world intrinsically mathematical and ultimately simple? Throughout much of the latter part of his life he sought to develop a 'Fundamental Theory'[8] which would tie together what we know of the large-and small-scale structure of the universe and, in so doing, enable the values of the dimensionless constants of nature to be calculated exactly. Sir Edmund Whittaker (1949) set out the guiding principle of Eddington's philosophy of Nature as follows:

> All the quantitative propositions of physics, that is, the exact values of the pure numbers that are constants of science, may be deduced by logical reasoning from qualitative assertions without making any use of quantitative data derived from observation.

Eddington took up the large number coincidences first recognised by Weyl (1919) in his famous textbook on general relativity (1923). He proposes that the characteristics of particles such as the electron should derive locally from the structure of the space-time in which they sit, so that he expects the equation

> *'radius of electron in any direction = numerical constant × radius of curvature of space-time in that direction'*

to hold as an extension of the general relativistic connection between gravity and the local curvature of space-time. He regards such

[8] This work was unfinished at the time of Eddington's death in 1944 but his manuscript was published posthumously as *Fundamental Theory* by Cambridge University Press in 1946 under the editorship of Eddington's former mentor E. T. Whittaker. A study of his unpublished work was undertaken by Slater who subsequently published an unsuccessful attempt to understand its logical basis (Slater 1957). For a review see Taub (1950).

'coincidences' as Weyl's as evidence for a rigorous relationship of this type.

Amongst the constants which Eddington regarded as of prime importance was the so-called 'Eddington number', N, which gives the number of protons and electrons in the universe. Eddington calculated this number to enormous precision (apparently on a transatlantic boat crossing) finding,

$$N = \tfrac{3}{2} \times 2^{256} \times 136 \sim 10^{79}. \tag{9}$$

Part of the fascination of this number for Eddington was the fact that it must be an integer and so could, in principle, be calculated *exactly*. He noticed an interesting coincidence that is similar to Weyl's – that the values of $N^{1/2}$ and the ratio of the electromagnetic to gravitational forces between a proton and an electron are of similar magnitude:

$$e^2/Gm_em_n \sim N^{1/2} \sim 10^{39}. \tag{10}$$

In the 1923 edition of *The Mathematical Theory of Relativity* (Eddington 1923), he wrote of the quantity on the left-hand side of this equation that

> It is difficult to account for the occurrence of a pure number (of order greatly different from unity) in the scheme of things; but this difficulty would be removed if we could connect it with the number of particles in the world – a number presumably decided by accident.

The fact that its magnitude was so different from unity convinced Eddington that this coincidence must again hide some unsuspected causal relationship. Later in his career Eddington would move away from any suggestion that N was 'decided by accident'. His 'Fundamental Theory' aimed to show that N was uniquely determined by combinations of microscopic constants whose values were determined purely combinatorically. Like many others after him he was attracted by the idea that the combination Gm_em_n/e^2 may be statistical in origin because it looks like $N^{-1/2}$, a random fluctuation in the number of particles in the observable universe.[9]

On the small-scale Eddington's efforts were directed towards explaining the value of the 'fine structure constant' $\alpha = e^2/\hbar c \sim 1/137$. His numerological efforts in this direction were mysterious and unconvincing to other physicists. Most found his arguments to be obscure, a criticism that exasperated Eddington who could not understand why the same criticisms were never levelled at his colleague Dirac. In a letter to Dingle (see Crowther 1952), he says: 'I am continually trying to find out why people find the procedure obscure . . . I cannot seriously believe that I ever attain

[9] A different form of the coincidence – that the mean free path of a proton in the universe is roughly equal to the radius of the observable universe – was noticed by Schrödinger (1938) in an investigation of (10) which he proposes be named 'Eddington's relation'. This relation also appears extensively in a series of papers and a book by A. Haas in the period 1932–8. See Barrow and Tipler (1986: 278) for details.

the obscurity that Dirac does.' The sceptical attitude of many contemporaries can be gauged from a spoof paper submitted to the German physics journal *Naturwissenschaften* by Beck, Bethe and Riezler in 1931. It parodies Eddington's attempts to derive the value of the fine structure constant (which he believed at that time to equal 1/137 exactly) by combinatorial arguments. The issue of 9 January 1931 carried the following paper (in German) from three young German post-doctoral workers at the Cavendish Laboratory in Cambridge:[10]

Concerning the Quantum Theory of Absolute Zero

> Let us consider a hexagonal crystal lattice. The absolute zero of this lattice is characterised by the fact that all degrees of freedom of the system are frozen out, i.e. all inner movements of the lattice have ceased, with the exception, of course, of the motion of an electron in its Bohr orbit. According to Eddington every electron has $1/\alpha$ degrees of freedom where α is the fine structure constant of Sommerfeld. Besides electrons our crystal contains only protons and for these the number of degrees of freedom is obviously the same since, according to Dirac, a proton is considered to be a hole in a gas of electrons. Therefore to get to the absolute zero we have to remove from the substance per neutron (= 1 electron plus 1 proton; our crystal is to carry no net charge) $2/\alpha - 1$ degrees of freedom since one degree of freedom has to remain for the orbital motion. We thus obtain for the zero point temperature
>
> $$T_0 = -(2/\alpha - 1) \text{ degrees.}$$
>
> Putting $T_0 = -273°$, we obtain for $1/\alpha$ the value 137, in perfect agreement within the limits of accuracy with the value obtained by totally independent methods. It can be seen very easily that our result is independent of the particular crystal lattice chosen.
>
> Cambridge, on 10 December 1930
>
> G Beck, H Bethe, W Riezler

Indeed, so convincing did this appear that Riezler was asked to present an exposition of the work in Munich at Sommerfeld's weekly seminar. Eddington, however, was not amused, and nor was the editor of the journal when he discovered that he had been made a fool of. Herr Berliner immediately published an 'erratum' on 6 March which pointed out that

> The Note by G. Beck, H. Bethe and W. Riezler, published in the 9 January issue of this journal, was not meant to be taken seriously. It was intended to characterise a certain class of papers in theoretical physics of recent years which are purely speculative and based on spurious numerical arguments. In a letter received by the editors from these gentlemen they express regret that the formulation they gave this idea was suited to produce misunderstandings.

[10] The translation from the German original is that of Delbrück (1972). Another, more transparent parody can be found in Born (1944: 37).

But George Gamow was not one soon to tire of a good joke and shortly afterwards he, Rosenfeld and Pauli wrote separate letters from different European addresses protesting to the editor that he had published another one of these disgraceful spoof articles and pointed the finger at another semi-numerological article, 'Origin of Cosmic Penetrating Radiation', in *Naturwissenschaften* by some poor unsuspecting author (A. V. Das) demanding that the editor obtain its withdrawal from the author in order to maintain the standards of the journal.

Another amusing sidelight which illustrates the difficulty many had in reconciling Eddington's work on fundamental constants with his monumental contributions to general relativity and astrophysics can be found in the biography of Kramers (Dresden 1987). It is a story told by Goudsmit himself:

> the great Arthur Eddington gave a lecture about his alleged derivation of the fine structure constant from fundamental theory. Goudsmit and Kramers were both in the audience. Goudsmit understood little but recognised it as farfetched nonsense. Kramers understood a great deal and recognised it as total nonsense. After the discussion, Goudsmit went to his friend and mentor Kramers and asked him. 'Do all physicists go off on crazy tangents when they grow old? I am afraid.' Kramers answered, 'No Sam, you don't have to be scared. A genius like Eddington may perhaps go nuts but a fellow like you just gets dumber and dumber.'

The cosmological constant,[11] introduced but soon abandoned by Einstein, played a pivotal role in Eddington's attempts to unify the macrophysical and microphysical constants and he laid great emphasis upon the significance of another coincidence amongst large dimensionless numbers involving the cosmological constant, Λ (Eddington 1935):

$$ch(m_n m_e/\Lambda)^{1/2} \sim 10^{39}. \qquad (11)$$

Far from following Einstein in discarding the cosmological constant, Eddington regarded it as the cornerstone of general relativity: 'if ever the theory of relativity falls into disrepute the cosmical constant will be the last stronghold to collapse. To drop the cosmical constant would knock the bottom out of space' (Eddington 1933).[12]

[11] Today the size of the quantity in (11) is still mysterious to cosmologists but it is conventional to consider a quantity like the reciprocal of that in (9), usually $\Lambda/\ell_{pl}^2 \leqslant 10^{-120}$, and to label the puzzle of its extreme smallness as the 'cosmological constant problem' (Barrow and Tipler 1986; Hawking 1983; Weinberg 1987). It is widely believed that this small number is evidence that $\Lambda \equiv 0$ for some undiscovered fundamental reason. Interestingly, many other 'large numbers' problems of the pre-war period are now cast as 'small numbers' problems by modern cosmologists (Barrow 1988a).

[12] Professor Bondi has suggested to me that the introduction of natural dimensional units by Johnstone Stoney and Planck played a role in moving thinking away from the notion that Nature was scale invariant and that Einstein's disenchantment with the cosmological constant was partially motivated by a desire to preserve scale invariance of the empty-space Einstein equations. In this connection I think it is interesting to compare with the earlier ideas of Laplace (1808), Delbœuf (1860, 1893) and Poincaré (1914) all of whom argued that if all length-scales in the universe were suddenly to increase by some constant factor then there would be no observational means of telling

Thus Eddington's writings during the 1920s and 1930s made the existence of large dimensionless ratios between cosmological and terrestrial constants familiar, if not notorious, to most physicists.

Beginning in the 1930s E. A. Milne had developed a kinematic theory of relativity in which there were to exist two fundamental timescales: one – t-time – to govern 'atomic' phenomena, and the other – τ-time – to govern 'gravitatitional' processes (Milne 1935, 1948). The two timescales were logarithmically related

$$\tau = \ln(t/t_o) + t_o \tag{12}$$

with t_o a constant. This development is relevant to our story in two ways. First, it introduced biological considerations into cosmology for the first time. Beginning in 1937 the biologist J. B. S. Haldane wrote several articles about the biological and geological consequences of Milne's proposal (Haldane 1937, 1944, 1955). He argued that the characteristic energies of mutation on biochemical activity which govern the pace of evolution would increase with respect to the τ timescale despite remaining constant on the t-timescale. Thus, for example, if radioactive decay occurred on the t scale with decay rate-constant λ then the apparent decay 'constant' on the τ scale would actually be $\lambda t/t_o$. In effect, there would exist a time-variation in the constants of physics governing chemical and biological processes. Haldane suggested that this could explain the discontinuous progress of evolution and produce a pattern of evolutionary change that we would now describe as the 'punctuated equilibrium' of Gould and Eldredge (1977):

> There was, in fact, a moment when life of any sort first became possible and the higher forms of life may only have become possible at a later date. Similarly a change in the properties of matter may account for some of the peculiarities of pre-Cambrian geology.

In view of these wide-ranging consequences he urged Milne and others to deduce how all microscopic physical quantities of relevance to biology

that this had occurred. Such a notion assumes (wrongly) that the structure of the universe is scale invariant. The earliest statement of this prejudice that the author has encountered is in Boscovich's remarkable *Theoria Philosophiae Naturalis* (1763). In paragraph 19 of Supplement II he writes: 'Moreover it might be the case that the whole Universe within our sight should daily contract or expand, while the scale of forces contracted or expanded in the same ratio; if such a thing did happen, there would be no change of ideas in our mind, and so we should have no feeling that such a change was taking place.' This statement is especially extraordinary since the entire mathematical argument of Boscovich's *Theoria* is proposing that the total force law acting upon all bodies in the universe is not just the scale-invariant inverse-square law of Newtonian gravity but a more general algebraic curve (which he approximates by a power series in inverse distance) which is alternately attractive and repulsive before asymptoting to the Newtonian inverse-square law. The force law is deliberately chosen to be of this *scale non-invariant* form by Boscovich in order to explain the finite sizes of bodies as equilibrium states between forces of attraction and repulsion. He was the first to introduce this notion but clearly failed to recognise that the very scale non-invariance of the force law required to explain objects with characteristic sizes guarantees that an expanded or contracted world could be observationally distinguished.

and chemistry would vary in his kinematic relativity theory. Such variations might also, he suggested, provide an escape from the conclusion that the universe is heading for a sorry end in the Heat Death foretold in the popular writings of Jeans and Eddington.

In Milne's theory the time-variation of traditional constants of Nature played a key part. Previously, this idea of varying 'constants' had surfaced only sporadically and eccentrically:[13] during the early 1930s there was always a background noise of arguments for an explanation of Hubble's cosmological redshift as a 'tired light' effect caused by the change of the speed of light or Planck's constant in time (although most of these looked to scattering of light by intervening material rather than varying constants for the reason), but Milne's papers ensured that the idea of a time-varying gravitation 'constant' was known to physicists such as Dirac by 1935.

5.4 Dirac: a radical departure

> That which has been believed by everyone, always and everywhere, has every chance of being false.
>
> Paul Valéry

Whilst on his honeymoon, in February 1937, Dirac wrote a brief letter to the editor of *Nature* in which he proposed a new explanation for the Large Numbers of Eddington and Weyl. This was followed, ten months later, by a longer exposition of his proposed 'new basis for cosmology' in the *Proceedings of the Royal Society* (Dirac 1937a, 1937b, 1938). He listed three large dimensionless numbers which could be created from fundamental constants and cosmological quantities like the density of the universe, ϱ and the age of the universe, then believed to be $t \sim 2 \times 10^9$ years, which, according to general relativity, are approximately related by $\varrho G t^2 \sim 1$. These were

$$N_1 = ct/(e^2/m_e c^2) \sim 10^{39}, \tag{13}$$

$$N_2 = e^2/Gm_n m_e \sim 10^{39}, \tag{14}$$

$$N = 4\pi(ct)^3 \varrho/3m_n \sim c^3 t/Gm_n \sim 10^{78}. \tag{15}$$

Of Eddington's deductions from such pure numbers Dirac was somewhat sceptical

> Eddington's arguments are not always rigorous . . . [but] . . . 10^{39} and 10^{78} are so enormous as to make one think that some entirely different type of explanation is needed for them.

[13] See Barrow and Tipler (1986: ch. 4) for a survey. Eddington never attempted to explain any Large Number coincidences by invoking varying constants. In fact, he was a fierce critic of all claims that the redshift was a 'tired-light' effect of some variation in constants of Nature like c and h which govern light propagation, dismissing such ideas as 'nonsensical' and 'self-contradictory' (Eddington 1946).

As an alternative to them he proposed his 'Large Numbers Hypothesis' (LNH), that

> Any two of the very large dimensionless numbers occurring in Nature are connected by a simple mathematical relation in which the coefficients are of order of magnitude unity.

According to this view the numbers N_1, N_2 and $N^{1/2}$ were actually *equal* up to small numerical factors of order unity. This hypothesis of equality between Large Numbers is not in itself original to Dirac. Eddington and others had written down such relations as (14), and also such expressions as (13) and (15), but they had suppressed the time-dependence of the radius of the obserable universe, ct, and its mass, $4\pi\varrho(ct)^3/3$, which appears in the numerator on the right-hand side of (15). Eddington had not distinguished between the number of particles in the entire universe – which might be infinite – and the number of particles in the *observable* universe which is defined to be a sphere about us with radius equal to the speed of light times the present age of the universe. This was a fundamental flaw in his entire methodology. The radical change precipitated by Dirac's LNH is that it requires us to believe that a collection of constants such as (14) must vary with the age of the universe, t

$$N_1 \approx N_2 \approx \sqrt{N} \propto t. \tag{16}$$

Now, because Dirac had included in his catalogue of Large Numbers two which possess an explicit dependence upon the age of the universe (N and N_1), the relations (16) require for consistency that

$$e^2/Gm_n \propto t. \tag{17}$$

Dirac chose to accommodate this requirement by concluding that Newton's gravitation constant is decreasing in value over cosmic timescales as

$$G \propto 1/t. \tag{18}$$

Thus one sees from (13)–(14)that $N_1 \propto N_2 \propto \sqrt{N} \propto t$ and the huge magnitude of the three Large Numbers is attributable to the great age of the universe. The conclusion $N \propto t^2$ subsequently led Dirac to conclude (Dirac 1973), quite wrongly, that this result required the continuous creation of protons. In fact all it is telling us is that as the universe ages we are able to see more and more protons coming within our horizon. Again, it is worth noting, as Dirac himself did, that the LNH leads to results in common with Milne's two-timing universe. Milne's theory requires that the mass of the observable universe, M_u, remains constant on the t-timescale. This requires $M_u \propto c^3G^{-1}t$ to be constant, hence $G \propto t$ in contrast to Dirac's result, but the gravitational radius of the universe, $R_g = 2GM_u/c^2$ is proportional to t in both Milne's theory and Dirac's because (18) gives $R_g \propto GM_u \propto t^{-1}Nm_n \propto t^{-1}t^2 \propto t$.

Dirac's first paper of 1937 and its sequel of 1938 created considerable interest and controvery.[14] Two responses are particularly interesting. The first was by Chandrasekhar (1937) who wrote a letter to *Nature* pointing out further astrophysical consequences of Dirac's LNH that all large dimensionless numbers of physics of order 10^{39} be directly proportional to the age of the universe. He noted that one could write down a one-parameter family of masses involving the proton mass m_n,

$$m(\xi) = (hc/Gm_n^2)^{\xi}\, m_n \tag{19}$$

involving the dimensionless combination $(hc/Gm_n^2) \sim 10^{39}$. Hence Dirac's LNH requires that

$$m(\xi) \propto t^{\xi} m_n. \tag{20}$$

Now, one observes the interesting coincidences that $m(3/2) \sim 10^{33}$ gm $\sim 1\,M_\odot$, $m(7/4) \sim 10^{11}\,M_\odot \sim 10^{44}$ gm, and $m(2) \sim 10^{21}\,M_\odot \sim 10^{55}$ gm are approximately equal to the masses of stars, galaxies and the observable universe respectively. Hence, according to Dirac's hypothesis Chandrasekhar predicts that the masses of stars and galaxies should be increasing with time as $t^{3/2}$ and $t^{7/4}$ respectively. Dirac took these to be observationally testable predictions of the LNH which he cited (1937b) in his defence against Dingle's stinging criticisms of the way in which Eddington, Milne and himself had departed from the scientific method in trying to force the universe into some philosophical strait-jacket divorced from observational motivation or support (Dingle 1937):

> This combination of paralysis of the reason with intoxication of the fancy is, shown, if possible, even more strongly in Prof. P. A. M. Dirac's letter in *Nature* of February 20th last, in which he, too, appears victim of the great 'Universe' mania ... Milne and Dirac ... plunge headlong into an ocean of 'principles' of their own making ... The criterion for distinguishing sense from nonsense has to a large extent been lost: our minds are ready to tolerate any statement, no matter how ridiculous it obviously is, if only it comes from a man of repute and is accompanied by an array of symbols in Clarendon type ... the most impressive garb of pseudo-profundity ... [Clarendon type is boldface]

Others, like Kothari (1938a, 1938b), Zwicky (1939) and Jordan (1938) proposed observational tests of these first 'continuous creation' cosmologies whilst Jordan derived a theory of gravitation with varying G from an action principle in which ideas like Dirac's could be evaluated self-consistently.

[14] Turing's biography (Hodges 1983) tells of his interest in Dirac's paper and his early fascination with Eddington's writings. He wondered whether the decrease in G with time could be tested if 'a palaeontologist could tell from the footprint of an extinct animal, whether its weight what was it was supposed to be'. A brief early discussion of the Dirac Large Number coincidences can be found in McCrea (1950).

Dirac's arguments were of course far from rigorous. Actually, he regarded the fact that they did not follow from any particular theory as a strength, indicative of their generality (Dirac 1937b), and not a sign of the lack of any solid foundation, as did Dingle (1937). He had implicitly used formulae, such as $G\varrho t^2 \sim 1$, derived from a gravity theory (general relativity) which only exists if G does not change with time, in order to conclude that it does. If one were to follow Dirac's arguments today then further Large Number coincidences can be given which over-determine the LNH in an inconsistent fashion. For example, the author first noticed that the ratio of the proton lifetime, $\tau_{pr} \sim 10^{32}$ yr, predicted by Grand Unified gauge theories to the Planck time, $t_{pl} \sim 10^{-43}$ s, given by (6) is of order 10^{79} (Barrow 1979); that is, we can create a new Large Number,

$$N_3 = \tau_{pr}/t_{pl} \sim \alpha^{-2} m_x^4 m_n^{-5} G^{-1/2} \sim 10^{79}, \tag{21}$$

where $m_x \sim 10^{15}$ GeV is the mass-energy scale at which 'Grand Unification' of the strong and electro-weak forces is supposed to occur. If this is added to Dirac's list and equated to N(t) then no single time-variation of G alone is consistent with the LNH because $N_3 \propto G^{-1/2} \propto t^{1/2}$ according to (16) so $G \propto t^{-1}$ but the LNH requires $N_3 \approx N \propto t^2$, which means $G \propto t^{-4}$.

Dirac's ideas filled a vacuum within theoretical cosmology at the time and were widely studied, criticised and extended. They also raised new cosmological questions which are still of great interest. In his 1938 paper Dirac argued that the LNH required that both the cosmological constant and the spatial three-curvature of the Friedmann universe be zero otherwise they would create further large dimensionless numbers. For example, the combination (11) involving Λ would need to vary with the age of the universe. Of this combination he writes,

> if it were not zero it would have to be very small not to be in disagreement with observation and its reciprocal would then provide us with a very large constant number, in contradiction to our fundamental principle.

It is apposite to notice that it was at this time that the anomalous size of the cosmological constant and the curvature radius of the universe were first noticed. Today, it is again fashionable to draw attention to these numbers although it has become conventional to use their reciprocals to display extremely small numbers which constitute what have become known as the 'flatness' and 'cosmological constant problems' (Guth 1981; Barrow 1988a).

An interesting response to Dirac's suggestion, which appeared in between the publication of Dirac's 1937 and 1938 papers was that of Arthur Haas (1938) who pointed out that Dirac's conclusion that G must vary with time was a consequence solely of his idiosyncratic taste in large numbers. If one supposed the universe to a closed Friedmann universe, and hence finite in size, then one could replace Dirac's Large Number N_1

which depends upon the *present* radius of the observed universe by N_1', the total number of particles in the entire universe. This depends upon the maximum size, ct_{max}, of the closed universe, so

$$N_1' = (ct_{max})m_e c^2/e^2. \tag{22}$$

Numerically N_1' should still lie close to 10^{39}, but there is a big conceptual change if one uses N_1' in place of N_1 because N_1' does *not* depend upon time. The maximum size is a time-independent quantity. Likewise in (15) we replace t, the present age of the universe, by t_{max} the age at its expansion maximum (which equals one half of its total age from beginning to end). As a result, we have the exact identity (Barrow and Tipler 1986)

$$N_1' \times N_2 = N \tag{23}$$

so there are now only two independent Large Numbers and no varying constants are required because the LNH just hypothesises relations between collections of *time-independent* quantities of similar magnitude.

Dirac's innovation created a new sub-culture within gravitation physics. After the war theorists proposed all manner of *ad hoc* time variations in physical constants traditionally regarded as constant. Following the early example of Jordan, historical evidence from geology and palaeontology was appealed to in an effort to determine whether gravity could have been so much stronger in the past. The most influential analysis was that of Teller (1948) who argued that the consequences of $G \propto t^{-1}$ would be in conflict with our own existence. Teller pointed out that the solar luminosity, $L_\odot$, would have been higher in the past because $L_\odot \propto G^7$ and, according to the conservation of angular momentum, the mean radius, R, of the Earth's orbit around the Sun would have been smaller as $R \propto G^{-1} \propto t$. Thus the temperature at the surface of the Earth, T_+, would have been significantly greater in the past

$$T_+ \sim (L_\odot/R^2)^{1/4} \propto G^{9/4} \propto t^{-9/4}. \tag{24}$$

Such rapid evolution in G implies that in the past the Earth's surface would have been far hotter than at present. Teller estimated that terrestrial oceans would have boiled away (taking the age of the universe to be 2×10^9 years he obtains $T_+ > 370\,°C$) in the pre-Cambrian era (400 million years ago) and the existence of tribolites would have been impossible. This adverse result later stimulated Gamow (1967, 1967a) to point out that the LNH does not necessarily require the result (18). Only the relation (17) is required by the LNH and Gamow pointed out that the deduction

$$e^2 \propto t \tag{25}$$

was equally acceptable.[15] However, the strategy of choosing (25) instead

[15] This variation was also deduced much earlier by Milne in the context of a discussion of Bohr's atomic theory and Eddington's relation (10) in the context of his two-time cosmology (Milne 1938). For an interesting account of Gamow's interest in Large Numbers see Alpher (1973).

of (18) ameliorates the 'boiling ocean' problem. A gradual temporal increase in the electron charge leaves the Earth's orbital motion unaffected but the structure of the Sun is affected through the effect upon the opacity which determines the rate at which energy is transported to the surface. The luminosity changes as $L_\odot \propto e^{-6}$ and hence $T_+ \propto L_\odot^{1/4} \propto t^{-3/4}$. This moves the epoch of boiling oceans so far into the past that any conflict with biological data is avoided.

Despite its novel aspects, Gamow's paper was actually just the first in a long line of incorrect studies of the cosmological and astrophysical consequences of the time variation of fundamental constants other than G. All these studies (see Dyson (1972) and Wesson (1978)) simply assume that the usual equations of physics hold in unchanged form when a constant is allowed to vary and then allow the constant quantity to be a time variable in the same equations in order to discover the consequences. In effect the world is being assumed to evolve through a family of different universes each with different but *constant* values of their fundamental constants.[16] Only in the case of the variation of gravity with time was something self-consistent actually done. New theories of gravitation in which Newton's gravitation constant, G, was replaced by a scalar field, able to vary in space and time, were first developed by Jordan but culminated in the elegant form developed by Brans and Dicke (1961). The formulation of this field theory of gravity, a natural generalisation of Einstein's, enabled Dirac's proposal to be confronted with observation in a self-consistent fashion.

In between Dirac's proposal and the development of Brans–Dicke theories of gravitation there emerged the steady-state theory of the universe. This theory offered no explanations for the Large Number coincidences[17] and Bondi, in his classic book, *Cosmology* (Bondi 1952) called Dirac's approach a 'counsel of despair' because it offered no guide as to how the subject should develop. Moreover, Bondi felt that the introduction of varying 'constants' by Dirac could only strengthen one's belief in the steady-state theory and the 'Perfect Cosmological Principle' of Bondi and Gold (1948) upon which it was based because it showed 'how limitless the variations are that may be imagined to arise in a changing universe'. The only answer was to outlaw them all. In retrospect this approach is not so transparent as first appears. The real question is at what logical level one applies the Perfect Cosmological Principle since all cosmological changes can be re-expressed as the time-invariance of some higher-order quantity. Bondi also went on to contrast Dirac's approach

[16] Self-consistent studies of the time-variation of constants other than G have only been made recently in Kaluza–Klein cosmological models with extra space dimensions (Barrow 1987; Marciano, 1984).

[17] This is somewhat ironic since the connection between microscopic physics and the large-scale structure of the universe that is always being sought in the studies of Large Numbers might have been expected to arise naturally only in the steady-state model. For in that theory microscopic physical constants must control the process of continuous creation which in turn necessarily determines the expansion rate and size of the observable universe.

sharply with that of Eddington, of which he was especially sceptical because of its complete absence of any forecasts which might be checked against observation, because whereas Eddington places huge significance upon the present values of the physical constants, Dirac endows them with none. For him they are ephemeral consequences of the cosmological time of day. Yet, curiously, it is Dirac's desire that the Large Number coincidences be found to be the same by observers at all cosmic times which motivates the LNH is in spirit just the Perfect Cosmological Principle of the steady statesmen.

5.5 From coincidence to consequence

> Accident: n. An inevitable occurrence due to the action of immutable natural laws.
>
> Ambrose Bierce

Robert Dicke was the first to realise that in an evolving universe there exists a preferred time period when observers like ourselves are most likely to exist.[18] Dicke first mooted this significant point in his *Reviews of Modern Physics* article of 1957 devoted to geophysical and palaeological evidences for the variation of physical constants:

> The problem of the large size of these numbers now has a ready explanation ... there is a single large dimensionless number which is statistical in origin. This is the number of particles in the Universe. The age of the Universe 'now' is not random but is conditioned by biological factors ... [because changes in the values of Large Numbers] would preclude the existence of man to consider the problem.

Four years later he elaborated it in detail, with reference to Dirac's Large Number coincidences, in a letter to *Nature* (Dicke 1961). Dicke argued that carbon-based life-forms owe their chemical basis to such elements as carbon, nitrogen, oxygen and phosphorus which are synthesised after billions of years of main-sequence stellar evolution. (The argument applies with equal force to any life-form based upon atomic elements heavier than helium.) When stars die these 'heavy' biological elements are dispersed throughout space by supernovae from whence they are incorporated into grains, planetesimals, planets, self-replicating 'smart' molecules like DNA and, ultimately, ourselves. Observers cannot arise until roughly the hydrogen-burning lifetime of a main-sequence star has

[18] Eddington had come close to employing this type of argument in 1930 when he remarked that there would need to be a remarkable coincidence in order that 'astronomers ... are just in time to observe [the spiral nebulae] this interesting but evanescent feature of the sky' in Lemaître's cosmology (Eddington 1930). However, he did not draw any connection between astronomical and evolutionary timescales. Dicke had previously applied Dirac's LNH to other fundamental interactions to claim that since the weak interaction could be characterised by a dimensionless number of order 10^{20} it should be found to vary as $t^{1/2}$ (Dicke 1957, 1959).

elapsed. This time-scale is determined essentially by fundamental constants of Nature to be

$$t_* \sim \left(\frac{Gm_n^2}{hc}\right)^{-1} \frac{h}{m_n c^2} \sim 10^{10}\,\text{yrs}. \qquad (26)$$

We would not expect to be observing the universe at times significantly in excess of t_* since all stable stars would have expanded, cooled and died. We seem strait-jacketed by the facts of biological life to gaze at the universe and develop cosmological theories after a time close to t_* has elapsed since the big bang. Thus the value of Dirac's N(t) that we observe is by no means random. It must have a value close to $N(t_*)$ where

$$N(t_*) = t_* c^3 G^{-1} m_n^{-1}. \qquad (27)$$

But if we combine (26) and (27) we see that this requires

$$N(t_*) = \left(\frac{Gm_n^2}{hc}\right)^{-2} \sim 10^{79}. \qquad (28)$$

This is precisely Dirac's famous coincidence (and Eddington's as well, in fact). It is a necessary property for an expanding big-bang universe to possess observers like ourselves that those observers witness this coincidence to hold. There is no need to give up Einstein's theory of gravitation, as Dirac implicitly required, nor is one warranted in deducing some quasi-statistical connection between the strength of gravity and the number of particles in the universe as Eddington and many others had done. Dirac's response (Dirac 1961), argued that 'On Dicke's assumption habitable planets could exist only for a limited period of time. With my assumption they could exist indefinitely in the future and life need never end.' However, this is incorrect since the decrease of G would eventually render all gravitationally-bound structures like planets unstable.

The result that we have just established can be restated in another more striking fashion. In order for a big-bang universe to contain the basic building blocks necessary for the subsequent evolution of biochemical complexity (by this we mean any elements heavier than helium) it must have an age at least as long as the time it takes for the nuclear reactions in stars to produce these elements. This means that the observable universe must be at least ten billion years old and therefore, since it is expanding, at least ten billion light years in size. We could not exist in a universe that was significantly smaller. The isolation of *necessary* conditions on fundamental constants and cosmological parameters in order that complex biochemistry can evolve has become known as the *Weak Anthropic Principle* (WAP), following the introduction of this terminology[19] by Brandon

[19] The term 'anthropic' was first used in an analogous connection in volume 2 of Tennant (1930: 83), where he argues that there is a relation between 'the intelligibility of the world . . . and the constitution and process of Nature' which arranges events into 'anthropic categories' which are consistent with the existence of rationality which need not be human.

Carter (1974).[20] As wielded by Dicke we see its use as a methodological principle that points out the existence of a selection effect upon our observations. Dirac assumed that all values of N(t) were in some sense equiprobable. Clearly they are not. Failure to appreciate the selection effect introduced by the fact that we exist led to entirely unwarranted and radical conclusions (since refuted by observation) regarding the gravitation constant.

This type of argument regarding the self-selection of the size of the observable universe by observers appears first in Mascall's Bampton Lectures[21] which were published in 1956 under the title *Christian Theology and Natural Science* where he attributes the basic idea to Whitrow. Stimulated by Whitrow's suggestions, he writes:

> if we are inclined to be intimidated by the mere size of the Universe, it is well to remember that on certain cosmological theories there is a direct connection between the quantity of matter in the Universe and the conditions in any limited portion of it, so that in fact it may be necessary for the Universe to have the enormous size and complexity which modern astronomy has revealed, in order for the earth to be a possible habitation for living beings.

Whitrow had employed this style of argument in a paper (Whitrow 1955) published at about that time which sought to provide an answer to the question 'why does space have three dimensions?' by showing that observers could only exist in three-dimensional worlds and hence three dimensions are a necessary pre-requisite for such problems to be posed. Similar stress upon the connection between the large-scale structure of the universe and the local conditions necessary for our own existence can be found in the work of the Soviet biophysicist and historian of science Grigori Idlis at about the same time (Idlis 1958) where he examines many of the features of the large-scale universe of stars and galaxies that are necessary for evolution to occur.

The fact that certain coincidences are necessary conditions for the existence of observers may not explain them causally. One way of proceeding from the fact of self-selection by observers to fuller causal

[20] Carter had presented similar ideas in a paper to the Clifford Centennial Meeting in Princeton 1970 and in 1967 had carried out a detailed analysis of the dependence of stellar structure and other necessary conditions for the evolution of life on the values of fundamental constants in an unpublished manuscript entitled 'The Significance of Numerical Coincidences in Nature'. Carter attributes his interest in the Large Number coincidences to the chapter on the subject in Bondi (1952) and his exasperation at the credence given to the completely unwarranted speculations of Jordan and Dirac about time-varying constants.

[21] The Bampton Lectures are a series of eight Divinity lectures to be delivered in the University Church at Oxford on Sunday mornings during the Lent and Act university terms. They were established by the will of the Reverend John Bampton, Canon of Salisbury (1690–1751), 'to confirm and establish the Christian Faith, and to confute all heretics and schismatics'. The first lectures were delivered in 1780 and continued annually until 1895 since when they have been biennial. This brief has traditionally been taken to include the subject of natural theology.

explanation is to appeal to an ensemble of different possible universes where fundamental constants take on all possible values at random, or to a variation in space of these constants within the actual universe. This latter course was also followed by those impressed by the array of remarkable coincidences between the relative values of dimensionless constants which then turn out to be necessary conditions for our existence (Carter 1974; Hoyle 1965; Pantin 1951, 1965). With regard to the existence of such life-supporting coincidences the Cambridge biologist Charles Pantin proposed (Pantin 1965):

> all we have is a speculation ... that the properties of the material Universe are uniquely suitable for the evolution of living creatures. To be of scientific value any explanation must have predictable consequences. These do not seem to be attainable. If we could know that our own universe was only one of an indefinite number with varying properties we could perhaps invoke a solution analogous to the principle of Natural Selection; that only in certain universes, which happen to include ours, are the conditions suitable for the existence of life, and unless that condition is fulfilled there will be no observers to note the fact.

5.6 Epilogue

> I have yet to see any problem, however complicated, which when you looked at it in the right way, did not become still more complicated.
>
> Poul Anderson.

The current legacy of the early speculations about Large Numbers is unexpectedly large. The Large Number coincidences still have no complete explanation but the past attempts to explain them have blossomed into answers to other questions. Varying constants have re-emerged in the context of Kaluza–Klein theories. The Anthropic Principle and the random variation of cosmic conditions throughout space have been found essential ingredients in the 'Inflationary Universe' of Guth and Linde (Barrow 1988; Guth 1981; Linde 1987). The explanation of the values of the fundamental constants is the challenge to be met by any new 'Theory of Everything', but whereas our story of the past has shown the astronomer persistently trying to strait-jacket the world of fundamental physics, today it is the particle physicist who seeks to encompass the universe. However, it is appropriate to end what is largely a story of magnificent failures with a caution as to how modern searches for a 'Theory of Everything' may fail even if they succeed. A fundamental Theory of Everything, whilst necessary to explain our observation of the universe today, is by no means sufficient. Initial conditions, random symmetry breakings and collective phenomena each play an irreducible role in determining the 'solutions' of the equations of any Theory of Everything and it is these that we observe (Barrow 1988a). We are on the wrong side of the tapestry. It is not the laws of Nature alone that determine what we see. Moreover, in a cosmological picture like the chaotic inflationary universe (Linde 1987) many features of the large-scale structure of the universe may have further *no* explanation save for the fact that they are the random symmetry breakings necessary for the evolution of astronomers. The 'Theory of Everything' on the other hand allows all sorts of other universes to arise. Whether the Large Numbers permit any deeper explanation than this remains a question that seeks an answer still.

Acknowledgements
I am indebted to Professor H. Bondi and Professor L. Kostro for helpful suggestions and to Professor W. H. McCrea for a number of invaluable discussions and references.

References

Alpher, R. A. (1973). Large Numbers, Cosmology and Gamow. *American Scientist*, **61**, 52.
Barrow, J. D. (1979). The Proton Half-Life and the Dirac Hypothesis. *Nature*, **282**, 698.
(1981). The Lore of Large Numbers: Some Historical Background to the Anthropic Principle. *Quarterly Journal of the Royal Astronomical Society*, **22**, 388.
(1983). Natural Units before Planck. *Quarterly Journal of the Royal Astronomical Society*, **24**, 24.
(1987). Observational Limits on the Time-Evolution of Extra Spatial Dimensions. *Physical Review*, **D35**, 1805.
(1988a). The Inflationary Universe: Modern Developments. *Quarterly Journal of the Royal Astronomical Society*, **29**, 101.
(1988b). *The World Within the World*. Oxford: Oxford University Press.
(1988c). Constants of Physics and the Structure of the Universe. In Saas Fee lectures on *Unités de Mesure et Constantes Physiques*, chapt 8 pp. 1–28, ed. M. Batato, R. Behn, J.-F. Loude & H. Weisen. Lausanne: Assoc. Vaudoise des chercheurs en Physique.
Barrow, J. D. and Tipler, F. J. (1986). *The Anthropic Cosmological Principle.* Oxford: Oxford University Press (rev. edn. 1988).
Beck, G., Bethe, H. and Riezler, W. (1931). Bemerkung zur Quantentheorie der Nullpunkstemperatur. *Naturwiss*, **19**, 39.
Bondi, H. (1952). *Cosmology*. Cambridge: Cambridge University Press.
Bondi, H. and Gold, T. (1948). The Steady State Theory of the Expanding Universe. *Monthly Notices of the Royal Astronomical Society*, **108**, 253.
Born, M. (1944). *Experiment and Theory in Physics*. Cambridge: Cambridge University Press.
Boscovich, R. J. (1753). *Theoria Philosophiae Naturalis* (English trans.). Cambridge: MIT Press, 1966.
Brans, C. and Dicke, R. H. (1961). Mach's Principle and a Relativistic Theory of Gravitation. *Physical Review*, **124**, 924.

Bridgman, P. W. (1920). *Dimensional Analysis*. New Haven: Yale University Press.
Carter, B. (1974). In *Confrontation of Cosmological Theories with Observation*, ed. M. S. Longair. Dordrecht: Reidel.
Chandrasekhar, S. (1937). Cosmological Constants. *Nature*, **139**, 757.
Crowther, J. G. (1952). *British Scientists of the Twentieth Century*. London: Routledge.
Delbœuf, J. R. L. (1860). *Prolegomènes philosophiques de la géométrie*. Paris.
(1893). L'Ancienne et les nouvelles géométries. *Revue Philosophique*, **36**, 449.
Delbrück, M. (1972). In *Cosmology, Fusion and Other Matters*, ed. F. Reines. Bristol: Adam Hilger.
Dicke, R. H. (1957). Principle of Equivalence and the Weak Interactions. *Review of Modern Physics*, **29**, 355.
(1959). Dirac's Cosmology and the Dating of Meteorites. *Nature*, **183**, 170.
(1961). Dirac's Cosmology and Mach's Principle. *Nature*, **192**, 440.
Dingle, H. (1937). Modern Aristotelianism: Deductive and Inductive Methods in Science: A Reply. *Nature*, **139**, 784 & 1011.
(1954). *The Sources of Eddington's Philosophy*. Cambridge: Cambridge University Press.
Dirac, P. A. M. (1937a). The Cosmological Constants. *Nature*, **139**, 323.
(1937b). *Nature*, **139**, 1001.
(1938). A New Basis for Cosmology. *Proceedings of the Royal Society*, **165**, 199.
(1961). *Nature*, **192**, 441.
(1973). Long-range Forces and Broken Symmetries. *Proceedings of the Royal Society*, **A333**, 403.
Dresden, M. (1987). *H. A. Kramers*. Berlin: Springer.
Duhem, P. (1954). *The Aim and Structure of Physical Theory*, trans. P. Weiner. Princeton: Princeton University Press.
Dyson, F. (1972). In *Aspects of the Quantum Theory*, ed. A. Salam and E. Wigner. London: Cambridge University Press.
Eddington, A. S. (1918). *Report on the Relativity Theory of Gravitation*. London: Physical Society of London (Fleetway Press).
(1920). *Space, Time and Gravitation*. Cambridge: Cambridge University Press.
(1923). *The Mathematical Theory of Relativity*. Cambridge: Cambridge University Press.
(1930). On the Instability of Einstein's Spherical World. *Monthly Notices of the Royal Astronomical Society*, **90**, 677.
(1935). *New Pathways in Science*. Cambridge: Cambridge University Press.
(1939). *The Philosophy of Physical Science*. Cambridge: Cambridge University Press.
(1946). *Fundamental Theory*. Cambridge: Cambridge University Press.
Einstein, A. (1911). Molekularbewegung in festen Körpern. *Annalen den Physik*, **35**, 687.
(1955). *The Meaning of Relativity*. London: Methuen 5th edn.
Fournier d'Albe, E. E. (1906). *The Electron Theory: A Popular Introduction to the New Theory of Electricity and Magnetism* (1st edn.). London: Longmans.
Gamow, G. (1967). Variability of Elementary Charge and Quasistellar Objects. *Physical Review Letters*, **19**, 913.

(1967a). Electricity, Gravity and Cosmology. *Physical Review Letters*, **19**, 759.
Gould, S. J. and Eldredge, N. (1977). Punctuated Equilibria: The Tempo and Mode of Evolution Reconsidered. *Paleobiology*, **3**, 115.
Guth, A. (1981). Inflationary Universe: A Possible Solution to the Horizon and Flatness Problems. *Physical Review*, **D23**, 347.
Haas, A. (1938). Relation Between the Gravitational Constant and the Hubble Factor. *Physical Review*, **53**, 207.
Haldane, J. B. S. (1937). *Nature*, **139**, 1002.
(1944). Radioactivity and the Origin of Life in Milne's Cosmology. *Nature*, **158**, 555.
(1955). In *New Biology No. 16*, ed. M. L. Johnson, M. Abercrombie and G. E. Fogg. London: Penguin.
Hawking, S. W. (1983). The Cosmological Constant. *Philosophical Transactions of the Royal Society*, **A310**, 311.
Heisenberg, W. (1958). *Physics and Philosophy*. New York: Harper and Row.
Hodges, A. (1983). *Alan Turing: The Enigma of Intelligence*. London: Hutchinson.
Hoyle, F. (1965). *Galaxies, Nuclei and Quasars*. London: Heinemann.
Idlis, G. (1958). Basic Features of the Observed Astronomical Universe as Characteristic Properties of a Habitable Cosmic System. *Izv. Astrophys. Inst. Kazakh*, **SSR7**, 39.
Joad, C. E. M. (1932). *Philosophical Aspects of Modern Science*. London: Unwin.
Johnstone Stoney, G. (1874). Report On the Physical Units of Nature. British Association, Belfast.
(1881). On the Physical Units of Nature. *Philosophical Magazine* (Series 5), **11**, 381.
(1891). On the Physical Units of Nature. *Transactions of the Royal Dublin Society*, **4**, 563.
(1894). *Philosophical Magazine* (Series 5), **38**, 418.
Jordan, P. (1938). Zur empirischen Kosmologie. *Naturwiss*, **26**, 417.
Kothari, D. S. (1938a). Cosmological and Atomic Constants. *Nature*, **142**, 354.
(1938b). The Theory of Pressure Ionization and Its Applications. *Proceedings of the Royal Society*, **A165**, 486.
Laplace, P. (1808). *Exposition du système du Monde*, book 4, chap. 16. Paris.
Linde, A. (1987). In *300 Years of Gravity*, ed. W. Israel and S. W. Hawking. Cambridge: Cambridge University Press.
Marciano, W. J. (1984). Time Variation of the Fundamental 'Constants' and Kaluza–Klein Theories. *Physical Review Letters*, **52**, 489.
Mascall, E. (1956). *Christian Theology and Natural Science*. London: Longmans.
McCrea, W. H. (1950). The Steady-State Theory of the Expanding Universe. *Endeavour*, **9**, 3 (Jan. issue no. 33).
Millikan, R. (1926). Under 'Electron' in *Encyclopaedia Brittanica*. London.
Milne, E. A. (1935). *Relativity, Gravitation and World Structure*. Oxford: Clarendon.
(1938). On the Equations of Electromagnetism. *Proceedings of the Royal Society*, **A165**, 313.
(1948). *Kinematic Relativity*. Oxford: Clarendon.
Pantin, C. F. A. (1951). Organic Design. *Advances in Science*, **8**, 138.

(1965). Life and the Conditions of Existence. In *Biology and Personality*, ed. I. T. Ramsey. Oxford: Blackwell.
Planck, M. (1899). Über irreversible Strahlungsvorgänge. *Sitzungsberichte der Preussisch–Akademie der Wissenschaften*, **5**, 479.
(1900). Über irreversible Strahlungsvorgänge. *Annalen der Physik*, **1**, 120.
(1906). *Theorie der Wärmestrahlung*. Leipzig: Barth. Engl. translation: *The Theory of Heat Radiation*, trans. M. Masius. NY: Dover, 1959.
Poincaré, H. (1914). *Science and Method* (trans. 1952). NY: Dover.
Rosenthal-Schneider, I. (1980). *Reality and Scientific Truth: Discussions with Einstein, von Laue and Planck*. Detroit: Wayne State Press.
Schrödinger, E. (1938). Mean Free Path of Protons in the Universe. *Nature*, **141**, 410.
Slater, N. B. (1957). *Development and Meaning of Eddington's Fundamental Theory*. Cambridge: Cambridge University Press.
Stebbings, S. (1944). *Philosophy and the Physicists*. Harmondsworth: Pelican.
Stewart, J. Q. (1931). Nebular Redshift and Universal Constants. *Physical Review*, **38**, 2071.
Taub, A. H. (1950). *Mathematical Reviews*, **11**, 144.
Teller, E. (1948). On the Change of Physical Constants. *Physical Review*, **73**, 801.
Tennant, F. R. (1930). *Philosophical Theology*, Vols 1 and 2. Cambridge: Cambridge University Press.
Turing, S. (1959). *Alan Turing*. Cambridge: Heffers.
Weinberg, S. (1972). *Gravitation and Cosmology*. NY: Wiley.
(1987). Anthropic Estimate of the Cosmological Constant. *Physical Review Letters*, **59**, 2607.
Wesson, P. (1978). *Cosmology and Geophysics*. Bristol: Adam Hilger.
Weyl, H. (1919). Eine neue Erweiterung der Relativitätstheorie. *Annalen der Physik*, **59**, 129.
(1934). Universum und Atom. *Naturwissenschaften*, **22**, 145.
(1949). *Philosophy of Mathematics and Natural Science* (rev. English trans. by O. Helmer). Princeton: Princeton University Press.
Whitrow, G. (1955). Why Physical Space has Three Dimensions? *British Journal for the Philosophy of Science*, **6**, 13.
Whittaker, E. T. (1949). *From Euclid to Eddington: A Study of Conceptions of the Physical World*. Cambridge: Cambridge University Press.
Witt-Hansen, J. (1958). *Exposition and Critique of the Conceptions of Eddington Concerning the Philosophy of Physical Science*. Copenhagen: Gads.
Yolton, J. W. (1960). *The Philosophy of Science of A. S. Eddington*. The Hague: Nijhoff.
Zwicky, F. (1939). On the Theory and Observation of Highly Collapsed Stars. *Physical Review*, **55**, 726.

Discussion

Hoyle:

Dirac's paper caused a good deal of discussion among us research students in Cambridge. The point was made that one cannot shift either the mass of bodies or the gravitational constant without running into difficulties with the orbits of close satellites such as Mars and Jupiter. (This is so long as one does not vary both in a conformal manner.)

Bondi:
1) There was a recent period when the geologists were very fond of a variable gravitational constant, in line with some of the theories you mentioned.
2) It was most interesting to hear you point out that as soon as the electron was identified, it was seen that this ended the Newton–Maxwell (–Einstein?) period of scale-free physics and yielded natural units. Incidentally, Einstein later disliked the cosmological constant because it destroyed the scale-free character of general relativity.

III
Geometrical and physical cosmology

6

Innovation, resistance and change: the transition to the expanding universe

G. F. R. Ellis

6.1 Introduction

In this paper, the overall view of the history of cosmology taken is that it can be seen as the interplay between shifts of viewpoint on the nature of cosmology and increasing observational evidence. These shifts of viewpoint comprise both major paradigm changes and more detailed changes of understanding, where by a *paradigm*, we mean here a basic viewpoint on the nature of cosmology as a topic of scientific study.

6.1.1 *Four major paradigms*

At the broadest level, the proposal is that we can use a framework of four paradigms (in general, inclusive rather than exclusive of the preceding paradigms) and three associated paradigm changes:

> **Paradigm 1:** a static or stationary unchanging universe. This necessitates introducing a cosmological constant Λ into Einstein field equations. The model can be tested by number counts of galaxies.
> **Paradigm 2:** an evolving universe model, characterised by geometrical properties and tested by a (redshift, distance) relation. This implies the possibility of an age problem arising from demanding consistency between the ages of objects in the universe and the age of the universe itself (determined from the Hubble constant). This is exclusive of the preceding paradigm, although the static and stationary models are limiting cases of the dynamic models.
> **Paradigm 3:** an evolving universe with a hot big bang initial phase whose evolution is determined by the laws of microscopic physics. This leads to predicted relations between primordial element abundances and the cosmic background radiation spectrum. Growth of small-scale inhomogeneity is controlled by these physical processes. This is inclusive of the preceding paradigm.
> **Paradigm 4:** a hot big bang universe with physical effects limited by causality horizons. This implies the possibility of field-theory

topological defects (domain walls, strings, and monopoles) and the probability of large-scale inhomogeneity. This is inclusive of the previous two paradigms except in so far as they are based on *a priori* symmetry assumptions (a 'cosmological principle').

The changes of paradigm have been rapid when they have taken place, but have in each case been preceded by forerunners and by resistance to the change taking place. The changes can be illustrated in a tabular form (see table 6.1). Note that the table is indicative rather than complete

Table 6.1. *Major paradigm changes in the history of cosmology*

Year	Paradigm 1: static or stationary	Change 1	Paradigm 2: evolving geometry	Change 2	Paradigm 3: non-equilibrium physics	Change 3	Paradigm 4: horizon limits
1917	**Einstein**						
	de Sitter						
1922	Einstein		Friedmann				
	Eddington						
1927			**Lemaître**				
	Robertson						
1929	Hubble						
	Tolman						
1930	⇒	*Eddington*					
			de Sitter				
			Tolman				
			Einstein				
1933			Robertson		Lemaître		
			Milne		Tolman		
1935			Walker				
			Heckmann				
1948	Bondi,		McVittie		**Gamow**		
	Gold		Bondi		Alpher,		
	Hoyle		Ryle		Herman		
1956			Sandage		Hayashi		Rindler
			Gunn				
					Zeldovich		Penrose
1965			⇒	*Peebles*			
					Fowler		**Misner**
			Tinsley		Wagoner		
					Thorne		
					Sachs, Wolfe		
			Tammann		Sakharov		
1972			Sandage		Weinberg		
					Schramm		
					Steigmann		Kibble
					⇒	*Guth*	
							Steinhardt
1982					Yoshimura		Albrecht
					Weinberg		Turner
1986							Linde

(many other names would be included in a fuller table[1]). Nevertheless, it clearly illustrates the main features.

In each case, there is a person or person who first discovers or understands the basic concept (I have indicated in bold type those who in my opinion first clearly presented the essence of the new paradigm). However, they have been ignored or resisted by the majority of their peers; they are forerunners of the change to come, but are before their time. Then there is a *transition figure* (indicated in italics) who triggers off the major paradigm change; this is not the person who first discovered or understood the new concept, but the first who put it in context convincing to the rest of their peers. This process is greatly aided by suitable observational evidence but by itself such evidence is not sufficient to make the change take place; indeed in each of the first two changes, observational evidence for the new view was available for a considerable time before its significance was understood. In the case of the last shift, the reasons for making the change are theoretical rather than observational; indeed there is still no substantial observational evidence supporting the new paradigm! In so far as the last three paradigms are inclusive of the previous, it is in these cases possible to effectively work in the old one, after the change has taken place, without denying the new one; such work will not specifically deal with the new concepts but will take cognizance of the new possibilities opened up by the new viewpoint (in contrast to work denying the new paradigm, which re-affirms the old viewpoint against the new).

In the remainder of this paper we will look in more detail at the first of these paradigm changes.

6.2 The transition to the expanding universe

Before looking at this transition, it is first necessary to explain briefly some of the sources of confusion in understanding the nature of the de Sitter universe as a world model.

6.2.1 *Redshift in the de Sitter universe*

It is important to realise that a cosmological model is specified only when a 4-velocity u^a representing the average motion of matter in the universe has been specified as well as the space-time metric g_{ab}; the observable relations in the model are determined by the choice of this 4-velocity, or equivalently of the associated fundamental world-lines. In particular, observed redshifts are determined by the formula

$$d\lambda/\lambda = (\dot{u}_a e^a + \dot{S}/S)dl, \qquad (1)$$

where $d\lambda$ is the change in wavelength of light emitted at wavelength λ

[1] See the bibliography in Ellis (1988).

Table 6.2. *Different cosmological descriptions of Minkowski, Einstein static, de Sitter, and FLRW space-times together with the first discoverers of each form*

	Static universe		FLRW universe		
Space-time	Geodesic	Accelerating	$k = +1$	$k = 0$	$k = -1$
Minkowski (C)	Minkowski	Rindler			Milne $S = t$
Einstein static (A)	Einstein		Einstein $S = const.$		
de Sitter (B)		de Sitter	Lanczos $S = cosh\, t$	Lemaître $S = exp\, t$	$S = sinh\, t$
FLRW (expanding)			Friedmann	Robertson	Friedmann

A blank space indicates no such form is possible. There are no redshifts in the geodesic static universes, but there are gravitational redshifts in the accelerating static universes.

observed by fundamental observers in a distance dl along the path of a light ray, $\dot{u}_a$ is the acceleration of the fundamental world-lines, $\dot{S}$ the rate of change of the average scale factor S, and e^a the direction of observation (see e.g. Ehlers 1961; Ellis 1971).

The basic problem then is that in the de Sitter universe, unlike general Friedmann–Lemaître–Robertson–Walker (FLRW) universe models, there is no unique choice of such a world-line (because the space-time is a space-time of constant curvature); so it can be written in many different ways as a cosmological model (see e.g. Hawking and Ellis 1973; Schrödinger 1950). In fact, it can be written as a *static universe* or as a *FLRW universe with* $k = 0, +1,$ *or* -1 (see row 3 of table 6.2, which for comparison shows also the Minkowski, Einstein static, and FLRW expanding models). In the static case, there are observed redshifts because the fundamental world-lines are non-geodesic; just as in the case of the observers in the Rindler version of a Minkowski universe, we can regard these as gravitational redshifts (corresponding to the first term in equation (1), as in the case of gravitational redshifts in a Schwarzschild solution). In the latter three cases there will be expansion ('Doppler') shifts as in any expanding FLRW universe (corresponding to the second term in equation (1), but the way these Doppler shifts vary with time will depend on the choice of fundamental observers (and so of orthogonal spatial sections). It is the second choice ($k = 0$) that corresponds to the 'steady state' form of the de Sitter universe. A further complication arises in that because the fundamental world-lines in the static case are not geodesic, a freely falling particle released by any static fundamental observer will be seen by him to accelerate away, generating local velocity (Doppler) redshifts corresponding to these 'peculiar velocities'.

Thus if a *static frame* is used, there are (a) gravitational redshifts for the fundamental galaxies plus (b) Doppler shifts for freely falling (non-fundamental) objects; the magnitude of the redshift depends on the extent to which observed objects move on 'fundamental world-lines'. It is (a) that became known as the 'de Sitter effect', reflecting the space-time curvature. What is particularly confusing is that the redshifts due to this effect were customarily expressed in km/sec (that is, in terms of an equivalent velocity) even though they did not represent a relative velocity. Thus for example Eddington wrote in 1924,

> de Sitter's theory gives a double explanation for this motion of recession; *first* there is a general tendency to scatter . . ., *second* there is a general displacement of spectral lines to the red in distant objects owing to the slowing down of atomic vibrations ... which would erroneously be interpreted as a motion of recession.

If an *expanding frame* is used, one obtains (c) standard FLRW expansion redshifts, but with their form depending on the FLRW frame used, that is, on the choice of fundamental observers in the space-time. This arbitrariness of redshift possibilities (caused by (a), (b), or (c)) arises because the 4-velocity of matter in the de Sitter universe can be chosen arbitrarily (the matter is dynamically negligible, and does not affect the space-time curvature).

A further issue that added to the confusion over the de Sitter universe was the existence of 'horizons' associated with some but not all coordinate systems. All coordinates except the $k = +1$ form cover only a part of the de Sitter hyperboloid, and additionally the past light cone of any point intersects only some of the world-lines in the universe model, so in each representation there are limits to the matter that can be seen. The nature of these horizons was not fully understood until the work of Rindler in 1956 and Penrose in 1963.

6.2.2 *1917 to 1933*

In the period from 1917 to 1930, everyone knew the universe was static. Consequently when the idea of the evolving universe was proposed, first by Friedmann and then by Lemaître, it was ignored or resisted: the scientific community was not ready for this concept. In fact after the initial discovery of the Einstein static universe (model A) and de Sitter static universe (model B), the period was dominated by the question: is the real universe better described by model A or B? (see table 6.3, classifying papers and comments of the period; sufficient references are given at the end to establish the main features of the table[2]).

After the publication of Einstein's and de Sitter's models, apart from a

[2] See also Berendzen et al. (1976), Smith (1982), and Ellis (1988) for reviews of this period.

paper by Levi Civita (1917), as far as I am aware no articles were published on relativistic cosmology for four years. This may have been because this time was needed for the rest of the physics community, lacking the requisite mathematical tools, to understand how to apply general relativity theory. It has also been suggested that it was because those who did understand the technical tools were applying general relativity to the solar system, and in particular were caught up by the 'hot topic' of the bending of light. In any case the resurgence of published theoretical work starts in 1922.

After Friedmann's path-breaking paper in 1922, Einstein published a brief note (1922) claiming an error in Friedmann's work; when it was pointed out to him it was his error, Einstein (1923) published a retraction of his comment, with a sentence that luckily was deleted before publication (Stachel 1986): '[Friedmann's paper] while mathematically correct is of no physical significance.' The Friedmann papers (1922, 1924) were largely ignored by the rest of the workers in the field, as was Lemaître's paper (1927). This paper was the first to put the idea of the expanding universe in the astronomical context by examining predicted redshifts in these universes; thus it went significantly further towards our current view of the universe than Friedmann's papers which were largely mathematical rather than physical in character (observational estimates of matter density had already been deployed by several people in the context of the Einstein and de Sitter universes, when Friedmann's papers were published; these papers, however, failed to make the crucial connection with systematic redshifts).

Einstein's rejection of the evolving universe idea was confirmed by comments to Lemaître in 1928 rejecting his expanding universe solution (Krugh 1987). Lanczos, Lemaître, and Weyl all found expanding versions of the de Sitter universe; Lemaître, Robertson, and Tolman each took the step of generalising to an arbitrary scale factor S(t), but only Lemaître understood the significance of this step. Indeed the Robertson (1929) paper, based on a previous paper in 1927 which determined the geometrical basis of the FLRW models but apparently only as a mathematical exercise, is remarkable in that he wrote down the general FLRW line element, understood both its conformal flatness and imbedding in 5-dimensional flat space-time, and referred to Friedmann's work – without realising the significance of the expanding metric. Rather, like Tolman (1929a), the purpose of the paper was to find all stationary FLRW universe forms – which turn out to be models A and B. Thus these papers are quite clearly written in the framework of paradigm 1, despite containing many of the ideas and results of paradigm 2.

Observational evidence supporting the expanding universe idea had been available at least since 1924, when Eddington summarised Slipher's redshift data in his book. This evidence did not trigger a change of

Table 6.3. *The first major paradigm change, showing major papers or comments by different workers according to year and paradigm implied*

Year	*Paradigm 1:* Static universe Issue: A or B?	Transition	*Paradigm 2:* Evolving geometry Issue: $k = +1, 0, -1$?
1917	**Einstein** A		
	de Sitter B		
	Levi Civita A		
1918	Eddington $\Rightarrow$ B		
	Einstein $\neq$ B		
1922	Lanczos B ($k = +1$)		Friedmann ($k = +1$)
	Einstein $\neq$ F		
	de Sitter		
1923	Lanczos B ($k = +1$)		
	Weyl B		
	Jeans $\Rightarrow$ B		
1924	Eddington $\Rightarrow$ A		Friedmann ($k = -1$)
	Einstein $\neq$ F		
	Wirtz		
	Silberstein		
	Lundmark		
1925	Lemaître (B) ($k = 0$)		
	Charlier		
	Lundmark		
1926	Hubble $\Rightarrow$ A		
1927		Robertson	**Lemaître**
1928	Robertson B ($k = 0$)	Freedericsz,	
	Tolman	Schechter	
1929	Hubble $\Rightarrow$ B		
	Robertson		
	Tolman		
	Zwicky		
1930	$\Rightarrow$	*Eddington*	
	Weyl B ($k = 0$)		de Sitter
			Lemaître
			Tolman
1931	Jeans		McVittie
			Einstein
			Tolman
			de Sitter
			Eddington
			McCrea

(continued)

Table 6.3. (*cont.*)

Year	*Paradigm 1:* Static universe Issue: A or B?	Transition	*Paradigm 2:* Evolving geometry Issue: $k = +1, 0, -1$?
1932	MacMillan		Heckmann
			Tolman
			McVittie
			Ward
			Einstein
			de Sitter
1933			Robertson
		Curtis	Milne
			Robertson
			de Sitter
			Lemaître
			Eddington

viewpoint. Indeed it is clear that Hubble in his famour paper setting the velocity–distance relation on solid ground (1929b) was working in paradigm 1. Referring to the possible theoretical interpretation of his work, he wrote:

> The outstanding feature however is that the velocity–distance relation may represent the de Sitter effect and hence that numerical data may be introduced into the discussion of the general curvature of space. In the de Sitter cosmology, displacements of the spectra arise from two sources, an apparent slowing down of atomic vibrations and a general tendency of particles to scatter. The latter involves a separation and hence introduces the element of time. The relative importance of the two effects should determine the form of the relation beween distances and observed velocities.

The interpretation seems to be based on Eddington's description of the static de Sitter universe (see section 6.2.1 above), in which the major redshift effect was not a velocity effect even though it is written in velocity terms.[3] The view of his work as an attempt to choose between A or B is confirmed by a contemporary popular article (Hubble 1929a), where he wrote: 'The necessary investigations are now under way with the odds, for the moment, favouring de Sitter.' The general acceptance of this point of view can be seen, for example, by the fact that after the publication of Hubble's paper, Tolman (1929b) wrote a major paper comparing de

[3] Incidentally, numerical data had already been introduced into the discussion of the general curvature of space by many people, including de Sitter (1917), Wirtz, Silberstein, and Hubble himself (1926).

Sitter's universe and the redshift observations. He decided the model did not adequately explain the observed redshifts.

The breakthrough occurred at a meeting of the Royal Astronomical Society in January, 1930. According to Eddington (1931),

> de Sitter propounded the dilemma that the actual universe apparently contained enough matter to make it an Einstein world and enough motion to make it a de Sitter world. This naturally called attention to the need for intermediate solutions.

Eddington then went on to make the crucial comment (1930a): 'One puzzling question is why there should be only two solutions. I suppose the trouble is that people look for static solutions.' He realised the need for expanding solutions, and with his student McVittie began to work on the problem. However, Lemaître, on reading Eddington's comments, wrote to him pointing out his own work; Eddington immediately saw its significance, and arranged publication of a translation in the *Monthly Notices of the Royal Astronomical Society*. The time was ripe for change; together with Eddington's proof of the instability of the Einstein static universe (1930b), this changed the viewpoint of the majority of workers in the field, e.g. de Sitter wrote (1931) that it was

> a solution of such simplicity as to make it appear self-evident . . . there cannot be the slightest doubt that Lemaître's theory is essentially true.

After the change of view, almost everyone now knew the universe must be expanding; this was the new orthodoxy. However, there was still some resistance, as expressed for example in papers by Curtis, Macmillan, Zwicky, and others. It is salutary to realise that even to the end of his life, Hubble did not particularly agree with the expanding universe interpretation of his data, as can be seen for example from his classic book *The Realm of the Nebulae* (1936) and his Halley lecture in Oxford (1953).

6.2.3 *A hidden history?*

The history of this change does not seem to be very widely known, and in particular the extent to which workers before the change believed in a static universe despite the evolving universe concept available to them and the observational evidence in favour of this alternative. I believe this comes from a combinaton of two causes: on the one hand we easily (with hindsight) read into the history what was not there, and on the other hand, the history has to some extent been hidden or distorted in documents that are widely read.

6.2.3.1 *Reading the present into the past*

As to the first effect, it is remarkably easy to misread the old literature because of our present standpoint. Particular examples are for instance Eddington's book (1924) and Hubble's book (1936). In each case discussions are presented of velocity–distance relationships where 'velocity' actually means equivalent velocity for a redshift which is not necessarily associated with relative motion; and I for one have read into this a support for the expanding universe concept which simply is not there. Closer examination makes the true situation apparent, but it is easy to fall into this trap.

6.2.3.2 *Misleading contemporary reviews?*

As to the second effect, some of the contemporary reviews are misleading to us now because of what they omit or emphasise. The easiest to focus on, because it is so clearly written and widely read, is Robertson's classic review (1933). This was of course written as a contemporary survey rather than a historical review, but nevertheless provides a viewpoint on the preceding history primarily through its excellent annotated survey of the previous literature. Taking into account the tendency to read into the past what was not there, this survey easily provides a wrong impression of the previous era, particularly the extent to which that era had been dominated by the question of whether model A or B was a better model for the universe, and the way the evolving models had been rejected.

This is for two reasons. Firstly a number of papers that throw a significant light on this development are omitted from Robertson's bibliography; they include papers by Charlier, Wirtz, and Silberstein, but particularly Einstein's (1922, 1924) rejection of Friedmann, Tolman's (1929b) attempt to fit Hubble's data to the de Sitter universe, and Hubble's paper (1929a) stating he was attempting to see if A or B was a better fit to the data. Secondly, Robertson's comments on various of the papers emphasise their contribution to the expanding universe idea but not that they were written in the previous paradigm. As an example, Hubble's (1926) paper is mentioned and that it established a uniform distribution of galaxies in visible portions of the universe; but not that it attempted to fit the data to the Einstein static universe. Particularly interesting is his comment on his own paper of 1929; he rightly says it presents the general FLRW line element but not that, as he did not realise the significance of this metric, he immediately specialised it to the two stationary cases. Thus if one takes his references at their face value, the picture that emerges is something like that in table 6.4, instead of table 6.3. This is a different view of the significance of the previous work than presented above; in effect Robertson emphasises the seeds of the new era to come, whereas table 6.3 emphasises the extent to which these germs of ideas were not developed. My contention is that without extra information, this view implied (as in table 6.4) would be bound to mislead one as to significant features of the previous era.

I should emphasise again here that Robertson's superb review was not

Table 6.4. *Alternative view of the first major paradigm change, as implied by Robertson's review (1933)*

Year	Static universe	Transition	*Evolving geometry* Issue: $k = +1, 0, -1?$
1917	**Einstein** A		
	de Sitter B		
1922	→	Lanczos B	Friedmann ($k = +1$)
	de Sitter		
1923	→	Lanczos B	
	Weyl B		
1924			Friedmann ($k = -1$)
1925	→	Lemaître	
1926	→	Hubble	
1927			**Lemaître**
1928	→	Robertson	
1929	→	Hubble	
	→	→	Robertson
	→	Tolman	
1930			Eddington
	→	Weyl	
			de Sitter
			Lemaître
			Tolman
1931			McVittie
			Einstein
			Tolman
			de Sitter
			Lemaître
			McCrea
1932			Heckmann
			Tolman
			McVittie
			Ward
			Einstein
			de Sitter

written as a historical document but rather as a contemporary presentation of the expanding universe idea seen from the view of paradigm 2. Nevertheless one has the impression the review reacts against the previous (paradigm 1) viewpoint to the extent of hiding, perhaps inadvertently, the central theme of that era.

The review contains one brief appendix commenting on Tolman's work which foreshadows paradigm 3. Robertson did not at the time realise the significance of these and other papers that were to be important later, as for example Kasner's vacuum anisotropic solution (1925) which on the face of it had little to do with relativistic cosmology. Only much later could one establish its significance as a key element in understanding the evolution of anisotropic universe models.

6.3 Conclusion

The view of the history of cosmology presented here is strictly a Whiggish view, i.e. the classification scheme used to separate the paradigms is clearly based on our current understanding of the nature of cosmology – the viewpoint that the curved space-time description of expanding universes[4] is 'correct'. A different classification into paradigms, and so different understanding of its history, might well result from a different view of the nature of cosmology. Given the present viewpoint, the analysis reveals substantial resistance to change preceding a rapid change of paradigm. It is also interesting how this resistance to change is somewhat hidden in later writings on this history.

The purpose of this paper is not in any way to denigrate the major contributions of the great figures of the past. Like all of us, they too have of necessity been imbedded in the knowledge and preconceptions of their times. The intention is rather to use this historical example as a warning to us today. Today's dominant view is not necessarily correct, and indeed there almost certainly is some new view waiting to be recognised; the precursors of that new view are probably already with us. On the other hand just because a new paradigm arises does not mean it is correct! The major message is that working scientists need to be aware of the pressures to conform, and the strength of the 'bandwagon' effect. There is a constant need to question and test the currently accepted foundations of cosmology and cosmological models.

References

References sufficient to establish the main features of the tables are as follows; for a fuller listing, see Ellis (1988).

Berendzen, R., Hart, R. and Seeley, D. (1976). *Man Discovers the Galaxies*. New York: Science History Publications.

Bondi, H. (1960). *Cosmology*. Cambridge: Cambridge University Press.

Charlier, C. V. L. (1925). On the Structure of the Universe. *Publications of the Astronomical Society of the Pacific*, **37**, 177–191.

Curtis, H. (1933). The Nebulae. In *Handbuch der Astrophysik*, Vol. II Part 1, Sections 69–79, ed. H. B. Curtis, B. Lindblad, K. Lundmark, and H. Shapley. Berlin: Springer, 891–908.

[4] An *expanding universe* is a space-time with invariant properties that change with time, not just a static space-time with particles expanding into it.

De Sitter, W. (1917a). On the Relativity of Inertia: Remarks concerning Einstein's Latest Hypothesis. *Proceedings of the Royal Netherlands Academy of Sciences (Amsterdam)*, **19**, 1217–1225.

(1917b). On Einstein's Theory of Gravitation and its Astronomical Consequences III. *Monthly Notices of the Royal Astronomical Society*, **78**, 3–28.

(1922). On the Possibility of Statistical Equilibrium of the Universe. *Proceedings of the Royal Netherlands Academy of Sciences (Amsterdam)*, **22**, 866–888.

(1930). The Expanding Universe: Discussion of Lemaître's Solution of the Equations of the Inertial Field. *Bulletin of the Astronomical Institute of the Netherlands*, **5**, No. 193, 211–218.

(1931). *Nature*. **128**, 706–709.

(1933). On the Expanding Universe and the Time Scale. *Monthly Notices of the Royal Astronomical Society*, **93**, 628–634.

Eddington, A. S. (1918). Letter (see Smith, 1982).

(1924). *The Mathematical Theory of Relativity*. Cambridge: Cambridge University Press.

(1930a). Remarks at the Meeting of the Royal Astronomical Society. *Observatory*, **53**, 39–40.

(1930b). On the Instability of Einstein's Spherical World. *Monthly Notices of the Royal Astronomical Society*, **90**, 668–678.

(1931). Council Note on Expansion of Universe. *Monthly Notices of the Royal Astronomical Society*, **91**, 412–416.

(1933). *The Expanding Universe*. Cambridge: Cambridge University Press.

Ehlers, J. (1961). Beitrage zur Mechanik kontinuerlichen Medien. *Abhandlungen der Akademie der Wissenschaft and Literatur, Mainz. Mathematik-Naturwissenschaften Klasse*, **11**, 1.

Einstein, A. (1917). Kosmologische Betrachtungen zur allgemeinen Relativitätstheorie. *Sitzungsberichte der Preussischen Akademie der Wissenschaften,* **7**, 142–152.

(1918). Kritisches zu einer von Hrn. de Sitter gegebenen Lösung der Gravitationsgleichungen. *Sitzungsberichte der Preussischen Akademie der Wissens chaften*, 270–272.

(1922). Bemerkung zu der Arbeit von A. Friedmann 'Ueber die Krümmung des Raumes'. *Zeitschrift für Physik*, **11**, 326.

(1923). Notiz zu der Arbeit von A. Friedmann 'Ueber die Krümmung des Raumes'. *Zeitschrift für Physik*, **16**, 228.

(1931). On the Cosmological Problem of General Relativity. *Sitzungsberichte der Preussischen Akademie der Wissenschaften*, **12**, 235–237.

Einstein, A. and de Sitter, W. (1932). On the Relation between the Expansion and the Mean Density of the Universe. *Proceedings of the National Academy of Science*, **18**, 213–214.

Ellis, G. F. R. (1971). Relativistic Cosmology. In *General Relativity and Cosmology*, ed. R. K. Sachs (Enrico Fermi School XLVIII). New York: Academic Press, 104–182.

(1988). The Expanding Universe: A History of Cosmology from 1917 to 1960. Einstein and the History of General Relativity, ed. D. Howard and J. Stackel. (Einstein Studies series.) Birkhauser: Boston 367–432.

Freederichsz, V. and Schechter, A. (1928). Notiz zur Frage nach der Berechnung der Aberration und der Parallaxe in Einsteins, de Sitters und Friedmanns Welten. *Zeitschrift für Physik*, **51**, 584–592.
Friedmann, A. (1922). Ueber die Krümmung des Raumes. *Zeitschrift für Physik*, **10**, 377–386.
(1924). Ueber die Möglichkeit einer Welt mit konstanter negativer Krümmung des Raumes. *Zeitschrift für Physik*, **21**, 326–332.
Gamow, G. (1948). The Origin of the Elements and the Separation of Galaxies. *Physical Review*, **70**, 505–506.
(1956). The Evolutionary Universe. *Scientific American* (Sept), 136–154.
Guth, A.H. (1981). Inflationary Universe: A Possible Solution to the Horizon and Flatness Problems. *Physical Review*, **D23**, 347–356.
Hawking, S.W. and Ellis, G.F.R. (1973). *The Large Scale Structure of Space-Time*. Cambridge: Cambridge University Press.
Heckmann, O. (1932). Die Ausdehnung der Welt und ihre Abhängigkeit von der Zeit. *Nachr. Ges. Gottingen Wiss.*, 97–102.
(1942). *Theories of Cosmology*. Berlin: Springer.
Hoyle, F. (1948). A New Model for the Expanding Universe. *Monthly Notices of the Royal Astronomical Society*, **108**, 372–382.
Hubble, E.P. (1926). Extra-Galactic Nebulae. *Astrophysical Journal*, **64**, 321–369.
(1929a). The Exploration of Space. *Harper's Magazine*, **58**, 732–738.
(1929b). A Relation between Distance and Radial Velocity among Extragalactic Nebulae. *Proceedings of the National Academy of Sciences*, **15**, 169–173.
(1936). *The Realm of the Nebulae*. New Haven: Yale University Press.
(1953). The Law of the Redshifts. *Monthly Notices of the Royal Astronomical Society*, **113**, 658–666.
Jeans, J. (1931). *Nature*, **128**, 703.
Kasner, E. (1925). Solutions of Einstein's Equations Involving Functions of Only One Variable. *Transactions of the American Mathematical Society*, **27**, 155–162.
Kragh, H. (1987). The Beginning of the World: Georges Lemaître and the Expanding Universe. *Centaurus*, **32**, 114–139.
Lanczos, C. (1922). Bemerkungen zur de Sitterschen Welt. *Physikalisches Zeitschrift*, **23**, 539–543.
(1923). On the Redshift in the de Sitter World. *Zeitschrift für Physik*, **17**, 168–189.
Lemaître, G. (1925). Note on de Sitter's Universe. *Physical Review*, **25**, 903.
(1927). A Homogeneous Universe of Constant Mass and Increasing Radius Accounting for the Radial Velocity of Extragalactic Nebulae. *Annales de la Société Scientifique de Bruxelles*, **A47**, 49–59. [In French.]
(1930). On the Random Motion of Material Particles in an Expanding Universe. *Bulletin of the Astronomical Institute of the Netherlands*, **5**, No. 200, 273–274.
(1931a). A Homogeneous Universe of Constant Mass and Increasing Radius Accounting for the Radial Velocity of Extragalactic Nebulae. *Monthly Notices of the Royal Astronomical Society*, **91**, 483–490 [translation of 1927].
(1931b). The Expanding Universe. *Monthly Notices of the Royal Astronomical Society*, **91**, 490–501.

(1933). L'Univers en expansion. *Annales de la Société Scientifique de Bruxelles* (A), **53**, 51–85.
Levi Civita, T. (1917). Realtá fisica di alcuni spazi normali del Bianchi. *Rendiconti Accademia dei Lincei*, **26**, 519–531.
Lundmark, K. E. (1924). On the Determination of the Curvature of Space-Time in de Sitter's World. *Monthly Notices of the Royal Astronomical Society*, **84**, 747–770.
MacMillan, W. D. (1932). Velocities of the Spiral Nebulae. *Nature*, **129**, 93.
McCrea, W. H. and McVittie, G. C. (1931). On the Contraction of the Universe. *Monthly Notices of the Royal Astronomical Society*, **91**, 128–133.
McVittie, G. C. (1931). The Problem of N Bodies and the Expansion of the Universe. *Monthly Notices of the Royal Astronomical Society*, **91**, 274–283.
(1932). Condensations in an Expanding Universe. *Monthly Notices of the Royal Astronomical Society*, **92**, 500–518.
Milne, E. A. (1933). World Structure and the Expansion of the Universe. *Zeitschrift für Astrophysik*, **6**, 1–95; 244; **7**, 185.
Misner, C. W. (1969). Mixmaster Universe. *Physical Review Letters*, **22**, 1071–1074.
Peebles, P. J. E. (1965). The Blackbody Radiation Content of the Universe and the Formation of Galaxies. *Astrophysical Journal*, **142**, 1317–1326.
(1971). *Physical Cosmology*. Princeton: Princeton University Press.
Penrose, R. (1963). Conformal Treatment of Infinity. In *Relativity Groups and Topology*, ed. C. de Witt and B. de Witt. New York: Gordon and Breach.
Rindler, W. (1956). Visual Horizons in World Models. *Monthly Notices of the Royal Astronomical Society*, **116**, 335–350.
Robertson, H. P. (1927). Dynamical Space-Time which Contain a Conformal Euclidean 3-space. *Transactions of the American Mathematical Society*, **29**, 481–490.
(1928). On Relativistic Cosmology. *Philosophical Magazine*, **5**, 835–848.
(1929). On the Foundations of Relativistic Cosmology. *Proceedings of the National Academy of Science*, **15**, 822–829.
(1932). The Expanding Universe. *Science*, **76**, 221–226.
(1933). Relativistic Cosmology. *Review of Modern Physics*, **5**, 62–90.
(1935). Kinematics and World Structure. *Astrophysical Journal*, **82**, 284–301; **83**, 187–201, 257–271.
Schrödinger, E. (1950). *Expanding Universes*. Cambridge: Cambridge University Press.
Silberstein, L. (1924). The Curvature of de Sitter's Space-Time Derived from Globular Clusters. *Monthly Notices of the Royal Astronomical Society*, **84**, 363–366.
Smith, R. (1982). *The Expanding Universe*. Cambridge: Cambridge University Press.
Tolman, R. C. (1928). *Proceedings of the National Academy of Science*, **14**, 348–353, 353–356.
(1929a). On the Possible Line Elements for the Universe. *Proceedings of the National Academy of Science*, **15**, 297–304.
(1929b). On the Astronomical Implications of de Sitter Line Element for the Universe. *Astrophysical Journal*, **69**, 245–274.

(1930): More Complete Discussion of the Time Dependence of the Non-Static Line Element for the Universe. *Proceedings of the National Academy of Science*, **16**, 409–420.
(1932). Models of the Physical Universe. *Science*, **75**, 367–373.
Tolman, R. C. and Ward, M. (1932). On the Behaviour of Non-static Models of the Universe when the Cosmological Constant is Omitted. *Physical Review*, **39**, 835–843.
Walker, A. G. (1936). On Milne's Theory of World Structure. *Proceedings of the London Mathematical Society*, **42**, 90–127.
Weinberg, S. (1972). *Gravitation and Cosmology*. New York: Wiley and Son.
Weyl, H. (1923). Zur allgemeinen Relativitätstheorie. *Physikalische Zeitschrift*, **24**, 230–232.
(1930). Redshift and Relativistic Cosmology. *Philosophical Magazine*, **9**, 936–943.
Wirtz, C. (1924). De Sitter's Kosmologie und die radial Bewegung der Spiralnebel. *Astronomische Nachrichten*, **222**, 21–26.
Zwicky, F. (1929). On the Redshift of Spectral Lines through Interstellar Space. *Physical Review*, **33**, 1077.

Discussion

Bondi:

(1) Perhaps the best illustration of your point of the persistence of the idea that the universe must be static is that the Milne–McCrea analysis of Newtonian models could not be carried out earlier because although Newtonian dynamics was so well known, a necessarily non-static solution simply could not be contemplated.

(2) I often like to imagine what would have happened if the Michelson–Morley experiment had been carried out in 1620. Surely people would have said 'Dear Mr Galileo, these ideas of Copernicus are beautiful and elegant but the *facts* show that the earth is at rest.'

Hoyle:

A question and a comment. The comment is that you have shown how strong conformist pressures can be, interestingly enough even when the greatest scientists are involved. What happens is that observations gradually pile up against populist views. Then there is a kind of quantum transition to a new view – usually the view that accommodates the observations with the minimum of change in theory. My question now: do you think the last quantum transition has now happened? If not then there are interesting implications!

Ellis:

Almost certainly we have not seen the last such transition.

Barbour:

1) I feel mention should be made in this conference of a beautiful little

paper by Karl Schwarzschild in 1900 in which he considered a model of a closed spherical universe (just a little bigger than our Galaxy). This paper was quoted by de Sitter in his 1917 MNRAS paper and I suspect had an important influence on de Sitter and possibly (through de Sitter) on Einstein.
2) With regard to the stimulus to Friedmann's papers, I think it is quite clear that they were basically mathematical exercises stimulated by the papers of Einstein and de Sitter. He says he aimed to find the most general solutions of the basic kind found by them, i.e. with a distinguished time coordinate and a homogeneous 3-space, whose scale, however, would be a function of the time. I think it is possible that Friedmann may have been prompted to such a study by a specific but almost 'throw away' remark at the end of Einstein's 1917 paper: 'It should however be emphasised that a positive curvature of space also results from the matter present in space even without the additional term [the cosmological term]; we only need that term to make possible a quasi-static distribution of matter as corresponds to the fact of the small stellar velocities.'

McCrea:
Does the ambiguity in the interpretation of the de Sitter model disappear if it is used to predict relations between observable quantities, e.g. the redshift-magnitude relation?

I know about the problem of fundamental observers in the de Sitter model, but it seems that the ambiguity of interpretation entered *after* fundamental observers had been identified (at any rate implicitly)

Ellis:
My view is that the observational relations are unique when and only when the fundamental observers have been identified.

Gibbons:
1) To support your explanation for the gap in the literature from 1917 to 1922, it is interesting to note that Einstein's visit to Paris in 1922 played an important role in creating interest in general relativity in France, and stimulated a lot of work on hierarchical cosmology by Borel, Selety, etc.

2) One can add a further paradigm – that associated with the idea that the universe was 'created from nothing'. In its various forms this has, since the inflationary paradigm, come to play an increasingly important part in cosmological discussion. It has its precursors – Boltzmann, Tryon, Fomin; and since 1982 a number of prominent proponents – Zeldovich, Vilenkin, Hawking.

7

Early inhomogeneous cosmological models in Einstein's theory

Andrzej Krasiński

7.1 Introduction

Just as everyone knew in the 1920s that the universe was static (Ellis 1988b), so almost everyone knew for sure into late 1970s that the universe was homogeneous. Therefore those works which explored the possibility of our universe being lumpy were ignored by mainstream cosmologists. However, a few early papers contain surprisingly powerful results. Here, an attempt is made to pull them out of the shade and give them the credit they deserve.

The investigation underlying this discussion was an offspring of a review of all exact solutions of Einstein's equations which generalize the Friedmann–Lemaître–Robertson–Walker (FLRW) cosmological models. The main review is not yet completed, so it is possible that other papers will be added at a later date to the present collection. In accordance with the framework of this symposium, the only generalizations of the FLRW models which will be discussed are those found before 1970. However, it is worth knowing that many more were found later, and the main review includes so far over seventy independently obtained (though not always independent) solutions.

Each of the sections 7.2 to 7.6 contains a concise presentation of one particular class of models, namely:

Section 7.2: the model found by Lemaître (1933), known under the name Tolman–Bondi, and the considerations of Tolman (1934) and Bonnor (1956) which were based on it.

Section 7.3: the model of McVittie (1933) which is a superposition of the Schwarzschild and the FLRW solutions.

Section 7.4: the spherically symmetric shearfree barotropic solution of Wyman (1946).

Section 7.5: the spherically symmetric shearfree perfect fluid spacetimes of Kustaanheimo and Qvist (1948).

Section 7.6: the general conformally flat expanding perfect fluid solution of Stephani (1967).

In section 7.7 note is taken of a few historical curiosities of a lesser physical significance.

7.2 The Lemaître–Tolman model

The solution of Einstein's equations underlying this model was found by Lemaître (1933). It became later known as the "Tolman–Bondi model," even though Tolman (1934) referred to Lemaître, and Bondi (1947) quoted Tolman, and none of the latter authors claimed priority. The model should thus properly be called after Lemaître, but in order to avoid confusion with the FLRW models I propose to call it Lemaître–Tolman (L-T).

In a convenient notation introduced by Zeldovich and Grishchuk (1989) the metric of the model is

$$ds^2 = dt^2 - \frac{r'^2 dR^2}{1+f(R)} - r^2(t,R)\,(d\theta^2 + \sin^2\theta\, d\phi^2), \tag{1}$$

where $f(R)$ is an arbitrary function, the prime denotes $\partial/\partial R$, $r(t,R)$ is determined by the equation

$$\left(\frac{\partial r}{\partial t}\right)^2 = \frac{F(R)}{r} + f(R) + \frac{1}{3}\Lambda r^2, \tag{2}$$

$F(R)$ is another arbitrary function and Λ is the cosmological constant. The matter density in the model is

$$\varkappa\varrho = \frac{F'}{r^2 r'}, \tag{3}$$

where $\varkappa = 8\pi G/c^2$ and the pressure is identically zero. The FLRW limit is obtained when

$$r = RS(t),\, f = -kR^2,\; F = CR^3, \tag{4}$$

where $S(t)$ is the FLRW scale function, k is the FLRW curvature index ($k = \pm 1, 0$), and C is a constant. In the limit, (2) becomes the Friedmann equation (with the mass-conservation integral already incorporated).

Using this solution Tolman (1934) showed that the Einstein static universe is unstable against a perturbation different from the one considered by Eddington (1930) (where the matter density was knocked off its Einstein value and allowed to vary in time, but forced to remain constant throughout space at each given moment). The L-T model can be thought of as such a perturbation of the Einstein universe in which the matter density is allowed to vary both in space and time. If the density is initially non-uniform in space, then the evolution determined by (2) will enhance rather than suppress the inhomogeneity. This is simultaneously an instability of the FLRW models against the growth of inhomogeneities. Tolman also observed that the universe could contain several homogeneous regions, each described by a FLRW metric and having a different matter-density, with transition zones interpolating between them described by L-T metrics. In such a universe, each FLRW zone would "then behave as in some particular completely homogeneous model without reference to the behavior of other parts of the model." Tolman concluded his paper as follows:

> Hence, it would appear wise at the present stage of theoretical development, to envisage the possibility that regions of the universe beyond the range of our present telescopes might be contracting rather than expanding and contain matter with a density and stage of evolutionary development quite different from those with which we are familiar. It would also appear wise not to draw too definite conclusions from the behavior of homogeneous models as to a supposed initial state of the whole universe.

The paper contains other statements that sound surprisingly modern:

> it is evident that some preponderating tendency for inhomogeneities to disappear with time would have to be demonstrated, before such models could be used with confidence to obtain extrapolated conclusions as to the behavior of the universe in very distant regions or over exceedingly long periods of time. ["such models" here means homogeneous] ... in regions where the density starts to increase it is evident from the full form of equation (18) that reversal in the process of condensation would not occur short of arrival at a singular state involving infinite density or of the breakdown in our simplified equations. [Tolman's equation (18) expresses $[\partial^2(\log \varrho)]/[\partial t^2]$ through ϱ and the metric functions.]

Bonnor (1956) used the L-T model to discuss the evolution of a localized condensation of matter in the universe. He assumed that the condensation is a sphere of a FLRW space surrounded by an L-T transition zone which is in turn surrounded by another FLRW space with a different density than in the central region. By investigating the evolution of such condensations Bonnor found that perturbations of homogeneous distribution of matter may produce galaxies by today only if they are initially several orders of magnitude larger than purely statistical fluctuations in density. This was a problem of much debate in the 1970s in the framework of the so-called theories of galaxy formation which were based on a linear approximation to the field equations. Bonnor had the result earlier and from the exact theory.

Note that from (1) and (4) we can write

$$f(R) = -k(R)R^2, \tag{5}$$

i.e. the curvature index k in the L-T model depends on position. It may thus happen that the L-T universe will be "open" ($k < 0$) in one part of the spacetime and "closed" ($k > 0$) elsewhere. This shows that the classification of cosmological models into open and closed according to the curvature of their spatial sections applies only to the FLRW class and is forced upon us by the strong *assumptions* about symmetry underlying this class. In more general models such a distinction may be sometimes impossible and so irrelevant in general (note also the last paragraph of section 7.6 below).

From (2) one can conclude that the big bang may have occurred at different times in different locations which is also quite an enlightening departure from the FLRW picture.

Although originally derived in a coordinate-dependent way, the L-T model can be uniquely defined by the following geometrical and physical properties: 1. The spacetime is spherically symmetric, 2. The source in the Einstein's equations is dust + the cosmological term, 3. The expansion and shear are nonzero. This characteristic follows from the work of Collins and Szafron (1979). The model is the limiting case of spherical symmetry and constant pressure ($p = \Lambda$) of the Szafron (1977) solutions. The subcase $\Lambda = 0$ is the spherically symmetric limit of one of the Szekeres (1975) solutions.

The L-T model is the only one among those described in this note that gained a limited recognition and is occcasionally used for discussing cosmological problems.

7.3 The McVittie (1933) model

This solution, called by McVittie "the mass particle in an expanding universe" is a superposition of the Schwarzschild solution and the FLRW models. The metric of the model (in a slightly modified notation) is

$$ds^2 = \left(\frac{1 - M(t,r)}{1 + M(t,r)}\right)^2 dt^2 - \frac{e^{\beta(t)}[1 + M(t,r)]^4}{c^2[1 + \frac{1}{4}kr^2]^2}[dr^2 + r^2(d\theta^2 + \sin^2\theta\, d\phi^2)], \tag{6}$$

where

$$M(t,r) = \frac{\mu(t)}{2r}(1 + \tfrac{1}{4}kr^2)^{1/2}, \tag{7}$$

k and c are constants, $\mu(t)$ is an arbitrary function, and $\beta(t)$ is related to μ by

$$\dot{\beta}\mu = -2\dot{\mu}. \tag{8}$$

In McVittie's original notation k was represented as $1/R^2$ which suggested (incorrectly) that the constant must be positive. The metric (6)–(8) fulfills the Einstein equations with the source being a shearfree expanding perfect fluid that has non-geodesic flowlines (the non-geodesic flow is forced by spatial gradients of pressure).

The Schwarzschild solution in the isotropic coordinates is obtained from (6)–(8) when μ and β are constant and $k = 0$. The FLRW models result when $\mu = 0$ (so $M = 0$), then (8) is fulfilled with arbitrary $\beta(t)$.

This solution looks appropriate for discussing, for example, the following problems:

1. The influence of the expanding universe on the mass of an isolated object. Note that $\mu(t)$ becomes the mass in the Schwarzschild limit, and $e^{\beta/2}$ becomes the scale factor in the FLRW limit. Consequently, (8) implies that the mass of an isolated object decreases with time ($\dot{\mu}<0$) when the universe expands ($\dot{\beta}>0$) which is a very neat Machian effect.
2. The influence of cosmological expansion on the orbits of gravitationally bound systems. The "Swiss cheese" model of Einstein and Strauss (1945) suggested that no such influence should occur. The McVittie model shows that the orbits should be expanding at least due to the decrease of the mass of the central body (see another reason two paragraphs below).
3. The description of black holes in the cosmological background (i.e. asymptotically nonflat black holes). Even the definition of such a black hole is a problem of debate (Ellis 1984) while here we have a rather obvious example for testing the (not yet formulated) theory.

Problem 3 is too modern to have been considered in the 1930s, but 1 and 2 could well have been attempted. McVittie did some preliminary work on problem 2. It was a qualitative discussion, without proper care being taken to define the radius of the orbit in an invariant or measurable way. Noerdlinger and Petrosian (1971) used the McVittie model to discuss, also only qualitatively, the general relativistic correction to the Newtonian effect of the expansion of an orbit due to the outflow of mass from within the orbit (in (6) the cosmic medium extends throughout the planetary system and streams outward; Noerdlinger and Petrosian were actually interested in the orbit of a galaxy in a cluster). No other studies of problem 2 and no studies at all of problem 1 are known to me.

The solution (6)–(8) was guessed rather than derived, and no derivation was presented until today. The model will be shown in section 7.5 to be a subcase of a more general class but the limiting transition has no physical interpretation.

McVittie observed that if the cosmological constant is nonpositive, then "at some time in the past the expansion started instantaneously with a finite velocity" and "there is a 'retarding force' slowing up the expansion which, obviously, cannot be the initial cause that started the latter." McVittie used this as an argument that λ must be positive, but today we would say that it was a prediction of singularity, apparently independent of the earliest singularity theorem by Tolman and Ward in 1932 (Ellis 1988a).

7.4 The Wyman (1946) solution

Wyman (1946) investigated the spherically symmetric solutions of Einstein's equations for which the source was a shearfree expanding perfect fluid obeying a barotropic equation of state. He found the follow- inhomogeneous solution:

$$ds^2 = \tfrac{1}{2}(\alpha C_2 C_3 t + \beta)^{-1}\mathscr{P}^{-2}\mathscr{P}_{,v}^{\,2}dt^2 - (a\mathscr{P}^2)^{-1}[dr^2 + r^2(d\theta^2 + \sin^2\theta\, d\omega^2)], \tag{9}$$

where C_2, C_3, α, β and c care arbitrary constants, $a = 144/c^2$,

$$v \stackrel{\text{def}}{=} t + \tfrac{1}{2}C_2 r^2, \tag{10}$$

and $\mathscr{P}(v)$ is determined by

$$\mathscr{P}_{v}^{\,2} = 4\mathscr{P}^3 - C_3. \tag{11}$$

The physical interpretation of the metric in unknown, but it is a general solution of a formal problem: under the assumptions stated,[1] (9)–(11) is the unique inhomogeneous model. Such a statement is contained in Wyman's paper, but the original proof made it difficult to see how (9)–(11) is singled out from among the other solutions (Krasiński 1988). A transparently complete proof by a different method was published later by Collins and Wainwright (1983).

In the limit of spatial homogeneity, $C_2 \rightarrow 0$, only the de Sitter model is obtained from (9)–(11), although the general FLRW models fulfill all the initial assumptions and should be expected to also show up as subcases. Thus the FLRW models are isolated in Wyman's class, with no inhomogeneous parent solutions. This is a consequence of the barotropic equation of state (Krasiński 1988, see also section 7.5).

7.5 The Kustaanheimo–Qvist (1948) spacetimes

Kustaanheimo and Qvist (1948) (abbreviated K–Q) considered the same problem as Wyman, but without the assumption $p = p(\varrho)$. The metric they obtained is

$$ds^2 = [\dot{F}/(A(t)F)]^2 dt^2 - F^{-2}[dr^2 + r^2(d\theta^2 + \sin^2\theta\, d\phi^2)], \tag{12}$$

where $A(t)$ is an arbitrary function, $x = r^2$ and $F(t,x)$ obeys

$$F_{,xx}/F^2 = f(x), \tag{13}$$

$f(x)$ being an arbitrary function. For two forms of $f(x)$ general solutions were given later: by Wagh (1955) for $f = \alpha x$ where $\alpha =$ constant and by Wyman (1978) for $f = (\alpha x^2 + bx + c)^{-5/2}$ where α, b and c are constants. Even with these special forms of f, the general solution for F is not an

[1] The assumptions were translated into modern language, the terms *shear* and *expansion* not being in use in 1946.

elementary function. K–Q discussed briefly the problem of solvability of (13) in elementary functions, and a more systematic investigation was done much later by Stephani (1983). The conditions given by K–Q and Stephani are sufficient, but not necessary, so they do not exhaust all the possibilities. The class defined by (12)–(13) includes as subcases, among others:

1. The FLRW models in which $f = 0$ and $F = (1 + \frac{1}{4}kr^2)/R(t)$.
2. The Wyman solution of section 7.4 in which $f = \frac{3}{2}C_2^2 = \text{const}$ and $F(t, x) = \mathscr{P}(v)$, $v = t + \frac{1}{2}C_2 x$.
3. The McVittie solution of section 7.3 in which $f = -\frac{3}{4}(\mu e^{\beta/2}/c)(x + \frac{1}{4}kx^2)^{-5/2}$ and $F = ce^{-\beta/2}(1 + \frac{1}{4}kx)[1 + M(t,x)]^{-2}$ (note that by (8), $\mu e^{\beta/2}$ is a constant).
4. The spherically symmetric limit of the Stephani model of section 7.6 in which $f = 0$ and F is the same as in FLRW, but $k = k(t)$ is an arbitrary function.

The limiting transition 2 is equivalent to imposing the barotropic equation of state on inhomogeneous solutions of (13). Transition 4 is equivalent to setting the conformal curvature to zero, and transition 1 results when pressure is homogeneous in addition. Transition 3 does not seem to have any clear interpretation.

The K–Q paper is a rare example of a problem satisfactorily treated already at first go. The authors reduced the Einstein's equations in their case to the simple (13). After forty years their result is still at the top of a large family of models, and no far-reaching generalizations were found (Barnes 1973 obtained the plane- and hyperbolically symmetric counterparts of (12)–(13), and Faulkes (1969) found the generalization of (13) in which a spherically symmetric electromagnetic field is present). Unfortunately, the paper was published in an unknown journal and the result did not reach anyone for a long time. Over the years, the K–Q equation (13) was rederived repeatedly by several authors (see list in section 7.7). Some of the particular solutions of (13) were discussed as stellar models. The only cosmological application of a subcase of (12)–(13) that I know about is the one discussed in section 7.3.

Note that, $f(x)$ given, the general solution of (13) will contain two arbitrary functions of time as integration "constants." Hence, with arbitrary f the family of solutions of (13) is labelled by three arbitrary functions, each of one variable. All this wealth of solutions shrinks to (9)–(11) plus the FLRW models when the barotropic equation of state is imposed. This shows how restrictive the assumption $p = p(\varrho)$ is.

7.6 The Stephani (1967) universe

Stephani (1967) found the following simple generalization of the FLRW models:

$$ds^2 = D^2(t,x,y,z)dt^2 - R^2(t)V^{-2}(t,x,y,z)(dx^2 + dy^2 + dz^2) \tag{14}$$

where

$$V = 1 + \tfrac{1}{4}k(t)[(x - x_o(t))^2 + (y - y_o(t))^2 + (z - z_o(t))^2], \tag{15}$$

$$D = F(t)\left(\frac{R}{V}\right)\left(\frac{V}{R}\right)_{,t}, \tag{16}$$

and $R(t)$, $k(t)$, $x_o(t)$, $y_o(t)$, $z_o(t)$ and $F(t)$ are arbitrary functions. The matter-density and pressure are given by

$$\varkappa\varrho = 3C^2(t), \tag{17}$$

$$\varkappa p = 3C^2 + 2C\dot{C}\frac{V}{R}\bigg/\left(\frac{V}{R}\right)_{,t}, \tag{18}$$

where $C(t)$ is a function related to k, F and R by

$$k = (C^2 - 1/F^2)R^2. \tag{19}$$

This is the most general conformally flat solution with an expanding perfect fluid source (Kramer et al. 1980). It was found in this generality only in 1967, but its spherically symmetric subcase (which is the limit $f = 0$ of the K–Q models and results from (14)–(16) when x_o, y_o and z_o are all constant) appeared already in the papers by Wyman (1946) and K–Q (1948), and was rediscovered many times more (see list in section 7.7). The FLRW models result when x_o, y_o, z_o and k are constant; then D becomes a function of t only and may be scaled to 1 by transforming t. After such a transformation $F = -R/\dot{R}$ and (19) coincides with the Friedmann equation.

Since $k = k(t)$, the model predicts that the universe can have its spatial curvature negative at one time t_1 (when $k(t_1) < 0$) and positive at another time t_2 (when $k(t_2) > 0$). This model is thus complementary to the L-T solution where the spatial curvature index varies in space (see the remark in the paragraph after (5)). This intriguing property of the Stephani model was pointed out in several earlier papers of the present writer, but as yet the solution has not been seriously considered as a viable model of our observed universe. The spherically symmetric subcase was repeatedly used as a model of stellar collapse.

7.7 Some historical curiosities

The previous sections presented only those papers which, in the present writer's opinion, were significant though not properly appreciated contributions to theoretical cosmology. The same solutions, or their subcases, were rediscovered later, however, more than once in each case. Such rediscoveries are listed here, but the lists are not guaranteed to be complete. The papers which make reference to previous discoveries are omitted deliberately.

The $\Lambda = 0$ subcase of the L-T model from section 7.2 was reobtained by Datt (1938). Four of the five special cases listed by Datt are trivial: (a) is the Minkowski metric in spherical coordinates, (b) is the Minkowski metric in a more clever disguise, (c) is the flat FLRW model and (d) is a simple coordinate transform of (b). Case (e) is the metric which later appeared in the paper by Oppenheimer and Snyder (1939) (see below). Datt's section 7 introduces one more solution which is a subcase of the Szekeres (1975) solutions and a generalization of the spherical Kantowski–Sachs (1966) solution to inhomogeneous matter-density. The L-T model with $\Lambda = 0$ was rederived once again by Carr and Hawking (1974).

Oppenheimer and Snyder (1939) investigated the gravitational collapse of a spherical cloud of dust. Their physical discussion of the process was a new result, but the metric they derived is the $\Lambda = 0 = f$ limit of the L-T solution and coincides with Datt's (1938) case (e) (they instantly specialized the metric further so that the flat FLRW model surrounded by Schwarzschild space resulted).

The Kustaanheimo–Qvist equation (13) was for the first time obtained (in different variables) by Wyman (1946), and then rederived by Wagh (1955, in Wyman's variables), Taub (1968, in still other variables), Cahill and McVittie (1970, in yet other variables), Barnes (1973), Wyman (1976), Glass (1979) and Banerjee and Chakravarty (1979). It should be noted that Barnes (1973) found *all* rotation-free shearfree perfect fluid solutions to Einstein's equations, among them the plane- and hyperbolically symmetric counterparts of the Kustaanheimo–Qvist spacetimes and the Stephani (1967) model. Wyman (1976) presented a wealth of solutions of the K–Q equation. Banerjee and Chakravarty (1979) reobtained the plane-symmetric models of Barnes (1973) (this may be difficult to reognize) and the Stephani (1967) model, but all under more restrictive assumptions. The papers dealing with various particular solutions of the K–Q equation (13), corresponding to different $f(x)$ are too numerous to be listed here (see Krasiński 1989 for a partial list). Also omitted are particular solutions with electric charge (see Sussman 1987 and 1988).

The spherically symmetric subcase of the Stephani (1967) model of section 7.6 was for the first time casually mentioned (and instantly dismissed because it does not admit $p = p(\varrho)$) by Wyman (1946), and then reobtained by Kustaanheimo and Qvist (1948, their case 1), Raychaudhuri (1955), Gupta (1959), Bondi (1967), Thompson and Whitrow (1967), Taub (1968), Cahill and McVittie (1970), Cook (1975), Glass (1979) and Pandey, Gupta and Sharma (1983). Some more authors discussed still simpler subcases of the model.

7.8 Conclusion

The foregoing sections show that inhomogeneous cosmological models have been since more than fifty years a subject of scientific activity which produced interesting and valuable results. Unfortunately, they were mostly ignored (with the possible exception of the L-T model) because most cosmologists *knew* that our actual universe was homogeneous. For this reason, the papers mentioned in this note did not really count in making history of cosmology. A symposium such as this one is perhaps the right place to recall them and in this way, at least, prevent further rediscoveries.

References

Banerjee, A. and Chakravarty, N. (1979). Irrotational Shearfree Motion of a Perfect Fluid in General Relativity. *Acta Physica Polonica*, **B10**, 3.

Barnes, A. (1973). On Shearfree Normal Flows of a Perfect Fluid. *General Relativity and Gravitation*, **4**, 105.

Bondi, H. (1947). Spherically Symmetrical Models in General Relativity. *Monthly Notices of the Royal Astronomical Society*, **107**, 410.

(1967). Bouncing Spheres in General Relativity. *Nature*, **215**, 838.

Bonnor, W. B. (1956). The Formation of the Nebulae. *Zeitschrift für Astrophysik*, **39**, 143.

Cahill, M. E. and McVittie, G. C. (1970). Spherical Symmetry and Mass-Energy in General Relativity. *Journal of Mathematical Physics*, **11**, 1382.

Carr, B. J. and Hawking, S. W. (1974). Black Holes in the Early Universe. *Monthly Notices of the Royal Astronomical Society*, **168**, 399.

Collins, C. B. and Szafron, D. A. (1979). A New Approach to Inhomogeneous Cosmologies: Intrinsic Symmetries. Parts I, II and III. *Journal of Mathematical Physics*, **20**, 2347, 2354 and 2362.

Collins, C. B. and Wainwright, J. (1983). Role of Shear in General Relativistic Cosmological and Stellar Models. *Physical Review*, **D27**, 1209.

Cook, M. W. (1975). On a Class of Exact Spherically Symmetric Solutions to the Einstein Gravitational Field Equations. *Australian Journal of Physics*, **28**, 413.

Datt, B. (1938). Über eine Klasse von Lösungen der Gravitationsgleichungen der Relativität. *Zeitschrift für Physik*, **108**, 314.

Eddington, A. (1930). Instability of the Einstein Spherical Universe. *Monthly Notices of the Royal Astronomical Society*, **90**, 668.

Einstein, A. and Strauss, E. G. (1945). Influence of the Expansion of Space on the Gravitation Field Surrounding the Individual Stars. *Reviews of Modern Physics*, **17**, 120.

Ellis, G. F. R. (1984). Relativistic Cosmology: Its Nature, Aims and Problems. In *General Relativity and Gravitation*, ed. B. Bertotti, F. de Felice and A. Pascolini, p. 215. Dordrecht: Reidel.

(1988a). The Expanding Universe: A History of Cosmology from 1917 to 1960. To appear in *Proceedings of Osgood Hill Conference*, ed. by D. Howard (Einstein studies series).

(1990). This volume.

Faulkes, M. C. (1969). Charged Spheres in General Relativity. *Canadian Journal of Physics*, **47**, 1989.
Glass, E. N. (1979). Shear-Free Gravitational Collapse. *Journal of Mathematical Physics*, **20**, 1508.
Gupta, P. S. (1959). Pulsating Fluid Sphere in General Relativity. *Annalen der Physik*, **2**, 422.
Kantowski, R. and Sachs, R. K. (1966). Some Spatially Homogeneous Anisotropic Relativistic Cosmological Models. *Journal of Mathematical Physics*, **7**, 443.
Kramer, D., Stephani, H., Herlt, E. and MacCallum, M. (1980). *Exact Solutions of Einstein's Field Equations*, p. 371 (theorem 32.15). Cambridge: Cambridge University Press.
Krasiński, A. (1988). Spherically Symmetric Shearfree Universes with a Barotropic Equation of State. Preprint, to be published.
(1989). Shearfree Normal Cosmological Models. *Journal of Mathematical Physics*, **30**, 433.
Kustaanheimo, P. and Qvist, B. (1948). A Note on Some General Solutions of the Einstein Field Equations in a Spherically Symmetric World. *Societas Scientiarum Fennica Commentationes Physico-Mathematicae*, **XIII** no. 16, 1.
Lemaître, G. (1933). L'Univers en expansion. *Annales de la Société Scientifique de Bruxelles*, **53A**, 51.
McVittie, G. C. (1933). The Mass-Particle in an Expanding Universe. *Monthly Notices of the Royal Astronomical Society*, **93**, 325.
Noerdlinger, P. D. and Petrosian, V. (1971). The Effect of Cosmological Expansion on Self-Gravitating Ensembles of Particles. *The Astrophysical Journal*, **168**, 1.
Oppenheimer, J. R. and Snyder, H. (1939). On Continued Gravitational Contraction. *Physical Review*, **56**, 455.
Pandey, S. N., Gupta, Y. K. and Sharma, S. P. (1983). Spherically Symmetric Conformally Flat Perfect Fluid Distributions of Class 1. *Indian Journal of Pure and Applied Mathematics*, **14**, 79.
Raychaudhuri, A. (1955). Perturbed Cosmological Models. *Zeitschrift für Astrophysik*, **37**, 103.
Stephani, H. (1967). Über Lösungen der Einsteinschen Feldgleichungen, die sich in einen fünfdimensionalen flachen Raum einbetten lassen. *Communications in Mathematical Physics*, **4**, 137.
(1983). A New Interior Solution of Einstein's Field Equations for a Spherically Symmetric Perfect Fluid in Shear-Free Motion. *Journal of Physics*, **A16**, 3529.
Sussman, R. A. (1987). On Spherically Symmetric Shearfree Perfect Fluid Configurations (Neutral and Charged), Part I. *Journal of Mathematical Physics*, **28**, 1118.
(1988). On Spherically Symmetric Shearfree Perfect Fluid Configurations (Neutral and Charged), Parts II and III. *Journal of Mathematical Physics*, **29**, 945 and 1177.
Szafron, D. A. (1977). Inhomogeneous Cosmologies: New Exact Solutions and their Evolution. *Journal of Mathematical Physics*, **18**, 1673.

Szekeres, P. (1975). A Class of Inhomogeneous Cosmological Models. *Communications in Mathematical Physics*, **41**, 55.
Taub, A.H. (1968). Restricted Motions of Gravitating Spheres. *Annales de l'Institut Henri Poincaré*, **A9**, 153.
Thompson, I. H. and Whitrow, G. J. (1967). Time-Dependent Internal Solutions for Spherically Symmetrical Bodies in General Relativity. *Monthly Notices of the Royal Astronomical Society*, **136**, 207.
Tolman, R. (1934). Effect of Inhomogeneity in Cosmological Models. *Proceedings of the National Academy of Sciences of the USA*, **20**, 169.
Wagh, R. V. (1955). On Some Spherically Symmetrical Models in Relativity. *Journal of the University of Bombay*, **24**, 1.
Wyman, M. (1946). Equations of State for Radically Symmetric Distributions of Matter. *Physical Review*, **70**, 396.
(1976). Non-Static Radially Symmetric Distributions of Matter (Jeffrey–Williams lecture 1976). *Canadian Mathematical Bulletin*, **19**, 343.
(1978). Nonstatic Spherically Symmetric Isotropic Solutions for a Perfect Fluid in General Relativity. *Australian Journal of Physics*, **31**, 111.
Zeldovich, Ya. B. and Grishchuk, L. P. (1984). Structure and Future of the 'New' Universe. *Monthly Notices of the Royal Astronomical Society*, **207**, 23p.

Discussion

Ehlers:

Could you please state again the assumptions underlying Wyman's solution?

Krasiński:

It follows uniquely from the following assumptions: 1) The spacetime is spherically symmetric, 2) The source is a perfect fluid, 3) Shear is zero, 4) Expansion is nonzero, 5) The barotropic equation of state holds, 6) The spacetime is inhomogeneous.

Ehlers:

But then it does not seem to reproduce the static spherically symmetric solutions.

Krasiński:

This is correct. In the process of integrating the Einstein equations for a spherically symmetric metric with a perfect fluid source, alternatives occur at which we must choose between mutually exclusive possibilities. One such choice is: either $g_{11,t} \neq 0$ and then $g_{oo} = f(t)[(\ln g_{11})_{,t}]^2$, or else a static solution results. From this point on, if we follow the first possibility, we lose control of the static case. Similarly, at a later point, another alternative occurs: either $(g_{11}^{-1/2})_{,uu} \stackrel{\text{deF}}{=} F = 0$ where $u = r^2$ (then the FLRW models result) or $F \neq 0$ and the Wyman solution follows. These possibilities are mutually exclusive, too: in the limiting case of spatial homogeneity, the

Wyman solution reproduces only the de Sitter spacetime, the general FLRW models are not contained in it, even though they fulfil assumptions 1 to 5.

Ellis:
1) De Sitter wrote down field equations for Lemaître–Tolman in 1917.
2) Guy Omer investigated inhomogeneous cosmologies in 1949.

Krasiński:
Omer investigated the Lemaître–Tolman solution. I did not take into account all papers written on inhomogeneous models, I only tried to present the different inhomogeneous solutions and a selection of papers in their interpretation.

8

Early work on "big bang" cosmology and the cosmic blackbody radiation

Ralph A. Alpher and Robert Herman

8.1 Introduction

It is the purpose of this review to describe briefly the early studies by the authors and by Gamow which have provided much of the basis for what has become the widely accepted canonical "big bang" model of the universe. We shall indicate which contributions were made by which researchers, since some confusion persists in the literature, and remark on some aspects of the sociology of science which have affected the acceptance both of the work described and of the model itself.

In preparing this review we have made liberal use of our earlier reviews (Alpher and Herman 1950, 1953, 1975, and 1988). For more details on the material we shall cover, we refer the reader to any and all of a number of recent and a few not-so-recent books and reviews; the following are only examples. On the big bang model, technical material will be found in Peebles (1971), Weinberg (1972), Reines (1972), Harrison (1981), Zee (1982), Zeldovich and Novikov (1983), Boesgard and Stiegman (1985), Bernstein and Feinberg (1986) and popularizations will be found in Weinberg (1977), Gribben (1986), and Kippenhahn (1987). A technical overview of recent work on the very early universe is given by Olive (1986). Also, there are a number of popularizations such as Zee (1986), Pagels (1985), and Davies (1984) regarding this very early epoch, particularly since it raises interesting questions about quantum mechanics and the early universe.

We take the origin of modern cosmology to be Einstein's general theory of relativity, which provided for the first time a conceptual physical/mathematical framework for dealing with problems on a cosmic scale (Einstein 1916, 1955). Einstein (1917) himself constructed the first cosmological model based on this theory of gravitation but modified his field equations with the inclusion of a cosmological constant so that the model would be time-independent, as was considered to be consistent with the state of observation at the time. Some years later Friedmann (1922, 1924) and independently Lemaître (1931; see review by McCrea 1970, and

Wagoner, this volume) developed nonstatic cosmological models based on the Einstein theory. It is interesting to note that Lemaître considered a closed universe containing both matter and radiation. Later Tolman (1931, 1934) described a nonstatic model containing only blackbody radiation. Meanwhile, de Sitter (1930, 1931) obtained analytic solutions for open, flat, and closed (or oscillating) models both with and without a cosmological constant.

All of the nonstatic models gained credence when during the late 1920s Hubble (1929, 1936) interpreted the redshifts of extragalactic nebulae which he, Humanson, and Slipher had been observing as indicating that the universe was expanding and therefore nonstatic. Again we note the special contribution of Lemaître (1950) who proposed a qualitative model of the universe which originated in the explosion of an all-encompassing primeval atom. While Einstein later characterized his introduction of the cosmological constant as a mistake, it has proved to be useful in cosmological modeling in several ways, including rationalizing the age of the universe in some models, and in providing an extra measure of freedom in model-building generally (see, for example, Weinberg 1972). Finally, then, the canonical model is based on the Friedmann–Lemaître solutions of Einstein's field equations, with a metric as given by Robertson and Walker, frequently referred to as the FLRW model (see appendix).

We note that there continues to be debate on whether the redshifts are in fact Doppler shifts, and on whether, as required in the FLRW model, the universe is homogeneous and isotropic on a scale sufficiently large to permit one to model the cosmic fluid as an ideal gas of matter and radiation. Moreover, there appears to be a significant discrepancy between the amount of luminous matter observed and the amount required by other considerations – the issue of dark or "missing" matter. We consider these questions to lie outside the scope of the present brief historical review.

Our contributions to "big bang" cosmology have their origin in the convergence of the concepts of nonstatic cosmological models with the realization that unusual physical conditions might be required for explaining the observed universal abundance distribution of the chemical elements (Gamow 1946; Alpher, Bethe, and Gamow 1948). During the same period that nonstatic models were being developed and the universal expansion was becoming an accepted phenomenon, scientists were becoming convinced that the observed relative abundances of the elements were indeed representative of the composition of the universe, in broad scope of course. From 1922 through the mid-1940s various attempts were made to understand cosmic abundances as the result of a rapid freezing in a state of thermodynamic equilibrium among nuclei (see reviews by Alpher and Herman 1950, 1953). Such an equilibrium had to have been at a high temperature and density in order for the abundances to develop according to the relative stability of nuclei (binding energies considered as Gibbs chemical potentials). It proved impossible to establish single values of temperature and density which were satisfactory over the entire range of the elements. In what proved to be one of the last and most detailed efforts in equilibrium theory, Chandrasekhar and Henrich (1942), developing earlier ideas of von Weizsäcker (1938), concluded that there was no single set of values of temperature and density which would do, but that the entire abundance distribution must nevertheless be involved in determining the conditions for nucleosynthesis. They went on to suggest that some nonequilibrium process was required. It is interesting that the contemporary view of nucleosynthesis in stellar interiors requires a mix of equilibrium and nonequilibrium considerations (see Fowler 1984, for a review of his and other contributions to stellar nucleosynthesis.)

8.2 Primordial nucleosynthesis by neutron-capture

It was in the context of the failure of earlier equilibrium theories that Gamow gave a talk at Ohio State University, later published in the relatively obscure *Ohio Journal of Science* (1935), in which he proposed that thermonuclear reactions, and in particular neutron-capture reactions which had then been recently described by Fermi, were basic to the formation of nuclei. Gamow's ideas presaged the so-called s-process of steller nucleosynthesis.

By 1942 Gamow was intrigued by the emerging picture of stellar energy generation as developed by Bethe, von Weizsäcker, Critchfield, Teller, Atkinson, Houtermans, and Gamow himself. Moreover, it had become clear that equilibrium theories of formation of the elements were not going to succeed; their failure was dubbed the "heavy element catastrophe" in that one could explain the light elements, but with the same physical conditions the calculated abundances of heavy elements were exceedingly insufficient. Gasmow gave a talk to the Washington Academy of Sciences in 1942 in which he proposed that the abundance distribution might be the result of a nonequilibrium breaking-up process, in which an original superdense object composed of nuclear matter would fragment, not unlike the primeval atom first proposed by Lemaître in 1931. Heavy elements would be built up by neutron-capture reactions, and fission fragments would cycle back as the limit of nuclear stability past uranium was exceeded, all in the context of expansion in the early stages of the universe.

The seminal viewpoint on nonequilibrium nucleosynthesis was contained in a later publication by Gamow (1946) in which he concluded that one must seek a nonequilibrium locale for nucleosynthesis and suggested that this locale might be the very early stages of the expanding universe. Using the FLRW model, he actually estimated the early rate of expansion of a matter-dominated model. He concluded, first, that at high densities such as might obtain during nucleosynthesis the rate of expansion would have been far too rapid for an equilibrium distribution of abundances to have been established, and second, that if neutron-capture reactions were the primary agent in establishing the abundance distribution, then there would have had to be a plentiful supply of free neutrons in the very early universe. Moreover, he suggested that the time available for thermonuclear reactions would have been short compared with the neutron lifetime. He then hypothesized a comparatively cold cloud of neutrons that somehow coagulated into heavier complexes, which in turn underwent beta-decay, moving into the nuclear stability region and thereby building up the heavier elements. The high abundance of hydrogen was taken to be the result of competition for neutrons between the coagulation process and neutron beta-decay.

Before we describe the next step it seems worthwhile to allude again to what actually stimulated all of these ideas, namely, the realization that the relative abundance distribution of the elements was universal. The story begins early indeed, with publications of Oddo (1914) and Harkins (1917) which proposed that the abundance distribution of the elements reflected their nuclear rather than their chemical properties. Then in the 1930s seminal papers by Goldschmidt (see, for example, 1938) synthesized the emerging view that stars were in lowest approximation all made of the same stuff in agreement with ongoing work on the composition of meteorites, the Earth's crust, and interstellar matter. This led to the first tabulations of cosmic abundance data as functions of atomic weight or atomic number. Shortly after World War II detailed corrections were made in the abundance distribution data by Brown Suess, Urey, and others (see Alpher and Herman 1950), so that there was a reasonable basis for studies of primordial nucleosynthesis. Figure 8.1 shows Brown's revision (1949) of Goldschmidt's data, which formed the basis of our work on this question. The superimposed curves are described below.

We note that the year 1946 represented a time of personal convergence, namely, that of the authors with Gamow. In 1944 Alpher, while a part-time graduate student pursuing evening courses at George Washington University, where Gamow was Professor of Physics, left the US Navy Department to join the staff of the Applied Physics Laboratory of The Johns Hopkins University (APL/JHU). Gamow later became a consultant there; Herman had joined APL/JHU several years before Alpher, after receiving his doctorate in physics at Princeton University and working for a time at the University of Pennsylvania and at the City College of New York. Herman's dissertation was in molecular spectroscopy; he studied relativity and cosmology with H. P. Robertson, who also headed his PhD, examining committee. We note that there is an interesting appreciation of Robertson by Zeldovich in *Physics Today* (1988). During the war years, Alpher worked on ship degaussing and proximity torpedo exploders,

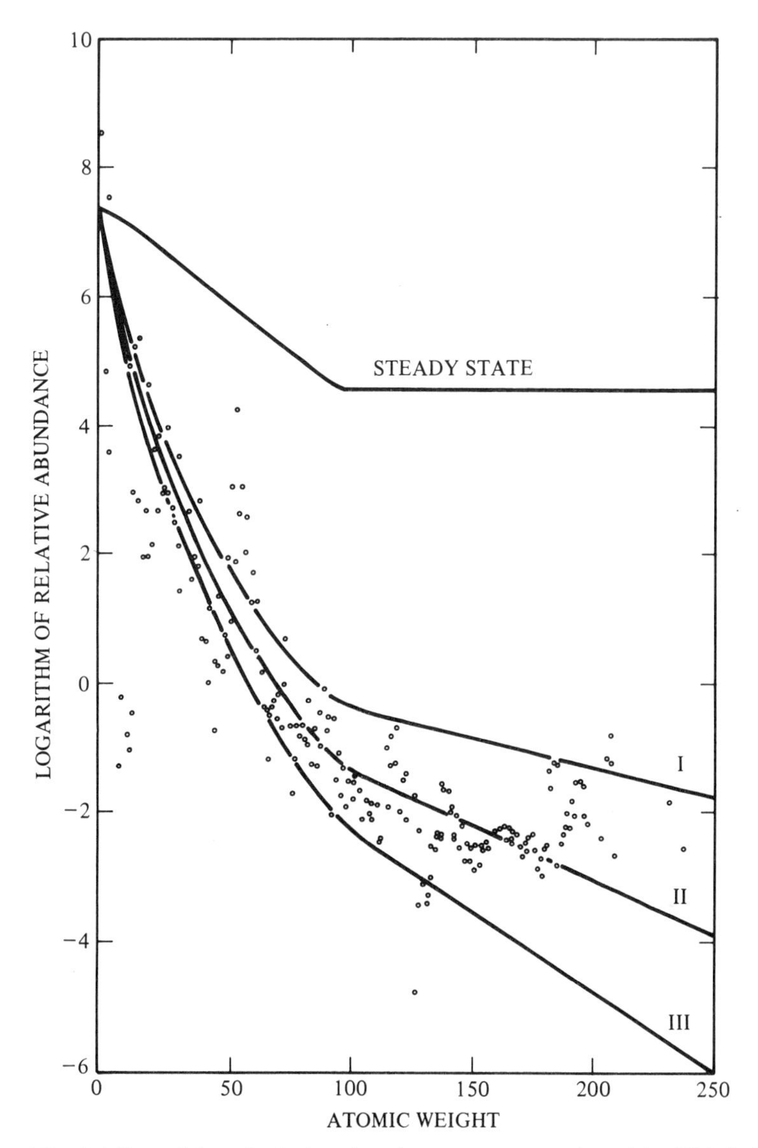

Fig. 8.1 Logarithm of relative abundance *versus* atomic weight. This is the basic cosmic elemental abundance data as compiled by Brown (1949) and used in our early primordial nucleosynthesis calculations. The superimposed curves marked I, II, III and Steady State are for various starting conditions in calculations based on the premises of the αβγ-paper (Alpher, Bethe, Gamow 1948) and are adapted from Alpher (1948b), as explained in the text. [Reprinted with permission from *Reviews of Modern Physics,* **22** (1950), 190.]

while Herman worked on the proximity fuze for naval antiaircraft ordnance. Immediately after the war Alpher switched to work in supersonic aerodynamics, as well as high-altitude rocket studies of primary cosmic rays with James Van Allen; Herman meanwhile pursued fundamental spectroscopic studies related to combustion.

With Gamow as his PhD adviser, Alpher began in 1946 to look at primordial nucleosynthesis as a topic for his doctoral dissertation. At that time he had already finished research begun in 1945 and had started to write up what was to be an ill-fated doctoral dissertation dealing with the rate of growth of adiabatic and isothermal condensations in a relativistic expanding medium dominated by matter. One day Gamow appeared with an issue of the *Journal of Physics of the USSR* he had just received, which contained a paper based on a dissertation by Lifshitz (1946) on the same subject, and incidentally with about the same results. And so it was that a new topic had to be found.

Gamow suggested to Alpher that he follow up on the ideas in Gamow's 1946 paper, namely, the possibility that the elements were synthesized in a nonequilibrium dynamical fashion in the early stages of an expanding universe. We – Alpher and Herman – were neighbors at APL/JHU, and Herman provided a sounding board for Alpher as work on the dissertation progressed. Eventually we became collaborators, beginning an exciting and productive association that has continued for over four decades. Our first joint paper (Alpher and Herman 1948a) included the prediction of the residual cosmic background radiation. That paper, as well as a joint paper with Gamow (Alpher, Gamow, and Herman 1948), were published in 1948 at very nearly the same time as Alpher's dissertation (Alpher 1948). Indeed, our first nucleosynthesis calculation (Alpher and Herman 1948b) was published in the same month as Alpher's dissertation.

We would be remiss in not acknowledging the excitement we experienced during our long collaboration with Gamow, who was known as Geo (Joe) to his friends. Gamow was certainly one of the great scientists of this century, a man with a deep love of science and an amazing intuitive feeling for physics. In our view he was not accorded anything comparable to the formal recognition he richly deserved for his many creative contributions (including such diverse topics as stellar energy sources, theory of alpha-decay, theory of beta-decay, model of nuclear fission, structure of DNA, structure of red giants, as well as cosmology). It is regrettable that Gamow's fun-loving and irrepressible approach to physics and to life in general led some scientists not to take his work seriously, and perhaps our work as well, because of our close identification with him.

What was the prevailing view of the universe at the time we began our work? The expansion of the universe was an accepted phenomenon, but many, including Einstein, still felt more comfortable with the idea of a static cosmos. Thus, the intellectual soil was fertile for the steady-state theory, which combined a steady-state cosmos with expansion maintained by matter creation, and which was the subject of active study principally by Hoyle, Bondi, and Gold (see Hoyle, this volume). It did not help the acceptance of an evolving, expanding model that Hubble's parameter, as measured in the late 1940s and early 1950s, gave an unacceptably short age for the universe, an age less than the age of the Earth. Despite this problem, Gamow and we tried to put physics into a nonstatic cosmological model and to explore the consequences. Even after new observations reduced the Hubble age problem, it took a long time for an evolving model to be accepted. We comment on this toward the end of this review, and at some length in a recent paper (Alpher and Herman 1988).

We note parenthetically that Alpher first introduced the use of the word "ylem" (derived from the Greek ΥΛΗ) to describe the primordial mix from which the elements were formed (Alpher 1948). Hoyle appears to have been responsible for introducing the expression "big bang" as a descriptor of an evolving, expanding universe. According to Gamow, Hoyle first used this phrase in a pejorative sense during a BBC radio debate with Gamow (see also Hoyle 1950, for the first published use). The appellation appears to have outlived the steady-state cosmology advocated by Hoyle, Bondi and Gold.

We return now to the problem of developing Gamow's 1946 ideas; it appeared to be a difficult and problematical task. There was a paucity of thermonuclear reaction cross section data; what data existed had not yet been declassified from wartime use, and were not available to us. However, there was a fortunate breakthrough for our research on cosmological nucleosynthesis with the publication of work by Hughes at Brookhaven (Hughes 1946). He had surveyed neutron-capture cross sections at 1 MeV for a large number of elements in an attempt to determine the suitability of various materials for reactor construction. What was exciting about Hughes' results was that the cross sections varied inversely with relative abundance data, as shown in fig. 8.2. Moreover, Hughes' data as well as other fragmentary results that could be related to his work, also shown in the figure, exhibited very low capture cross sections for the "magic number" nuclei, whose abundances were significantly elevated relative to their neighboring nuclei. This was further corroboration of a strong inverse relationship between abundance and neutron-capture cross section.

Hughes' results were then used by Alpher in his dissertation calculations under Gamow's guidance. It was assumed that at an appropriate early time the universe consisted of a gas of neutrons. As protons became available from neutron decay, sequential neutron-capture reactions commenced, building up the element abundance distribution. In this first attempted model of primordial nucleosynthesis, in which the expansion was not taken into account, the neutron-capture cross sections were represented as showing a simple exponential increase up to atomic weight 100, with a constant value thereafter for the heavier elements, all as shown in fig. 8.2. The nuclear reactions were assumed to begin when the universal temperature had fallen below the binding energy of the deuteron, about 0.1 MeV; the cross sections were corrected using a $1/E$ relation. It was

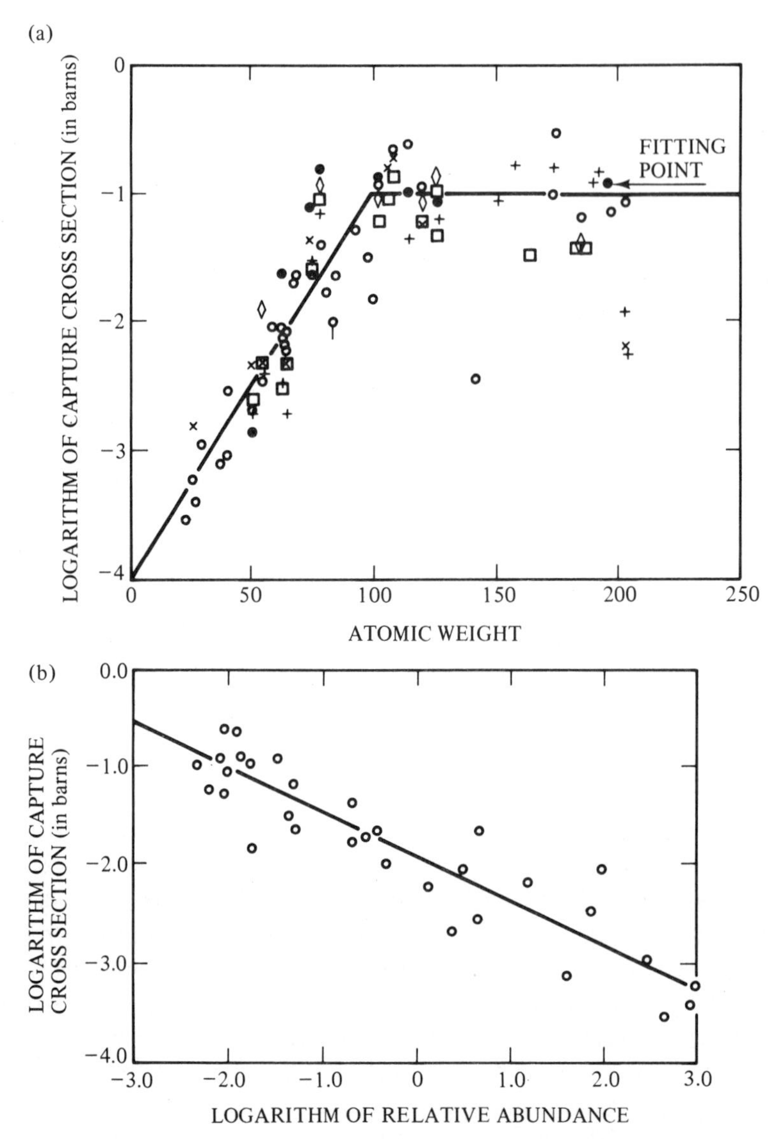

Fig. 8.2 In (a) the measurements of Hughes (1946) of radiative neutron-capture cross sections at 1 MeV are plotted *versus* atomic weight. In (b) is shown a correlation of these measurements with relative abundance. Data by other investigators were added to Hughes' data where appropriate. The straight lines represent the smoothed values used in our calculations. Details are given in Alpher (1948b) and Alpher and Herman (1950). [Reprinted with permission from *Reviews of Modern Physics*, **22** (1950), 184 and 185.]

further assumed that nucleosynthesis was completed in a time short compared both to the time scale of the universal decrease in temperature resulting from expansion and also the neutron decay lifetime.

This first very simplified static calculation, described in what has become known as the "αβγ-theory," did not explicitly consider the radiation domination of the universe at early times, although this was discussed in Alpher's detailed dissertation (Alpher 1948), and led to what seemed at first glance to be a very high value of the density of matter during nucleosynthesis. Simplified as this approach may have been, it did nevertheless map the general nature of the abundance distribution, which had not been done before, provided a rationale for the relatively high abundance of the light elements, and appeared to require an early universe that was both hot and dense.

These results confirmed Gamow's 1946 ideas, and also gave the first intimation of a hot, dense origin for the universe. The results of the αβγ-theory as described in more detail by Alpher (1948b) in a paper based on his dissertation, are shown in fig. 8.1, superimposed on the basic cosmic abundance data referred to earlier. Curves I, II, and III represent different values of the product of the starting neutron concentration and the assumed duration of nucleosynthesis. For Curve II, this product is 8×10^{17} sec cm^{-3}; for Curve I it is 1.6 times this value; and for Curve III, 0.6. The "steady state" is the result achieved by allowing the process to run indefinitely. Even before the appearance of this paper, Alpher and Herman, as discussed below, were occupied with improved calculations, taking the universal expansion into account explicitly.

8.3 Prediction of the cosmic background radiation

We emphasize here that the αβγ-paper did not consider any current consequence of a hot radiation-dominated universe at early times, contrary to a recent statement by Hawking (1988). The conceptual and quantitative step to a present cosmic background radiation was made by Alpher and Herman, later in 1948, when they calculated the ~5 K blackbody relict radiation temperature. Moreover, contrary to statements by Wilkinson and Peebles (1983), Gamow made no reference in his 1948 *Nature* paper to a 10 K background; in addition, in their 1948 *Nature* paper Alpher and Herman had not "repeated Gamow's calculation with greater accuracy, using a computer," in the context of the background blackbody radiation. They did re-do Gamow's calculations with respect to galaxy formation but the prediction and calculation of a cosmic blackbody relict radiation of 5 K had nothing to do with the subject of Gamow's paper.

In their developments beyond the αβγ-paper, it was immediately apparent to the authors that sequential build-up of the elements was not adequate, that one would have to look in detail at all of the thermonuclear reactions among light nuclei and not just at a smoothed neutron-capture cross section function, and that there would be a major problem resulting from the absence of stable nuclei at masses 5 and 8. Another problem to which we have already alluded was the short age for the cosmological model, given the then-current value of the Hubble parameter. Furthermore, one had to consider a cosmological model with a mixture of matter and radiation as the working fluid, and with radiation dominating the behavior of the model at early times. It was also evident, as assumed in the $\alpha\beta\gamma$-paper, that sequential build-up could not have been significant until the temperature had fallen below a value equivalent to the binding energy of the deuteron, where photodisintegration would have become unimportant.

Following up these points in one of his characteristically insightful but "quick and dirty" calculations, Gamow tried to get at the conditions in the early universe by looking only at the synthesis of deuterons and at the same time see what he could do about the tantalizing problem of the formation of galaxies. Two publications by Gamow in 1948 (1948a, 1948b) covered these ideas in some detail. Alpher and Herman found some errors in the second of these papers; in so doing they came coincidentally to their prediction of the relict blackbody radiation in the universe, as recounted below.

Gamow supposed that the time dependence of the temperature and of the density of matter was given by an approximation, valid at early times, of the relativistic description of an expanding model (see appendix). He further supposed that one could start with neutrons only and use simple rate equations for neutron and proton concentrations, with neutrons being lost to beta-decay and appearing as protons, and with neutrons and protons being used up in the formation of deuterons. He then employed a theoretical n(p,γ)d reaction cross section given by Bethe (1947) and integrated the rate equations subject to the constraint that the final concentration of protons be 1/2. The free parameter in his calculations was the density of matter, ϱ_o, at a particular early epoch. Gamow's calculation gave a value of $\varrho_o \simeq 7 \times 10^{-3}\,\mathrm{g\,cm^{-3}}$ at one second, which was consistent with the primordial density of matter calculated in the $\alpha\beta\gamma$-paper and in later calculations by Alpher and Herman. He then used these quatities together with Jeans' gravitational instability criterion to derive a typically interesting Gamowian relation between the diameter and mass of galaxies on the one hand, and the binding energy of the deuteron on the other hand. He did suggest that galaxies might be expected to have formed at the epoch when the densities of matter and radiation were equal, a crossover or decoupling time (decoupling because at this time the universe changed from being opaque to being transparent to the radiation), a time between the control of the expansion by radiation earlier, and by matter later. However, his numerical estimates for this epoch and of the consequent matter and radiation densities were substantially incorrect because he carried the early time approximations for the time-behavior of matter and radiation densities too far.

During the summer of 1948 Gamow had sent Alpher and Herman a copy of the manuscript dealing with this calculation which he had already submitted to *Nature*. Alpher and Herman pointed out errors in the manuscript in a telegram sent to Gamow at Los Alamos where he was spending the summer. However, Gamow felt it was too late for him to make the needed corrections, and he urged the authors to submit a paper to *Nature* with these comments. We sent him a proposed manuscript. Gamow advised *Nature* that a paper by Alpher and Herman was coming, and asked that it be published as soon as possible after his paper. It was in the course of these considerations that Alpher and Herman found it possible to integrate the full relativistic equations of the expansion, not only finding a correct crossover or decoupling time, but also realizing that they were in a position to examine the time dependence of all the relevant physical variables over the entire evolution of the universe. In particular they were able for the first time to predict the existence of a cosmic background blackbody radiation and to calculate its temperature and contribution to the density in the present universe. The calculation (see appendix), based in a relatively simple way on the then-known values of the pertinent parameters, yielded a present cosmic background blackbody radiation at a temperature of ~5 K. This result was stated explicitly in the Alpher–Herman 1948 *Nature* paper, and was discussed in considerable detail in three later publications by Alpher and Herman (1949, 1950, 1951; see also the further elaboration in our 1975 and 1988 reviews).

The authors' 1948 calculation of the residual blackbody radiation temperature as well as their subsequent calculations depends on the simple relation between the densities of matter and radiation which must hold during the expansion if there is no interconversion of matter and radiation, namely,

$$\varrho_r \varrho_m^{-4/3} = \text{constant}.$$

This relation is used in the following way; if one knows the density pair $[\varrho_\tau, \varrho_m]$ at any one epoch, then the relation defines possible values of the pair of densities at any other epoch, and, if three of the densities are known, the fourth can be determined. In the first 1948 calculation, the following data were used: a matter density as required for nucleosynthesis of $\sim 10^{-6}\,\mathrm{g\,cm^{-3}}$, and a radiation density calculated from the early time approximation (see appendix) of $\sim 1\,\mathrm{g\,cm^{-3}}$, each at about 640 seconds into the expansion; the present mean density of the universe was given by Hubble in 1948 as $\sim 10^{-30}\,\mathrm{g\,cm^{-3}}$. The above density pair relation then contained the present radiation density as the only unknown, and the calculated value for a background blackbody radiation temperature was found to be ~5 K or a radiation density of $\sim 5 \times 10^{-33}\,\mathrm{g\,cm^{-3}}$.

It is interesting to consider the various times at which we republished this type of calculation. The original prediction of ~5 K, as already noted, was in 1948, approximately six months after the αβγ-paper; the next, which was the first detailed discussion, was one year precisely after the αβγ-paper; the next, in 1950, was also after a one-year period. The following paper, in 1951, was once again after yet another one-year period. This last calculation was in a way unfortunate, because in the interim a new value for the universal mean density had become current (long since revealed to be in error) and this led to an estimate of a background temperature of ~28 K. Scholarly research unfortunately being what it was and still is, this particular result has not infrequently been quoted as though it were a failure of the model or of the work of the authors Alpher and Herman, rather than being properly considered in its dependence on what the community of scientists at the time considered to be reasonable cosmological parameters.

8.4 More on primordial nucleosynthesis

The year 1951 was not a great year for advocates of the big bang model. The same data just mentioned which gave far too high a value of the background radiation temperature also led to a very short age for the model. This provided much encouragement to devotees of the steady-state model, and despite their not having faced up to the problem of finding a source of the cosmic abundance of helium in their model, or for doing much more than handwaving about the formation of the rest of the elements, the steady-state model was in its ascendancy. It was not until 1964 that Hoyle and Tayler formally recognized the formation of helium as an insuperable problem for the steady-state theory, although by then the theory of the formation of the heavier elements in stellar interiors was in reasonably good shape. Since that time the abundance of deuterium has become a more firmly known quantity, and is also well represented by primordial nucleosynthesis (see Wagoner, this volume).

Subsequent to the αβγ-paper there were a number of improved calculations of primordial nucleosynthesis based on neutron-capture cross sections. As an example, Smart (1948, 1949) considered shielded isobars, as well as the even–odd atomic weight effect in abundances, in the context of a neutron-capture sequence. In 1951 Alpher and Herman considered again the straightforward neutron-capture sequence. They used the smoothed neutron-capture cross sections in rate equations which explicitly included neutron decay and the universal expansion. Figure 8.3 shows the evolution with τ (where $\tau = \lambda t$, and λ is the neutron decay constant) of relative abundances for selected atomic weights. After a $\tau = 0.13$ the reactions die out and the curves become parallel as the remaining neutrons decay and the mix of nuclei dilutes and cools in the expansion. Note the relatively high abundance at the atomic weight of helium. In fig. 8.4 the final relative abundance values are plotted against atomic weight and compared with the cosmic abundance distribution. The best fit was obtained with a matter density chosen to be $\sim 1.5 \times 10^{-3}\,\mathrm{g\,cm^{-3}}$ at one second into the expansion, or $\sim 9 \times 10^{-4}\,\mathrm{g\,cm^{-3}}$ at a value of $\tau = 0.13$ in fig. 8.3.

The relative abundances calculated with smoothed neutron-capture cross sections were of interest, but it was recognized that as soon as

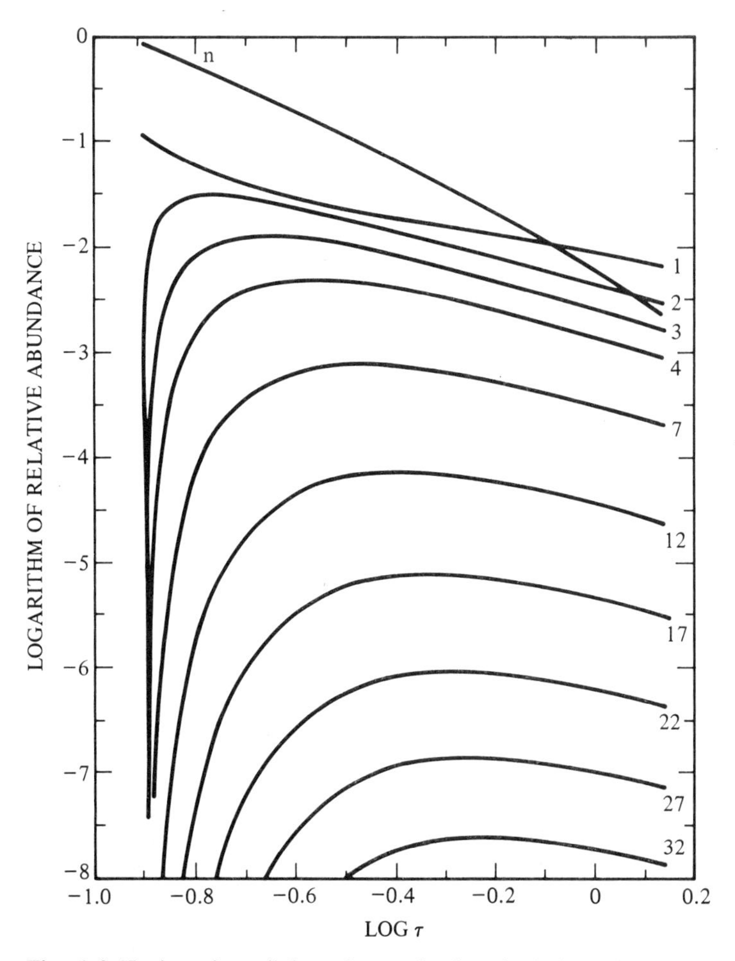

Fig. 8.3 Early primordial nucleosynthesis calculations based on a neutron-capture reaction sequence, including free neutron decay and the universal expansion, according to Alpher and Herman (1951). The time scale is in units of the neutron decay constant. [Reprinted with permission from *Physical Review*, **84** (1951), 63.]

detailed thermonuclear reaction rates among the light elements in particular became available, they should be explored in this context. As the result of attending a colloquium by Alpher in late 1948, Enrico Fermi, in collaboration with Anthony L. Turkevich, collected or estimated as required the cross sections for 28 thermonuclear reactions among nuclei up to atomic weight 7. They used these cross sections, together with the early-time approximations for the cosmological model, to solve coupled rate equations, including the universal expansion, among the 28 reactions considered. The calculations were subject to the conditions that the reactions began at about 300 seconds into the expansion, when the temperature had fallen to about 0.07 MeV and photodissociation reactions

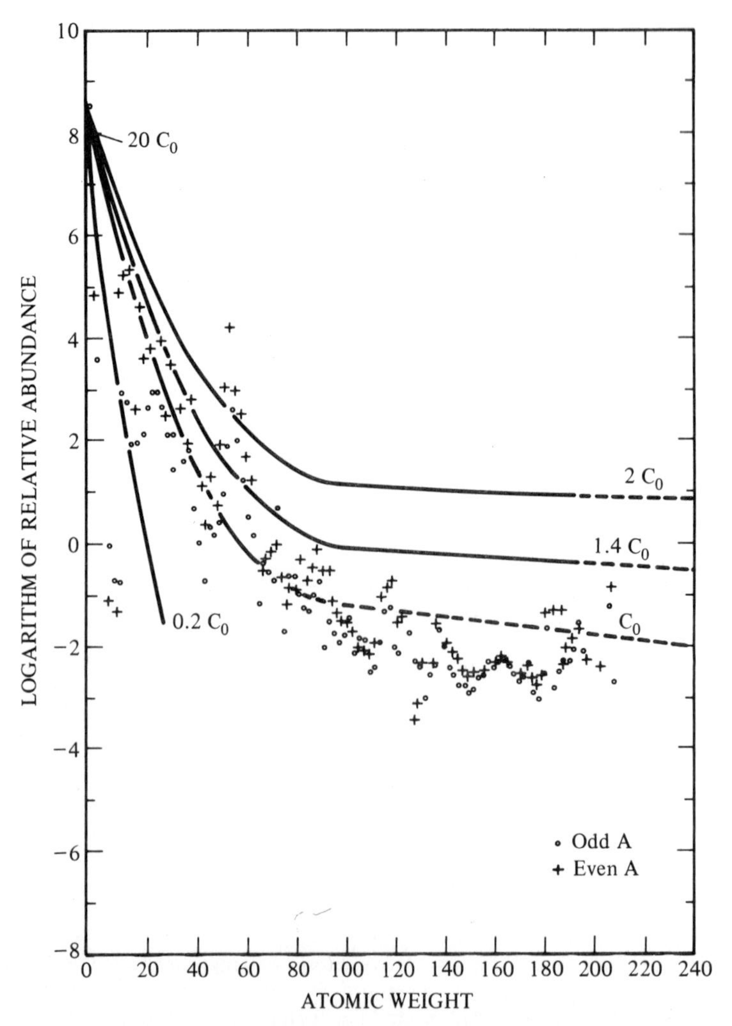

Fig. 8.4 Final relative abundance values calculated in Fig. 3, plotted *versus* atomic weight and compared with cosmic abundances. The best fit is represented by the C_0 curve, which represents a starting matter density of $\sim 1.5 \times 10^{-3}\,g\,cm^{-3}$ at one second into the expansion. See Alpher and Herman (1951). [Reprinted with permission from *Physical Review*, **84** (1951), 64.]

would no longer be important, with neutrons and protons initially in the ratio 7:3, as would result from starting with neutrons only at 1 second, and, finally, with a neutron decay constant of $10^{-3}\,sec^{-1}$. Their results for a starting matter density of $1.7 \times 10^{-3}\,g\,cm^{-3}$ are shown in fig. 8.5, as reported by Alpher and Herman (1950). Again note the high abundance of primordial helium. Needless to say, no way was found at that time to continue a building up process past atomic weights 5 and 8. Suggestions for bridging the atomic weight gaps at 5 and 8 were made by Fermi, Wigner, Hayashi, and others over the next few years, but none of these resolved the

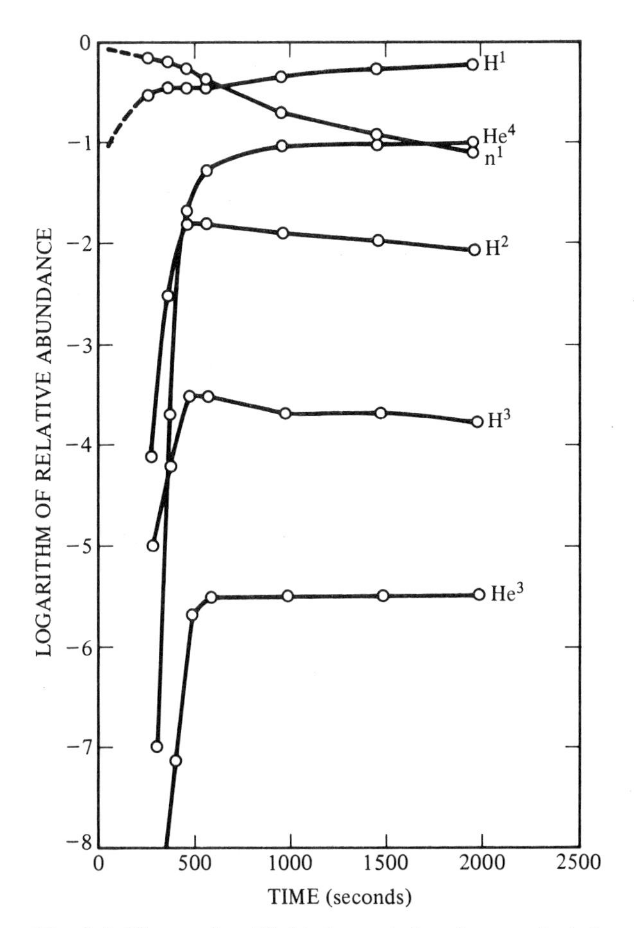

Fig. 8.5 The results of light element abundance calculations performed by Fermi and Turkevich, as reported by Alpher and Herman, 1950. These were based on consideration of 28 reactions, through Li and Be. The initial nucleon concentration at 300 seconds was taken to be $10^{21}\,cm^{-3}$, a matter density of $\sim 1.7 \times 10^{-3}\,g\,cm^{-3}$. Expansion of the universe was taken into account. [Reprinted with permission from *Reviews of Modern Physics*, **22** (1950), 195.]

difficulty at the time. Several researchers more recently have considered the possibility that density or compositional inhomogeneities in the early universe would provide a mechanism for bridging these gaps and allow primordial nucleosynthesis of the heavier elements (Hayashi and Nishida 1956; Applegate, Hogan and Scherrer 1987; Malaney and Fowler 1988).

An additional concern in the early work was the lack of detailed knowledge of the initial conditions for the period of nucleosynthesis. In late 1950 Hayashi proposed that instead of starting with an ylem of pure neutrons, one should calculate the initial relative concentrations of neutrons and protons for nucleosynthesis by considering the outcome of reactions among elementary particles and radiation starting at a temperature in the expansion just less than that equivalent to the meson rest mass. He performed calculations along these lines and derived an abundance ratio of neutrons to protons which was too small to be useful as a starting condition for the simple neutron-capture sequence to represent nucleosynthesis properly. Thereafter, Alpher and Herman, together with colleague James W. Follin, Jr. of APL/JHU (Alpher, Follin, and Herman 1953a) undertook the most detailed calculations they could manage in the spirit of Hayashi's approach, under the constraints of the state of physics in 1952. Major differences between the calculations of Alpher, Follin, and Herman and those of Hayashi were in that the former used relativistic quantum statistics appropriate to the prevailing conditions, and also used a then-current value for the neutron decay constant which was significantly different from that employed by Hayashi.

The Alpher–Follin–Herman calculations were begun at a temperature of 100MeV, $\sim 10^{12}$ K, with the assumption that thermodynamic equilibrium prevailed for all species except baryons and antibaryons, for which a temperature of 100 MeV was too low for interconversion. The predominance of baryons over antibaryons had been established prior to the time when the temperature had dropped to $\sim$100 MeV by physical processes then unknown and still somewhat problematical. The universe had cooled to about 100 MeV by $\sim 6 \times 10^{-5}$ sec. Using relativistic quantum statistics and the temperature-time relation appropriate for early times in the expansion, Alpher, Follin, and Herman then calculated the contributions of baryons, leptons, and radiation to the total mass density which controlled the expansion. Two Dirac neutrino cases were considered, namely, distinguishable and indistinguishable neutrinos, which gave significantly different results for the ratio of the total density to the radiation density. While the neutron–proton ratio obtained was indeed different from that obtained by Hayashi, it also was not appropriate for a successful neutron-capture sequence for all the elements. Alpher, Follin, and Herman continued their collaboration until 1960, using their earlier work as a basis for further consideration of the formation of the light elements, but their results were reported only in a series of contributed papers at American Physical Society meetings (see Alpher, Follin, and Herman 1953b). There seemed to be little interest in the results at the time.

In one additional utilization of the results of Alpher, Follin, and Herman, the authors (1958) showed that the baryon–antibaryon asymmetry in the universe could not be explained as a residue of a statistical fluctuation from the early time when the baryon concentration was of the same order as the photon concentration.

Another interesting result of this collaboration was the realization that the neutrinos would freeze in when the temperature had dropped to $\sim$10 MeV and would then continue to expand adiabatically to the present. With the methodology of the Alpher, Follin, and Herman paper, and given the knowledge about neutrinos at the time of its preparation, one would expect there to be a background of neutrinos today at a temperature of about 2 K, with the present background blackbody radiation at 2.73 K. Similar calculations in the same paper also suggested a residual background today of gravitons, if indeed they exist, with a temperature of about 1.2 K.

The authors note that Matzner (1968) found an algebraic error in the Alpher, Follin, and Herman paper which affected the graviton decoupling temperature they used, and recalculated the temperature of any present background gravitons to be about 1.6 K, while Weinberg (1972) has estimated about 1 K for the graviton background. These numbers are most interesting in the context of recent work by Rawley, Taylor, Davis, and Allan (1987) who estimated upper limits to the present density of gravitational radiation. Their results, based on studies of the stability of the "clock rate" of the millisecond pulsar PSR 1937+21 over a four-year period, enabled them to put an upper limit on the graviton density of $\sim 1.7 \times 10^{-34}$ g cm^{-3} for a frequency of 0.8 cycles/year and an upper limit of $\sim 7 \times 10^{-36}$ g cm^{-3} for a frequency of 0.23 cycles/year. We note that if one assumes a number density for the massless residual gravitons given by the Planck distribution, 1 K corresponds to a graviton density of $\sim 3 \times 10^{-36}$ g cm^{-3}, while 1.6 K corresponds to a graviton density of $\sim 2 \times 10^{-35}$ g cm^{-3}.

In 1975, Alpher and Herman had occasion to prepare an extensive review of their work on the big bang model and their prediction of the background radiation. By this time the relevant cosmological parameters had changed significantly. In particular they used a present matter density of 3.4×10^{-31} g cm^{-3}, a value consistent with the work of Gott, Gunn, Schramm, and Tinsley (1974), but still well below the closure density for a Hubble parameter of 55 km sec^{-1} Mpc^{-1}, a matter density at 1 second into the expansion of $\sim 6.9 \times 10^{-2}$ g cm^{-3}, close to that obtained by Wagoner (1973) in his detailed study of light element nucleosynthesis, and a radiation density of 4.48×10^{-5} g cm^{-3} at 1 second from the early-time approximation (see appendix). These values yield 2.7 K for the present cosmic background temperature. The authors show in fig. 8.6 a plot of the densities, temperatures, proper distances, and redshifts for these particular parameters of the model. Gamow called such plots "Divine Creation Curves" and used one as a logo on his personal letterhead for a considerable period in his later years. Clearly some concepts will have to give if the assumed present matter density needs to be corrected upward by a factor of 10 or more because of the presence of dark or faint matter, or to meet the closure requirement of inflationary models. These considerations are beyond the purview of the present discussion.

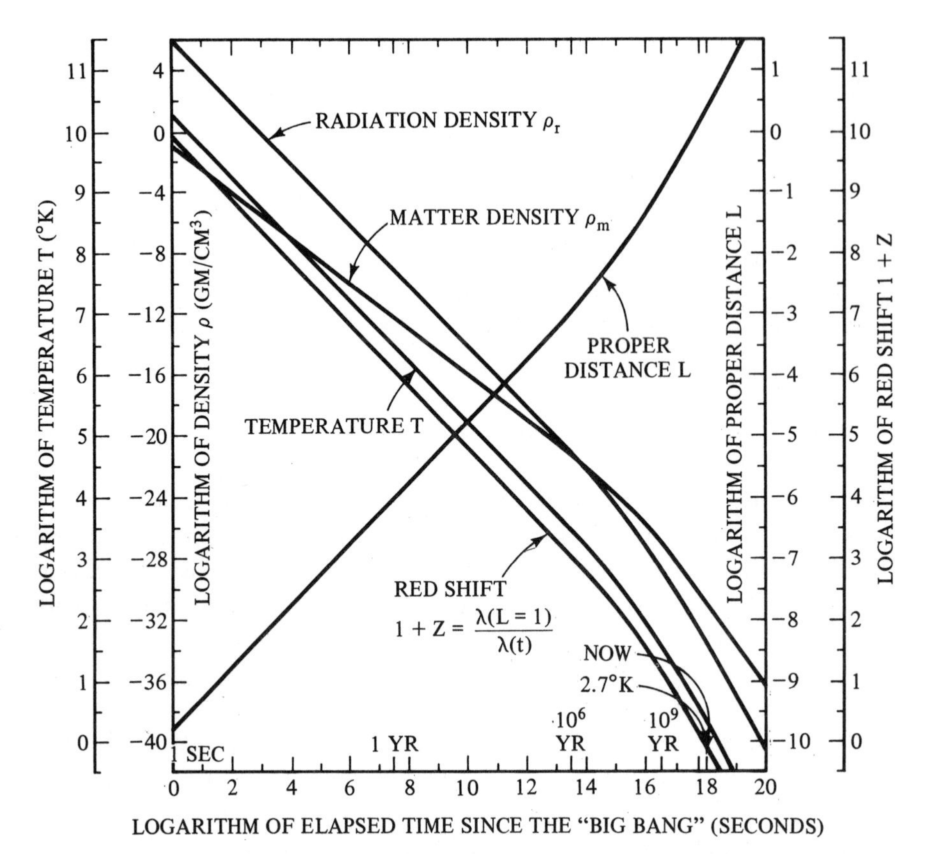

Fig. 8.6 A plot of various cosmological parameters as a function of the epoch, for a standard big bang open model, with matter density 0.06 times the critical density, a present value of the Hubble parameter of 55 km sec^{-1} Mpc^{-1}, and a present background radiation temperature of 2.7 K. This "Divine Creation Curve" (so-named by Gamow) is described by Alpher and Herman (1975). [Reprinted with permission from *Proceedings of the American Philosophical Society*, **119** (1975), 333.]

8.5 Some comments on the history and sociology of science

There are some curious aspects of our interactions with Gamow on the specific question of the background radiation. In 1948 and 1949 he argued with the authors personally and in correspondence that even if the concept of a relict cosmic blackbody radiation was real, it was not useful, because of the presence of starlight at the Earth at about the same energy density. Despite this, in 1950 Gamow made a remarkable statement in an article in *Physics Today*, namely, that there is a 3 K background radiation, with not the slightest hint in the article or in Gamow's correspondence to the authors on how the 3 K figure was produced or where it came from. He was, of course, fully aware of our calculations at the time, and may have "rounded off" in his own inimitable way!

Three years later in an article in a Danish journal (Gamow 1953) he estimated a 7 K background temperature by means of a strange linear extrapolation of matter and radiation densities in the expansion, despite the availability to him of our approach. Again three years later (Gamow 1956) he persisted with yet another arcane calculation, obtaining 6 K. The authors became aware of these papers of Gamow only after they appeared in print; they chose for a variety of reasons not to publish their disagreement with his approaches to estimating a background radiation temperature. The fact that Gamow made no reference to the earlier calculations of Alpher and Herman in any of his three papers has made for considerable confusion in the literature and added to the problem of correct attribution by later workers to the authors' original calculations.

By 1956 there were in the archival literature four papers by Alpher and Herman (1948b, 1949, 1950, 1951) and three by Gamow (1950, 1953, 1956) discussing a residual background radiation. Our ~5 K prediction was also described in a news release from the American Physical Society in January 1949 on the occasion of a paper we gave at the Annual Meeting. A number of science writers around the U.S. based articles on this press release and included the prediction in their articles. Finally, in 1964 Doroshkevich and Novikov discussed a cosmic background radiation. Their paper, which was submitted by Zeldovich, referred to a 1–10 K range for such radiation, curiously attributing the result to a 1949 paper by Gamow in which, in fact, he had not alluded to the possibility of such radiation.

Penzias and Wilson (1965) reported their observation of a background radiation at 7.35 cm wavelength in 1965 without their own interpretation, but referred instead to an explanation given in an accompanying paper by a group at Princeton, viz., Dicke, Peebles, Roll, and Wilkinson (1965). The latter interpreted the radiation as a remnant background from what they called the "primeval fireball." It is reported that when they made this identification, the Princeton group had nearly completed setting up a radiometer to look for a background radiation reflecting a prior "big crunch." In a short time they confirmed the Penzias and Wilson result at their radiometer wavelength of 3.2 cm. There is a brief historical discussion of these matters, including the authors' considerably earlier prediction, by Gloria Lubkin in *Physics Today* (1978), whose article was prepared on the occasion of the awarding of the Nobel Prize to Penzias and Wilson. The authors cannot help noting that the Princeton group made no references in 1965 or for some years thereafter to the many prior publications in the archival literature on the cosmic background radiation (however, see Peebles 1971).

In view of the foregoing, one can only wonder how Hawking, in his recent best-seller (1988), found it possible to state "For this, Penzias and Wilson were awarded the Nobel prize in 1978 (which seems a bit hard on Dicke and Peebles, not to mention Gamow!)."

The verification of a cosmic background radiation, which is widely acknowledged as among the most important cosmological findings of modern times, stimulated a veritable explosion of activity in the scientific community, on both theory and observation, as well as an enormous burgeoning of public interest. Some of the theoretical work for several years thereafter plowed ground already thoroughly worked over. For example, there were numbers of contributions in *Nature* working out the details of a matter-radiation universe which had long since been well documented in the archival literature.

A most interesting facet of the background radiation story is that there was information in the literature published during the World War II years (Adams 1941; McKellar 1941) regarding the rotational temperature of interstellar CN radicals, the interpretation of which was at least suggestive of a background radiation field at several degrees K. Adams observed interstellar CN absorption towards ζ Ophiuchi, and McKellar interpreted the observations as indicating an excitation temperature of 2.3 K at a wavelength of 2.64 mm, with no obvious source of excitation. The result is the more interesting not only because it was overlooked, but also because it lay close to the blackbody peak, while the ground-based measurements of the background radiation were all at longer wavelengths. This mode of observation of the background has recently given one of the best measures of the radiation temperature, namely, 2.796 (+0.014; −0.039) K, as reported by Crane, Kutner, Hegyi, Blades, Palazzi and Mandolesi (1989), for a wavelength of 2.64 mm. These events cause us to wonder again about the efficiency of the scientific enterprise.

Primordial nucleosynthesis of the lightest elements in the early universe, and stellar nucleosynthesis of the heavier elements are by now both reasonably well understood. There is a considerable literature in both areas, with the classical paper on stellar nucleosynthesis being that of Burbidge, Burbidge, Fowler, and Hoyle (1957), and that on primordial nucleosynthesis being Wagoner, Fowler, and Hoyle (1967) (see Wagoner 1973, and this volume). We note with some satisfaction that work on primordial nucleosynthesis uses the methodology of the 1953 paper by Alpher, Follin, and Herman, and that the physical conditions required for primordial nucleosynthesis are not radically different from those we deduced earlier. We certainly appreciate the statement made by Weinberg (1977) referring to the Alpher, Follin, and Herman paper as "the first thoroughly modern treatment of the early universe," as well as his commending the authors together with Gamow for being the first to take the big bang model seriously.

Some cosmologists suggest that the very early universe exhibited an inflationary period which appears to require that from that epoch to the present the universe should exhibit a mean density quite exactly equal to that required for closure. The standard big bang model of primordial nucleosynthesis on the other hand requires a density of matter well below the closure density. If the inflationary model is correct, and if there is indeed "missing mass" in galaxies, then this requirement acts as a constraint either on the form of the unseen or "missing mass," or on the conditions surrounding primordial nucleosynthesis. For example, in expanding on the work of Applegate, Hogan, and Scherrer (1987), Malaney and Fowler (1988) suggest that by assuming early inhomogeneities one can have primordial nucleosynthesis, a universal closure density, and a nonexotic baryonic "missing mass."

It is interesting to note that the conditions for primordial nucleosynthesis are providing useful constraints on the current activity among particle physicists and theorists trying to unify the forces of nature (see, for example, the early paper by Steigman, Schramm, and Gunn 1977). Thus, cosmology places limits on the allowable number, type and degeneracy of neutrinos, the number and properties of weakly interacting particles, the mean density of luminous matter and the entropy per baryon, or the photon-to-baryon ratio. For many years that ratio was the one "free" parameter in the canonical big bang model, although we, with Gamow (1967) and others, had long since suggested that it should not be considered "free," but rather should be amenable to explanation as a natural consequence of the physics of the very early universe.

In this context, but again reminding the reader that the present paper is not a review, a very considerable amount of research has been and is being done on the early universe. While the Alpher–Follin–Herman paper pushed the time frontier of the big bang model back to a few microseconds, this could be considered a rather late time compared to the 10^{-43} sec horizon of the Planck time being invoked in current research. Research along these lines during the last ten years or so is incidentally addressing some of the conceptual problems of the canonical big bang model, such as the origin and significance of the high entropy per baryon, baryon–antibaryon asymmetry, the horizon problem, etc. A recent paper by Prigogine, Geheniau, Gunzig, and Nardone (1988) presents a most interesting approach to the question of the generation of entropy.

There are many remaining problems. They range from understanding the formation of galaxies, given the very high level of isotropy of the cosmic microwave background, to the "missing mass" problem. There is also the possibility of some source of energy addition to the short wavelength region of the 2.8 K spectrum, perhaps in the pregalactic universe, should recent measurements of the submillimeter background radiation be confirmed (see recent summary by Andersen 1988). There are also continuing attempts underway to explain the microwave background as other than a relict radiation, including thermalization of starlight by grains (see, for example, Hoyle, this volume; Barnothy 1988).

The temperature of the radiation background now appears to be the most precisely known of all cosmological parameters, and its use as an "absolute frame of reference" (see Peebles and Wilkinson 1968) is helping to provide insights into deviations from a uniform Hubble flow (see Partridge 1988). We look forward with great anticipation to the cosmological insights which are certain to result from new theoretical studies and from observations with the Hubble Space Telescope and with the Cosmic Background Explorer satellite (COBE).

Let us now return for a brief comment on the period after the first observation of the background radiation by Penzias and Wilson. By early 1967 Gamow and we had become very distressed by how our early work continued to be ignored. Gamow took the initiative of urging a joint publication of an historical nature to attempt to set the record straight. The resulting paper was published during the same year in the *Proceedings of the National Academy of Sciences* (Alpher, Gamow, and Herman 1967) in the interest of speed. The main emphasis of the paper was to document once more our work on the big bang model, the Alpher and Herman prediction of the cosmic background radiation, and the joint work on galaxy formation with Gamow. Correct attributions began to appear elsewhere after this paper, and our contributions were more widely recognized. We have had the opportunity of publishing several subsequent review papers. We believe that the publication in 1977 of Weinberg's popularized account *The First Three Minutes* had a major influence on the acceptance of contemporary cosmological ideas by the scientific community, as well as of our role in developing those ideas. It is clear that Weinberg consulted original source material in preparing his book. Nevertheless, accounts continue to appear, both in the archival literature and in popularizations, that deal incorrectly with the early work by us and by Gamow. Too many authors evidently rely on more recent publications, many of which continue to propagate errors, rather than consulting original source material.

One wonders about the forces that shape the activities of some scientific authors. A number of questions have been raised in this paper which beg to be looked at with proper objectivity by historians and sociologists of science, particularly in terms of what such matters say about scholarship and integrity in science, which is one of the most important of human endeavors. We do not accept the argument of some that correct attribution does not matter, but that only the furtherance of science matters. This view does not reflect the ideals and realities of the scientific enterprise. A correct history of science as a human endeavor does matter, both for the present and for the future.

The authors have contemplated some of the advantages and disadvantages of living for a relatively long time. It has been amusing to note that there are on occasion colleagues who discuss this subject area as though the authors are no longer living or scientifically productive. For many years we have contemplated the possibility that the reason for a lack of acceptance of the worth of our early work lay in our both being employed by large industrial research laboratories at the time when the background radiation was first observed, and for some years thereafter. We have derived enormous pleasure from the creative process, pain from the lack of appreciation of our work and of Gamow's work, and a measure of satisfaction and pleasure in realizing that at long last some scientific colleagues view these early contributions as meritorious. It was very pleasant to have a distinguished cosmologist remark to one of us during one of the Texas Symposia that "you guys did it all."

It has been exciting to be involved in the early development of the big bang model and to have witnessed the explosive development of research in this area, addressing as it does the fundamental concern to understand the mystery of existence, and to have seen cosmology evolve from a discipline considered skeptically, as an area not worked in by sensible scientists, to one of the most popular and profound areas in contemporary physics and astronomy. The big bang model has survived for almost two generations – a very short time on any cosmic scale.

What marvels of human imagination and understanding does the future hold?

Acknowledgements

We are grateful to the American Institute of Physics for permission to use material from our article in *Physics Today*, **41**(8) (1988), 24.

Appendix

For a universe containing an ideal fluid of matter and radiation which are not interconverting, the Friedmann–Lemaître nonstatic solution of Einstein's field equations with the Robertson–Walker metric leads to the following relativistic energy equation (the FLRW model), which expresses the time rate of change of proper distance ℓ:

$$\frac{d\ell}{dt} = +\left[\left(\frac{8\pi G}{3}\right)(\varrho_m + \varrho_r)\ell^2 - \frac{c^2\ell_o^2}{R_o^2}\right]^{1/2} \text{cm sec}^{-1},$$

where the positive sign indicates expansion, ϱ_m and ϱ_r are the densities of matter and radiation, respectively, and c, G, ℓ_o, and R_o are the velocity of light, the universal gravitational constant, a unit proper distance, and the radius of curvature, respectively. The equation is readily recast in terms of $\ell/\ell_o = L = (1 + z)$, where z is redshift. Conservation of matter requires

$$\varrho_m\ell^3 = \text{constant},$$

while requiring the adiatic expansion of radiation leads to

$$\varrho_r\ell^4 = \text{constant}.$$

If radiation dominates at early times in the expansion, then

$$\varrho_r \gg \varrho_m, \text{ and } c^2\ell_o^2/R_o^2 \text{ is small,}$$

so that one can write

$$\varrho_r = 4.48 \times 10^5\, t^{-2}\, \mathrm{g\, cm^{-3}},$$

and consequently

$$T = 1.52 \times 10^{10}\, t^{-1/2}\, \mathrm{K},$$

where the numerical coefficients in ϱ_r and T contain only universal constants.

The energy equation can be integrated to yield the age of the expansion as a function of the other variables, viz.,

$$t = K_1 + K_2^{-1}(\gamma\varrho_{r'} + \gamma\varrho_{m'} L + K_2L^2)^{1/2}$$
$$-[(\gamma\varrho_{m'})/(2K_2^{3/2})] \ln [(\gamma\varrho_{r'} + \gamma\varrho_{m'} L + K_2L^2)^{1/2}$$
$$+ K_2^{1/2}L + \gamma\varrho_{m'}/(2K_2^{1/2})] \sec,$$

where

$$K_1 = [\gamma\varrho_{m'}/(2K_2^{3/2})] \ln [(\gamma\varrho_{r'})^{1/2} + \gamma\varrho_{m'}/(2K_2^{1/2})]$$
$$- (\gamma\varrho_{r'}/K_2^2)^{1/2},$$
$$\gamma = 8\pi G/3,$$

and

$$K_2 = c^2/|R_o^2|.$$

The quantity $\ell/\ell_o = L = (1 + z)$ is a dimensionless proper distance, with ℓ_o chosen so that $L = 1$ now, while $\varrho_{m'}$ and $\varrho_{r'}$ are the densities of matter and radiation when $L = 1$, and are not to be confused with the running variables. Note that in models of this type energy is not conserved. Evaluation of the constant of integration requires that one specify the radius of curvature.

The above expression for the time since the big bang can be simplified if one wants to calculate the age of the present epoch, which is matter-dominated. One defines the critical total density (the density for closure of the universe) by taking the radius of curvature to be infinite and the present value of Hubble's parameter as $H_o = [d\ell/(\ell dt)]$ now, whence the energy equation yields:

$$\varrho_c = 3H_o^2/(8\pi G).$$

It is customary to work with Ω, defined as

$$\Omega = (\varrho_{m'} + \varrho_{r'})/\varrho_c \simeq \varrho_{m'}/\varrho_c.$$

Then the present age of the big bang, t_o, can be written in relatively simple terms as a function of present values of Ω and the Hubble parameter as follows (see, for example, Weinberg 1972):

$$\Omega > 1 \quad H_o t_o = -1(\Omega - 1) + \Omega/[2(\Omega - 1)^{3/2}] \times \operatorname{arc\,cos}[(2 - \Omega)/\Omega],$$

$$\Omega = 1 \quad H_o t_o = 2/3,$$

$$\Omega < 1 \quad H_o t_o = 1/(1 - \Omega) - \Omega/[2(1 - \Omega)^{3/2}] \times \operatorname{arc\,cosh}[(2 - \Omega)/\Omega].$$

References

Adams, W. S. (1941). Results with the Coudé Spectrograph of the Mt. Wilson Observatory. *Astrophysical Journal*, **93**, 11.

Alpher, R. A. (1948a). Origin and Relative Abundance Distribution of the Elements. Dissertation, George Washington University, Washington, DC.

(1948b). A Neutron-Capture Theory of the Formation and Relative Abundance of the Elements. *Physical Review*, **74**, 1577.

Alpher, R. A., Bethe, H., and Gamow, G. (1948). The Origin of Chemical Elements. *Physical Review*, **73**, 803.

Alpher, R. A., Follin, J. W. Jr., and Herman, R. (1953a). Physical Conditions in the Initial Stages of the Expanding Universe. *Physical Review*, **92**, 1347.

(1953b). Initial Conditions in the Expanding Universe and Element Synthesis. *Physical Review*, **91**, 479A; Follin, J. W. Jr., Alpher, R. A., and Herman, R. (1959). Light Element Formation During Early Stages of the Expanding Universe. *Bulletin of the American Physical Society*, Ser. II(4), 476; (1960). Formation of D, Li, Be and B in an Expanding Universe. Ser. II(5), 287.

Alpher, R. A., Gamow, G., and Herman, R. (1967). Thermal Cosmic Radiation and the Formation of Protogalaxies. *Proceedings of the National Academy of Sciences*, **58**, 2179.

Alpher, R. A. and Herman, R. (1948a). Evolution of the Universe. *Nature*, **162**, 774.

(1948b). On the Relative Abundance of the Elements. *Physical Review*, **74**, 1737.

(1949). Remarks on the Evolution of the Expanding Universe. *Physical Review*, **75**, 1089.

(1950). Theory of the Origin and Relative Abundance Distribution of the Elements. *Reviews of Modern Physics*, **22**, 153.

(1951). Neutron-Capture Theory of Element Formation in an Expanding Universe. *Physical Review*, **84**, 60.

(1953). The Origin and Abundance Distribution of the Elements. *Annual Review of Nuclear Science*, **2**, 1.

(1958). On Nucleon-Antinucleon Symmetry in Cosmology. *Science*, **128**, 904.

(1975). Big Bang Cosmology and the Cosmic Black-Body Radiation. *Proceedings of the American Philosophical Society*, **119**, 325.

(1988). Reflections on Early Work on "Big Bang" Cosmology. *Physics Today*, **41**(8), 24.

Andersen, P. H. (1988). Short Wavelength Distortions Seen in the Cosmic Background Radiation. *Physics Today*, **41**(8), 17.

Applegate, J. H., Hogan, C. J., and Scherrer, R. J. (1987). Cosmological Baryon Diffusion and Nucleosynthesis. *Physical Review*, **D35**, 1151.
Barnothy, J. M. (1988). Private communication.
Bernstein, J. and Feinberg, G. (ed.) (1986). *Cosmological Constants: Papers in Modern Cosmology*. New York: Columbia University Press.
Bethe, H. A. (1947). *Elementary Nuclear Theory*. New York: John Wiley.
Boesgard, A. M. and Stiegman, G. (1985). Big Bang Nucleosynthesis: Theories and Observations. *Annual Review of Astronomy and Astrophysics*, **23**, 319.
Brown, H. (1949). A Table of Relative Abundances of Nuclear Species. *Reviews of Modern Physics*, **21**, 625.
Burbidge, E. M., Burbidge, G. R., Fowler, W. A., and Hoyle, F. (1957). Synthesis of the Elements in Stars. *Reviews of Modern Physics*, **29**, 547.
Chandrasekhar, S. and Henrich, L. R. (1942). An Attempt to Interpret the Relative Abundance of the Elements and their Isotopes. *Astrophysical Journal*, **95**, 288.
Crane, P., Kutner, M. L., Hegyi, D. J., Blades, J. C., Palazzi, E., and Mandolesi, N. (1989). Cosmic Background Radiation Temperature at 2.64 mm, 1.32 mm, and 0.6 mm. *Proceedings of the Moriond Astrophysics Conference.* (in press).
Davies, P. C. W. (1984). *Superforce*. New York: Simon and Schuster.
De Sitter, W. (1930). The Expanding Universe. Discussion of Lemaître's Solution of the Equations of the Inertial Field. *Bulletin of the Astronomical Institute of the Netherlands*. **5**(193), 211.
(1931). Some Further Computations Regarding Non-Static Universes. *Bulletin of the Astronomical Institute of the Netherlands*, **6**(223), 141.
Dicke, R. H., Peebles, P. J. E., Roll, P. G., and Wilkinson, D. T. (1965). Cosmic Black-Body Radiation. *Astrophysical Journal*, **142**, 414.
Doroshkevich, A. G. and Novikov, I. D. (1964). Mean Density of Radiation in the Metagalaxy and Certain Problems in Relativistic Cosmology. *Soviet Physics-Doklady*, **9**, 111.
Einstein A, (1916). Die Grundlage der allgemeinen Relativitätstheorie. *Annalen der Physik* **49**, 769.
(1917). Kosmologische Betrachtungen zur allgemeinen Relativititätstheorie. *Preuss. Akad. Wiss. Berlin* **6**, 142
1955). *The Meaning of Relativity*. Princeton University Press.
Friedmann, A. (1922). Über die Krümmung des Raumes. *Zeitschrift für Physik*, **10**, 377.
(1924). Über die Möglichkeit einer Welt mit konstanter Negativer Krümmung des Raumes. *Zeitschrift für Physik*, **21**, 326.
Fowler, W. A. (1984). The Quest for the Origin of the Elements. *Science*, **226**, 922.
Gamow, G. (1935). Nuclear Transformations and the Origin of the Chemical Elements. *Ohio Journal of Science*, **35**, 406.
(1942). Concerning the Origin of Chemical Elements. *Journal of the Washington Academy of Sciences*, **32**, 353.
(1946). Expanding Universe and the Origin of Elements. *Physical Review*, **70**, 572.
(1948a). The Origin of Elements and the Separation of Galaxies. *Physical Review*, **74**, 505.

(1948b). The Evolution of the Universe. *Nature*, **162**, 680.
(1949). On Relativistic Cosmogony. *Reviews of Modern Physics*, **21**, 367.
(1950). Half an Hour of Creation. *Physics Today*, **3**(8), 16.
(1952). *Creation of the Universe*. New York: Viking Press.
(1953). Expanding Universe and the Origin of Galaxies. *K. Danske Vidensk. Selsk., mat.-fys. Medd.* **27**, 1.
(1956). The Physics of the Expanding Universe. In *Vistas in Astronomy*, ed. A. Beers. New York: Pergammon Press.
Goldschmidt, V. M. (1938). Geochemische Verteilungsgesetze der Elemente. IX. Die Mengenverhältnisse der Elemente und der Atom-Arten. *I. Matematisk-Naturvidenskapelig klasse*, No. 4, 1937.
Gott, J. R. III, Gunn, J. E., Schramm, D. N., and Tinsley, B. M. (1974). An Unbound Universe? *Astrophysical Journal*, **194**, 543.
Gribben, J. (1986). *In Search of the Big Bang*. New York: Bantam Books.
Harkins, W. D. (1917). The Evolution of the Elements and the Stability of Complex Atoms. *Journal of the American Chemical Society*, **39**, 856.
Harrison, E. R. (1981). *Cosmology: The Science of the Universe*. New York: Cambridge University Press.
Hawking, S. W. (1988). *A Brief History of Time*. New York: Bantam Books.
Hayashi, C. (1950). Proton–Neutron Concentration Ratio in the Expanding Universe at the Stages Preceding the Formation of the Elements. *Progress in Theoretical Physics (Japan)*, **5**, 224.
Hayashi, C. and Nishida, M. (1956). Formation of Light Elements in the Expanding Universe. *Progress in Theoretical Physics (Japan)*, **16**, 613.
Hoyle, F. (1950). *The Nature of the Universe*. New York: Harper. (See pp. 119, 124.)
Hoyle, F. and Tayler, R. J. (1964). The Mystery of the Cosmic Helium Abundance. *Nature*, **203**, 1108.
Hubble, E. P. (1929). A Relation Between Distance and Radial Velocity Among Extra-Galactic Nebulae. *Proceedings of the National Academy of Sciences*, **15**, 168.
(1936). *The Realm of the Nebulae*. New Haven: Yale University Press.
Hughes, D. J. (1946). Radiative Capture Cross Sections for Fast Neutrons. *Physical Review*, **70**, 106A.
Kippenhahn, R. (1987). *Light from the Depths of Time*. Berlin: Springer-Verlag.
Lemaître, G. (1931). Homogeneous Universe of Constant Mass and Increasing Radius. *Monthly Notices of the Royal Astronomical Society*, **91**, 483.
(1950). *The Primeval Atom. An Essay on Cosmogony*. New York: Van Nostrand.
Lifshitz. (1946). On the Gravitational Stability of the Expanding Universe. *Journal of Physics of the USSR*, **10**, 116.
Lubkin, G. (1978). Nobel Prizes to Kapitza for low-temperature studies and to Penzias and Wilson for 3 K Cosmic Background. *Physics Today*, **31**(12), 17.
Malaney, R. A. and Fowler, W. A. (1988). The Transformation of Matter After the Big Bang. *American Scientist*, **76**(5), 472.
Matzner, R. A. (1968). On the Present Temperature of Primordial Black-Body Gravitational Radiation. *Astrophyscial Journal*, **154**, 1123.
McCrea, W. H. (1970). Cosmology Today. *American Scientist*. **58**(5), 521.

McKellar, A. (1941). Molecular Lines from the Lowest States of Diatomic Molecules Composed of Atoms Probably Present in Interstellar Space. *Publications of the Dominion Astrophysical Observatory (Victoria, B. C. Canada)*, **7**, 251.

Oddo, G. (1914). Die Molekularstruktur der Radioaktiven Atome. *Zeitschrift für Anorganische Allgemeine Chemie*, **87**, 253.

Olive, K. A. (1986). Cosmology and Particle Physics: A General Review. *Annals of the New York Academy of Sciences*, **470**, 1.

Pagels, H. R. (1985). *Perfect Symmetry*. New York: Bantam Books.

Partridge, R. B. (1988). The Angular Distribution of the Cosmic Background Radiation. *Reports on Progress in Physics*, **51**, 647.

Peebles, P. J. E. (1971). *Physical Cosmology*. Princeton: Princeton University Press.

Peebles, P. J. E. and Wilkinson, D. T. (1968). Comments on the Anisotropy of the Primeval Fireball. *Physical Review*, **174**, 2168.

Penzias, A. and Wilson, R. W. (1965). A Measurement of Excess Antenna Temperature at 4080 Mc/S. *Astrophysical Journal*, **142**, 419. (See also A. Penzias [1979]. The Origin of the Elements. *Science*, **205**, 549.)

Prigogine, I., Geheniau, J., Gunzig, E., and Nardone, P. (1988). Thermodynamics of Cosmological Matter Creation. *General Relativity and Gravitation* (in press).

Rawley, L. A., Taylor, J. H., Davis, M. M., and Allan, D. W. (1987). Millisecond Pulsar PSR 1937+21: A Highly Stable Clock. *Science*, **238**, 761.

Reines, F. (ed.) (1972). *Cosmology, Fusion and Other Matters: George Gamow Memorial Volume*. Boulder: Colorado Associated University Press.

Smart, J. S. (1948). On the Formation of Shielded Isobars by the Nuclear Photo-Effect. *Physical Review*, **74**, 1882.

(1949). Effects of Nuclear Stability on the Formation of the Chemical Elements. *Physical Review*, **75**, 1379.

Stiegman, G., Schramm, D. N., and Gunn, J. E. (1977). Cosmological Limits to the Number of Massive Leptons. *Physics Letters*, **B66**, 202. (See also Yang, J., Schramm, D. N., Stiegman, G., and Rood, R. T. [1979]. Constraints on Cosmology and Neutrino Physics from Big Bang Nucleosynthesis. *Astrophysical Journal*, **227**, 697.)

Tolman, R. C. (1931). On the Theoretical Requirements for a Periodic Behaviour of the Universe. *Physical Review,* **38**, 1758.

(1934). *Relativity, Thermodynamics and Cosmology*. Oxford: Clarendon Press.

Von Weizsäcker, C. F. (1938). Über Elementumwandlungen im Innern der Sterne. *Physikalische Zeitschrift,* **38**, 176; **39**, 633.

Wagoner, R. V., Fowler, W. A., and Hoyle, F. (1967). On the Synthesis of Elements at Very High Tempertures. *Astrophysical Journal*, **148**, 3. (See also Wagoner, R. V. [1973]. Big Bang Nucleosynthesis Revisited. *Astrophysical Journal*, **179**, 343; in addition see Wagoner, R. V., in this volume.)

Weinberg, S. (1972). *Gravitation and Cosmology*. New York: Wiley.

(1977). *The First Three Minutes*. New York: Basic Books.

Wilkinson, D. T. and Peebles, P. J. E. (1983). Discovery of the 3 K Radiation. In *Serendipitous Discoveries in Radio Astronomy*, ed. K. Kellerman and B. Sheets. Green Bank: NRAO.

Zee, A. (1982). *Unity of Forces in the Universe*. Singapore: World Scientific.
(1986). *Fearful Symmetry*. New York: MacMillan.
Zeldovich, Y. B. (1988). Cosmology from Robertson to Today. *Physics Today*, **41**(3), 27.
Zeldovich, Y. B. and Novikov, I. D. (1983). *The Structure and Evolution of the Universe*. Chicago: University of Chicago Press.

9

Deciphering the nuclear ashes of the early universe: a personal perspective

Robert V. Wagoner

9.1 Introduction

Even before the development of nuclear physics, the idea that the pattern of element abundances reflects the history of matter was a compelling one (Tolman 1922; Suzuki 1928). Below we shall trace how this "cosmic archeology" has been applied to the universe as a whole, obtaining clues to the nature of our deep past from its nuclear debris. Throughout we shall stress how the evolution of concepts of the primeval universe was related to the development of the relevant physics, in the absence of many relevant observations.

The scope of this historical review will be limited to the period through 1973, in keeping with the wishes of the organizers of the symposium from which this volume was developed. A list of important contributions to the development of cosmological nucleosynthesis during this period is presented in table 9.1. That particular year has been chosen for this contribution because by then the power of the (still accepted) theory of primordial nucleosynthesis as a sensitive indicator of conditions in the early universe had been realized, and dramatically illustrated via the detection of interstellar deuterium (Rogerson and York 1973). This date also happens to coincide with the end of the eight-year period during which this subject was a major area of research of the author.

We shall focus in most detail on developments since 1950, when Hayashi's demonstration of the proper initial conditions made possible the construction of the modern self-consistent model of the era of nucleosynthesis. The previous history has been authoritatively reviewed by two of its major participants, Alpher and Herman (1950), and their latest update appears in this volume. Although the perspective that governs this paper is more limited, we hope that the history will be vitalized by some personal recollections and points of view.

Table 9.1. *Milestones*

1922	A. Friedmann	Models of expanding, matter-filled universes.
1928	S. Suzuki	Within equilibrium theory, observed He/H abundance ratio requires a temperature of about 10^9 K.
1931	H. C. Urey C. A. Bradley, Jr.	Observed relative abundances of isotopes cannot have resulted from nuclear equilibrium at any single temperature.
1931	G. Lemaître	Introduction of big bang via "primeval atom."
1942	S. Chandrasekhar L. R. Henrich	Detailed equilibrium calculation approximately matches observed abundances only for mass numbers in the range $A = 12 - 53$.
1946	G. Gamow	Suggested neutron capture in (cold) big bang as origin of the elements.
1948	R. A. Alpher G. Gamow R. C. Herman	First calculation of neutron-capture process; subsequently included radiation density, neutron decay, and deuterium production.
1949	R. A. Alpher R. C. Herman	Time-dependent treatment of expanding matter and radiation allowed prediction of $T_o = 5$ K from conditions necessary for primordial element production.
1950	E. Fermi A. Turkevich	Extensive nuclear reaction network up to $A = 7$ indicated that lack of stable nuclei at $A = 5$ (and 8) prevents significant production of heavier nuclei in big bang.
1950	C. Hayashi	Assumption of initial neutron gas invalid; equilibrium at $T > 10^{10}$ K fixed neutron–proton ratio.
1953	R. A. Alpher J. W. Follin, Jr. R. C. Herman	Complete formulation of evolution of physical conditions in hot big bang.
1957	E. M. Burbidge G. R. Burbidge W. A. Fowler F. Hoyle	Explanation of most abundances as due to synthesis within stars.
1963	Y. B. Zel'dovich	Inferred $T_o = 20$ K if observed helium was produced in a hot big bang; thus rejected the model.
1964	F. Hoyle R. J. Tayler	From first detailed calculation, argued that helium was produced within expansion (of universe or massive objects) from very high temperatures, and noted that more than two types of neutrino would increase its abundance.
1965	A. A. Penzias R. W. Wilson	Discovery of $T_o = 3$ K background radiation.
1966	P. J. E. Peebles	Calculation of abundances of ^{2}H, (^{3}He), and ^{4}He

Table 9.1. (*cont.*)

		within standard model and those with other expansion rates.
1966	S. W. Hawking R. J. Tayler K. S. Thorne	Calculated effect of anisotropic expansion on helium abundance.
1967	R. V. Wagoner W. A. Fowler F. Hoyle	Detailed calculation of all elements produced within big (and little) bangs, including possibility of universal neutrino degeneracy.
1970	H. Reeves W. A. Fowler F. Hoyle M. Meneguzzi J. Audouze H. Reeves	Showed that spallation of interstellar gas by cosmic rays produces observed abundances of those light nuclei not made in the early universe.
1972	L. Searle W. L. W. Sargent	Discovered blue dwarf galaxies with deficient heavy elements but normal helium abundance.
1973	H. Reeves J. Audouze W. A. Fowler D. N. Schramm	Argued that no galactic process was capable of producing a significant abundance of deuterium.
1973	R. V. Wagoner	Improved calculation implies that most deviations from standard model (but not necessarily baryon inhomogeneities) must have been small during nucleosynthesis, and present baryon density most likely insufficient to close universe if deuterium is primordial.
1973	J. B. Rogerson D. G. York	Determination of the interstellar deuterium abundance from Lyman lines observed by the Copernicus satellite.

9.2 Early history

It was very natural that the first class of theories of the origin of the elements was based on the hypothesis that matter had evolved from a state extreme enough to allow all the reactions which determined their abundances to achieve equilibrium. If the density and/or temperature had fallen sufficiently fast, the resulting abundance pattern would then have been "frozen" in its equilibrium distribution. This hypothesis was the simplest to work with because the only information needed was nuclear binding energies, rather than reaction cross sections. In the more modern versions (after the discovery of the neutron in 1932), the abundance distribution was then a uniquely specified function of the temperature and the densities of protons and neutrons.

However, as early as 1922 Tolman determined that if the observed helium–hydrogen ratio was a consequence of equilibrium (with electrons), temperatures of at least 10^6 K would have been required; while Suzuki (1928) foreshadowed what was to come with his calculation indicating that an amount of helium no greater than the amount of hydrogen would be produced "if the cosmos had, at the creation, the temperature higher than 10^9 degrees." But soon thereafter Urey and Bradley (1931) provided the first indication of the inadequacy of the theory with their detailed calculation showing that the relative abundances of the isotopes of various elements could not be reproduced by equilibrium at any single temperature.

Although both stellar and universal sites for this process were considered, in the years following the publication of the first comprehensive abundance compilation by Goldschmidt (1938) the focus shifted more to a primordial universal origin. The reason appears to be twofold: the apparent uniformity of the observed abundances was then hard to understand as arising from individual stellar events, and the physical conditions within stars were considered to be not extreme enough to produce the required temperatures, densities, and eventual explosions on the scale required.

The first detailed, physically correct equilibrium calculation was carried out in 1942 by Chandrasekhar and Henrich, who showed that the guiding principle – abundance should be correlated with how tightly a nucleus is bound – held for a range of nuclei at densities of about $10^6\,g\,cm^{-3}$ and temperatures of about 10^{10} K. However, hopes for explaining the origin of most elements in this way were rapidly dimming because of three glaring discrepancies. The most serious was the fact that the observed abundances reached a roughly constant level at mass numbers $A > 70$, whereas the equilibrium abundances fell throughout this range (because the binding energy per nucleon reaches a maximum at iron). In addition, the calculations overproduced the rare light elements Li, Be, and B, while they underproduced the iron-peak elements. This is illustrated in fig. 9.1, showing the results of a subsequent calculation (for somewhat different conditions) by Klein, Beskow, and Treffenberg (1946).

Curiously, the development of the cosmological model which would provide the physical scenario necessary for any such theory of element production does not appear to have been significantly influenced by the equilibrium theory. The framework was provided by Friedmann (1922), who derived the evolving universe models from the Einstein field equations. But the first person to consider the conditions that might have existed in the remote past within those models that expanded from high densities was Abbé Georges Lemaître, the "father of the big bang." In contrast to Eddington, whose opposing view motivated his proposal, Lemaître felt that "it is a very happy circumstance that relativity provides for a natural beginning." His picture of the state of matter as a "primeval

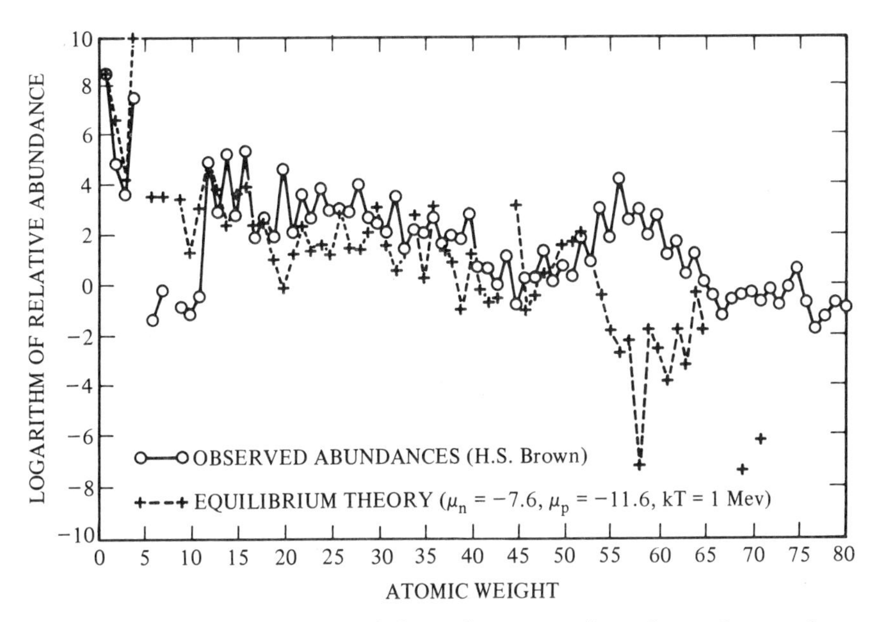

Fig. 9.1 Abundances calculated from the assumption of complete nuclear equilibrium by Klein, Beskow and Treffenberg (1946) are compared with those observed. The values of the chemical potentials chosen correspond to a density of $4 \times 10^8\,\mathrm{g\,cm^{-3}}$ and n/p = 55. There is a computed point for A = 78 at −11.73; the neutron concentration has not been included in the normalization point at A = 1. [From Alpher and Herman (1950).]

atom" (1931, 1933, 1950), although of necessity incomplete, breathed life into the early universe. The primeval atom, eventually considered to be a giant isotope of the neutron filling all of the (closed) universe, underwent "superradioactive disintegration" into the observed cosmic rays as well as the rest of matter. This general idea was later resurrected in a stellar context by Mayer and Teller (1949) in their polyneutron fission theory of the origin of the heavy elements.

The seeds of the presently accepted theory of cosmological nucleosynthesis were planted by George Gamow (1946), like Georges Lemaître (but even more so) a daring, flamboyant physicist. Although he also was misdirected in believing that *all* the elements had to emerge from the early universe, he fully recognized that some nonequilibrium process was required and was more natural. His key assumption of an initially pure gas of neutrons was unfortunate, but did follow the tradition of previous investigations. However, it remains mysterious that Gamow did not realize the implications of the existence of neutrinos, as we shall discuss below.

The processes that Gamow considered in his initial investigations were neutron captures and subsequent beta decays within a cold big bang emerging from a density high enough to provide an expansion rate (via the Friedmann equations) fast enough to postpone neutron decay, yet slow enough to allow the neutron captures time to build up all the elements. In order to test this proposal, Gamow began his collaboration with Ralph Alpher, which led to the first calculation (Alpher, Bethe, and Gamow 1948). (Some details concerning the inclusion of Bethe (for aesthetic reasons) and other manifestations of Gamow's *joie de vivre* have been collected by Reines 1972.) They were motivated by the fact that the products of the known cross sections and the corresponding abundances were roughly the same, but lack of extensive neutron-capture data forced them to employ a simple dependence on mass number. Encouraged by their reproduction of the gross dependence of the abundances on nuclear mass, they brought Robert Herman into the collaboration and explicitly included the effects of the (dominating) radiation density and the expansion, as well as the details of the reactions involving the neutrons and protons (Alpher 1948; Alpher and Herman 1948, 1951; Gamow 1948). Examples of their results are shown in figs. 9.2 and 9.3. But it can be

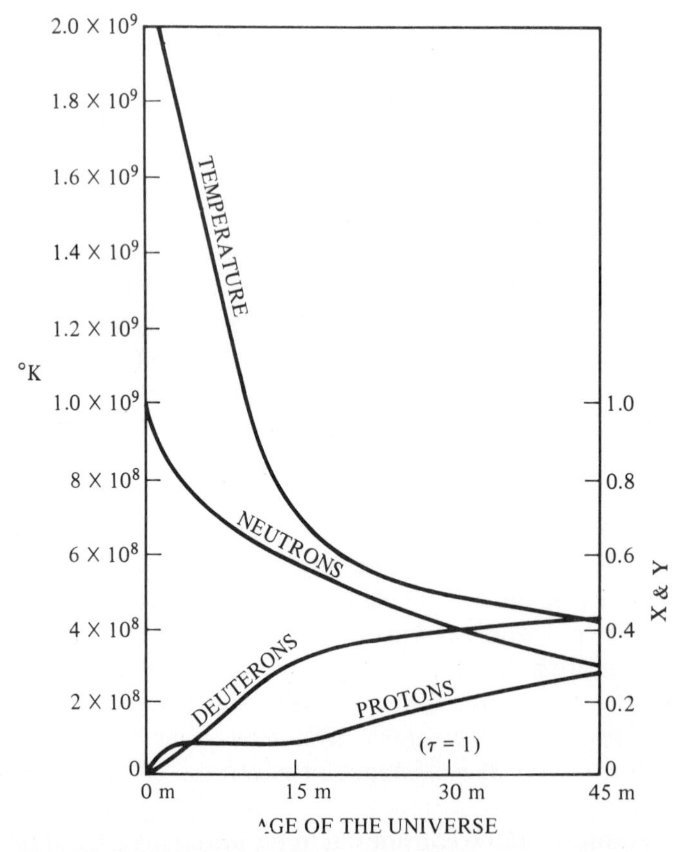

Fig. 9.2 The first calculation of the evolution of abundances in a hot big bang model, in which the contribution of the radiation density to the expansion rate was taken into account. The computations were carried out by R. Alpher, and appeared in the paper of Gamow (1948).

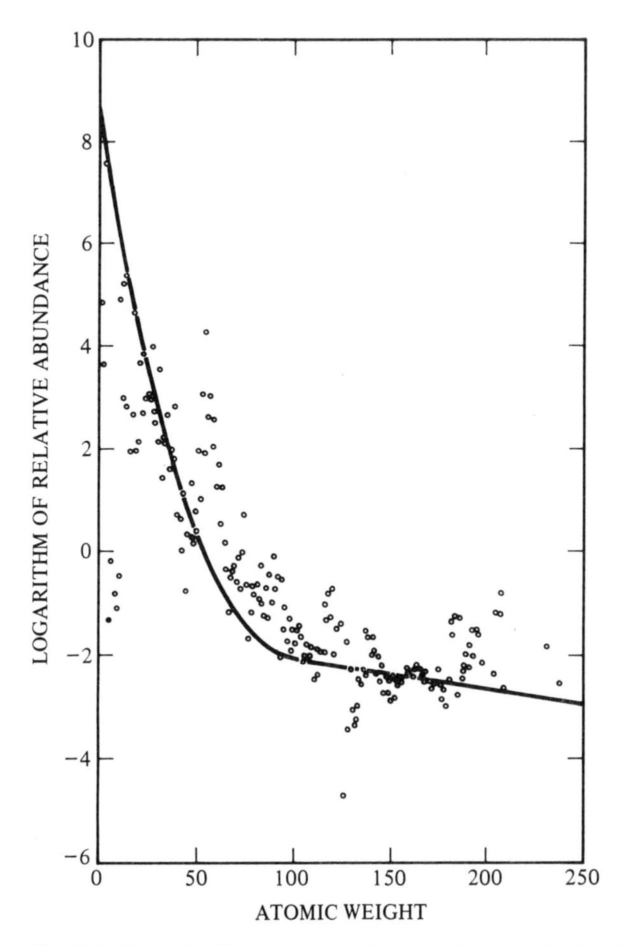

Fig. 9.3 The solid line represents the abundances calculated from the (smoothed) neutron-capture theory (with neutron decay but without universal expansion) by Alpher and Herman (1948). The density was chosen to be $4 \times 10^8\,\mathrm{g\,cm^{-3}}$. The observed abundance points are again those given by Brown (1949). [From Alpher and Herman (1950).]

argued that the most important result of their calculations was the prediction of a present radiation temperature of 5 K (Alpher and Herman 1949), which followed from the primordial conditions required to match the abundances.

However, even the qualitative success of this theory was in grave danger, as Alpher, Gamow, and Herman knew all too well. The lack of stable nuclei at mass numbers 5 and 8 were barriers to the production of heavier nuclei via neutron captures. This blockage was shown to operate even if charged-particle reactions were included, in a groundbreaking calculation by Fermi and Turkevich (presented by Alpher and Herman 1950). A network of 28 nuclear reactions up to mass 7 was employed, leading to the results shown in fig. 9.4. Under physical conditions similar to

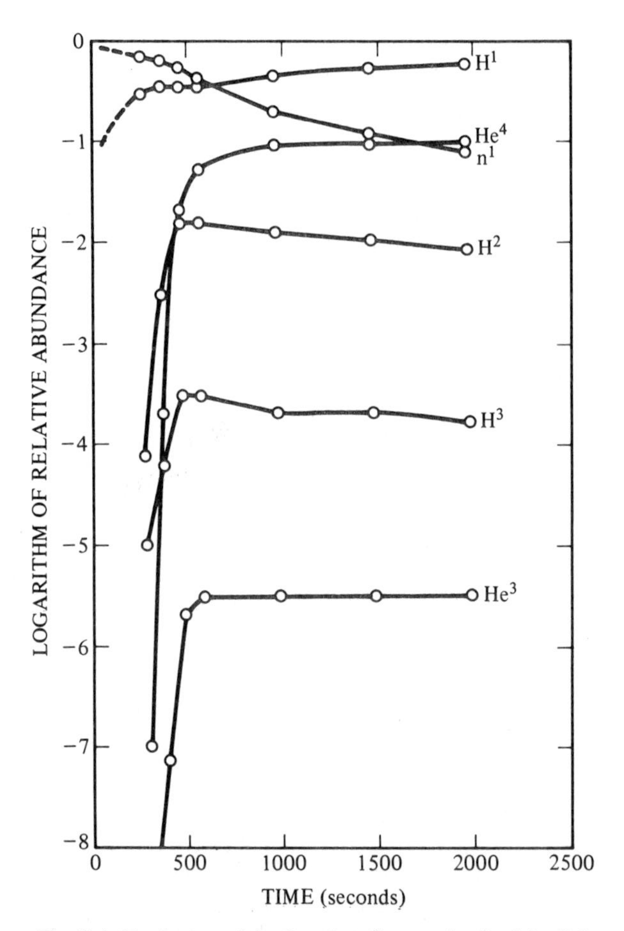

Fig. 9.4 Evolution of the fraction (by number) of the lightest nuclei in a radiation-dominated expanding universe, according to an unpublished calculation of E. Fermi and A. Turkevich presented by Alpher and Herman (1950). The initial neutron density (at 1 second) was taken to be 10^{21} cm^{-3}.

those employed by Alpher, Gamow, and Herman, only about 10^{-7} of the mass was converted into nuclei heavier than helium. Even though there were grounds for hoping that more extensive nuclear cross section measurements might lead to increased production of ^{7}Li, ^{7}Be, and therefore the heavier nuclei, these results must have dampened the enthusiasm of those pursuing this grand scheme of producing all the elements in one event. There was not yet a reason to focus on the lightest nuclei.

Although a scenario for cosmological nucleosynthesis was laid out by Gamow, Alpher, and Herman, it was a realization by Hayashi (1950) which allowed it to be transformed into a self-consistent physical theory. It is interesting to note an historical coincidence: the paper by Hayashi, which initiated the "modern era" of these investigations, appeared in the same month (April, 1950) as the classic review of the origin and abundance of the elements by Alpher and Herman, which summarized the previous era.

Hayashi pointed out that at temperatures $T > 10^{10}$ K, only ten times higher than those at which nucleosynthesis could begin (because the destruction of deuterium by photons had been reduced sufficiently), the neutrons and protons were brought into equilibrium by their reactions with electrons and neutrinos (and their antiparticles). Pair production and annihilation also kept the neutrinos and antineutrinos in equilibrium with the electrons and positrons, which likewise were kept in equilibrium with the photons. This meant that the ratio of neutrons to protons was a uniquely specified function of temperature. The initial conditions for nucleosynthesis were no longer arbitrary; but fixed, and independent of the previous history of the universe. One question remains: why did Fermi, a key developer of the theory of these weak interactions, not realize this profound consequence?

9.3 Development of the standard model

Motivated primarily by the implications of Hayashi's discovery for nucleosynthesis, Alpher and Herman joined forces with James Follin, Jr. to produce the first detailed description of what is now called the "standard model" of this phase in the evolution of the universe (Alpher, Follin, and Herman 1953). This model is based upon a homogeneous, isotropic universe whose expansion is governed by general relativity, within which the net lepton number per photon is small enough to keep the neutrinos nondegenerate. It is is remarkable that, except for some details such as the number of types of neutrinos, their calculation remains a valid quantitative description of the major constituents: photons, electrons and positrons, and neutrinos and antineutrinos. They were thus also able to follow in detail the evolution of the neutron–proton ratio, shown in fig. 9.5. The only parameter of the standard model, the baryon number per photon, does not play a role in such calculations because it contributes insignificantly to the total density and because nuclear reactions were not included. The basic reason why their calculation has stood the test of time is because they were careful to include all the relevant physical processes, whose properties were known at the relevant energies, $kT < 100$ MeV. It remains fortunate that the calculation can thus be begun from a state of complete statistical equilibrium, making the basic results independent of the uncertain physics of earlier epochs. These properties are depicted in fig. 9.6.

They showed how the neutrinos (and antineutrinos) decoupled from the other particles at temperatures of a few MeV, and correctly determined the evolution of their energy distribution function. They also followed the subsequent electron–positron annihilation and the conversion of their mass-energy into that of photons, raising the photon temperature above

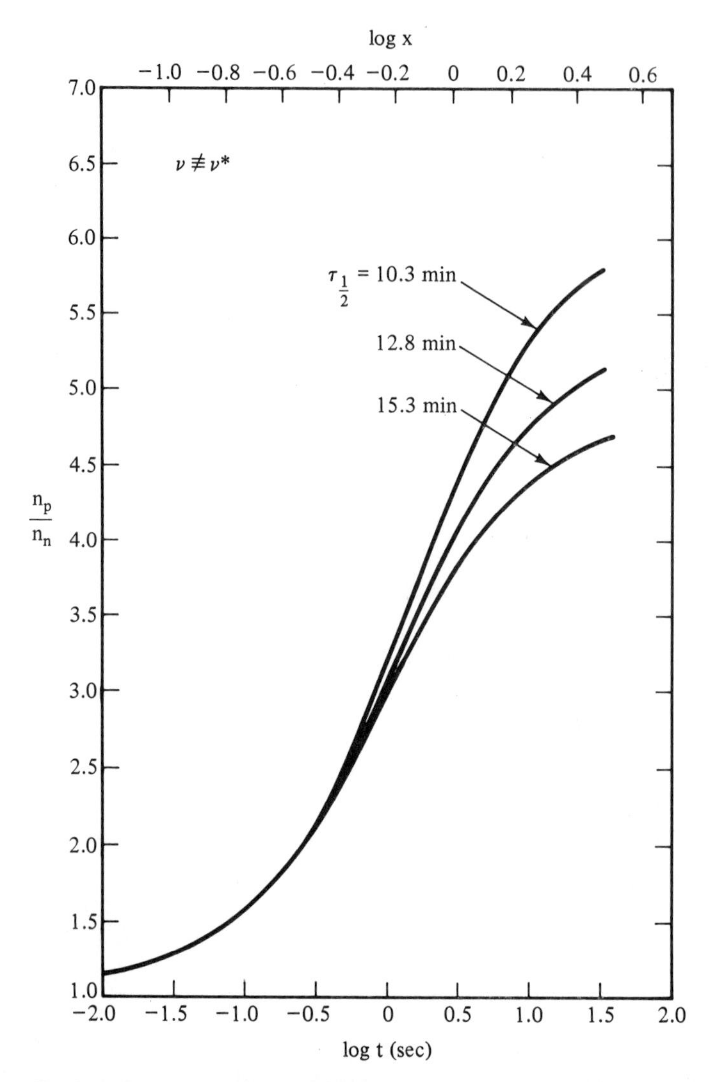

Fig. 9.5 The first detailed calculation of the evolution of the proton-neutron ratio in the standard big bang model, from Alpher, Follin and Herman (1953). A single type of Dirac neutrino was assumed in this case, and various values of the neutron half-life were employed. ($x = m_e c^2/kT$)

that of the neutrinos. Afterwards, both again were cooled adiabatically by the expansion. Finally, they realized that their computed decrease in the neutron–proton ratio with time would have important implications for nucleosynthesis (as did Hayashi). Indeed, they had redone their neutron-capture calculation, beginning with the conditions specified by Hayashi's model at kT = 0.1 MeV, where they believed nucleosynthesis would begin (Alpher and Herman 1953). But they left "for future study to re-examine the formation of the elements by thermonuclear reactions as a subsequent part of the picture developed here."

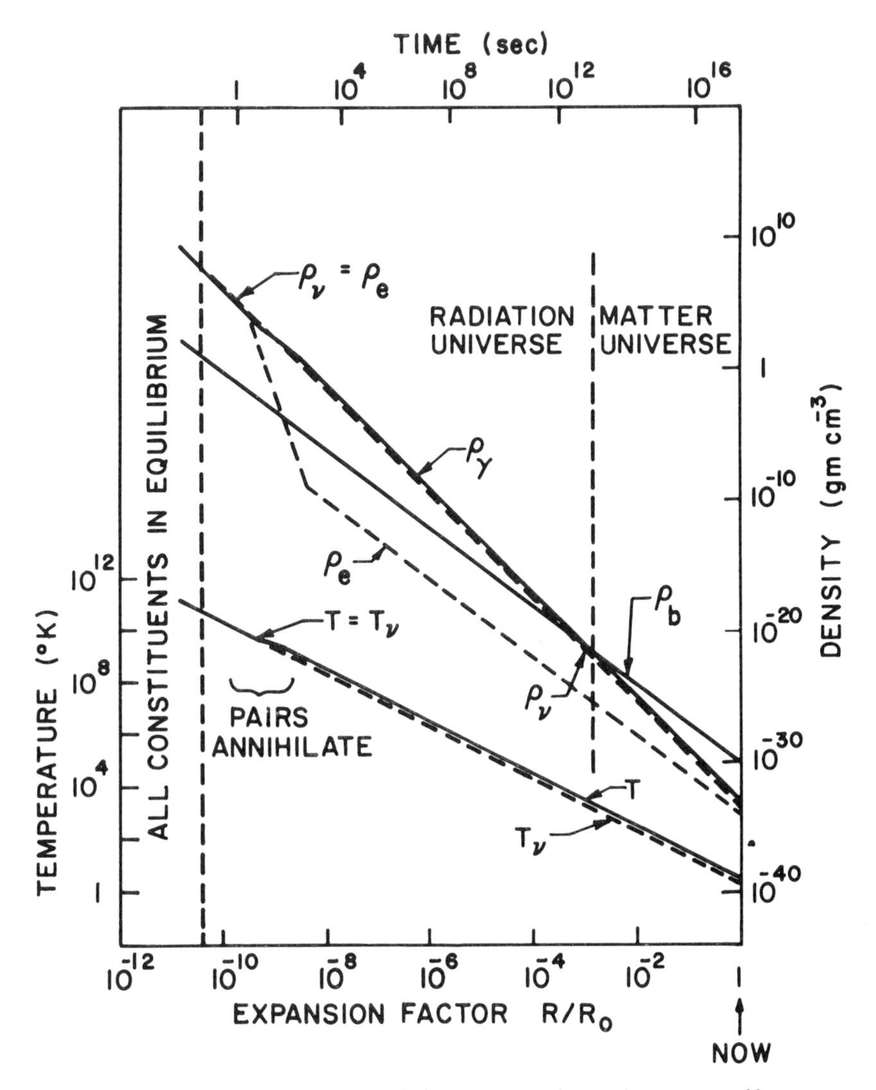

Fig. 9.6 Evolution of the densities of photons, neutrinos, electrons, and baryons, as well as the photon and neutrino temperatures, during the expansion of the standard big bang. [This is a version of a figure presented by Dicke et al. (1965), and generalized by Wagoner, Fowler and Hoyle (1967).]

At this point another intriguing question emerges. The framework for cosmological nucleosynthesis had been constructed. Why wasn't the nuclear reaction network developed by Fermi and Turkevich employed to calculate the resulting abundances? Among the possible reasons, one seems to stand out. The seeds of a revolution in nucleosynthesis were in the wind. The first harvest appeared in 1957, in the monumental paper of Burbidge, Burbidge, Fowler, and Hoyle and that of Cameron (1957). Their demonstration that conditions within stars were indeed sufficient to synthesize most, if not all, of the elements must have dramatically reduced the motivation to consider other sites.

The field of cosmological nucleosynthesis thereby lay fallow for a decade, until Zel'dovich (1963) and Hoyle and Tayler (1964) reconsidered the production of the most abundant nucleus, ^{4}He. Zel'dovich gave few details of his calculation (and did not refer to the results of Alpher, Follin, and Herman), but obtained a present radiation temperature of 20 K from the conditions necessary to produce 10% helium (by mass?), which he felt was "a satisfactory composition." Since he felt that "such a density of radiation is obviously in conflict with both radio-astronomical observations and the indirect data of cosmic-ray theory," he rejected the hot big bang (later to become one of its strongest proponents).

In a paper of remarkable prescience, Hoyle and Tayler carried out the first detailed calculation of the helium abundance, within a simplified model of the evolution of the dominant constituents which (as they noted) agreed very well with the results of Alpher, Follin, and Herman. Since most of the neutrons still present when nucleosynthesis commenced (accurately estimated by Hoyle and Tayler) were quickly processed into helium, it was sufficient to know the neutron–proton ratio then in order to obtain the resulting helium abundance. The abundance they obtained was only slightly higher than that observed. But they showed that conditions within exploding supermassive stars could be similar to that in the early universe, and thus the same amount of helium could emerge. They had first noted that the observed amount of starlight only required the production of 10% of the observed amount of helium. Thus they concluded "that most, if not all, of the material of our everyday world . . . has been 'cooked' to a temperature in excess of 10^{10} K." Although the incompatibility between the age of the universe and that of the Earth had been resolved, Hoyle still favored the steady-state theory and therefore the supermassive stellar site for helium synthesis.

Finally, it should be emphasized that Hoyle and Tayler were the first to point out the important relationship between the number of neutrino families and the helium abundance. They noted that "if there were more than two kinds of neutrino the expansion would have to be faster in order to overcome the gravitational attraction of the extra neutrinos, and . . . the larger the ratio He/H turns out to be." They (like Zel'dovich) also noted the dependence of the helium abundance on the nucleon (baryon) density via its small effect on the temperature (and therefore time) at which nucleosynthesis could commence.

9.4 Post microwave background

With the discovery of the microwave background radiation (Penzias and Wilson 1965; see also the contribution of Wilson in this volume) and the rapid verification of its blackbody spectrum (Field and

Hitchcock 1966; Roll and Wilkinson 1966; Thaddeus and Clauser 1966), the early universe was revealed to us, and cosmological nucleosynthesis as a probe of an even earlier epoch was placed on a firm foundation. Almost immediately, two new assaults on the problem were initiated at Princeton and Caltech.

Jim Peebles had played a major role in developing the rationale for the Princeton group's search for the primordial radiation (Dicke, Peebles, Roll, and Wilkinson 1965), and became aware of the arguments of Hoyle and Tayler (1964) regarding the helium abundance as well as the work of Fermi and Turkevich (1950) and Alpher, Follin, and Herman (1953). It was therefore natural that he produced a computer calculation which followed the evolution of all the major constituents, including five nuclear reactions governing the buildup through helium as well as those weak reactions governing the interconversion of neutrons and protons. His results appeared the year Lemaître died (Peebles 1966a, 1966b).

He obtained abundances of ^{2}H, ^{3}He, and ^{4}He, both within the standard model and within universes having different expansion rates. Possible causes of a nonstandard expansion rate which he mentioned were additional energy density provided by gravitational radiation or "a new kind of neutrino field," a violation of general relativity, or anisotropy (although he considered this unlikely because of the observed present isotropy). His results for ^{4}He (shown in fig. 9.7 for the standard model) were fairly accurate, since as indicated previously its production is mainly dependent on the neutron–proton ratio. The importance of the helium predictions was immediately recognized, although the observations of the helium abundance at that time shed little direct light on its primordial value. On the other hand, the lack of a sufficient number of nuclear reactions and cross section data affected his predictions for deuterium and, to an even greater extent, ^{3}He (for which no numerical results were presented within the standard model, presumably because of its low abundance).

The Caltech assault was initiated by Willy Fowler, as a natural extension of his development of the field of nuclear astrophysics, and Fred Hoyle, who had continued his frequent interactions with Willy. Building upon the foundation in experimental and theoretical nuclear physics laid by Charles Lauritsen and his son Thomas in the Kellogg Radiation Laboratory, Willy had developed a (the!) center for theoretical astrophysics as well. Arriving from Stanford as a fresh PhD in 1965, I was soon caught up by the infectious enthusiasm of Willy and the marvelous group he had assembled. Being near the center of the emerging story of the quasars also contributed greatly to the overwhelming excitement of that unique time and place. Among the participants in our weekly SINS seminar, also led by the (initially) frightening John Bahcall and Bob Christy, were Dave Arnett, Don Clayton, John Faulkner, Jim Gunn, Bruce Peterson, Vahe Petrosian, Giora Shaviv, Peter Strittmatter, and Kip Thorne. Willy taught me the joy

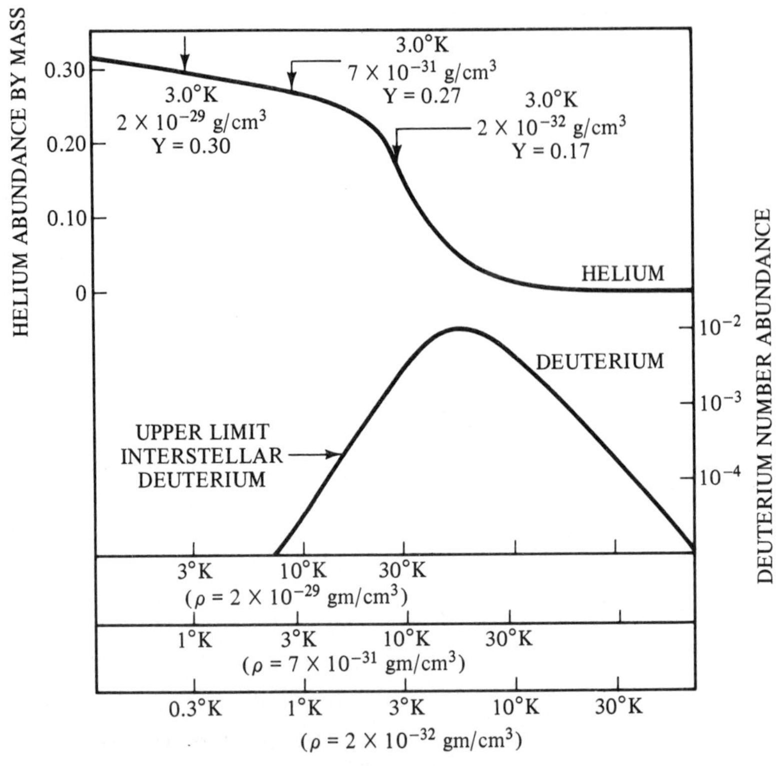

Fig. 9.7 The amount of ^{4}He and ^{2}H produced in the standard big bang, according to Peebles (1966a). The final abundances are presented as a function of the present radiation temperature, for various choices of the present baryon density.

of research, through almost daily discussions in the lab (and memos on hundreds of nuclear reaction rates, one of which is reproduced in fig. 9.8), two summers at Fred Hoyle's new Institute of Theoretical Astronomy (with virtually unlimited access to the computer), and the famous parties hosted by him and Ardie or Tommy and Marge Lauritsen. And it was especially thrilling to work with Fred as well, since it was his Messenger Lectures on cosmology at Cornell in 1960 that had almost instantly converted me from engineering to theoretical astrophysics. Although my thesis research had been in general relativity, Willy and Fred quickly convinced me of the potential power of nuclear physics as a probe of the hot big bang. (Figure 9.9 is a photograph of us with Don Clayton at Caltech then.)

My major role in our collaboration was the development of a computer program that could accurately follow the synthesis of 41 nuclei connected by 79 nuclear and weak reactions. I also learned how to convert nuclear data into at least estimates of the reaction rates that we needed from Willy, Georgeanne Caughlan, and Barbara Zimmerman, who had begun their classic compilations of rates of astrophysical importance (Fowler,

CALIFORNIA INSTITUTE OF TECHNOLOGY — INTER-OFFICE MEMO.

To RVW — From WAF — Date 4/13/67

Subject

Bob,

I prefer the treatment of threshold reactions in which

$$\sigma \propto \left(\frac{E_{th}}{E}\right)^{1/2}\left(1-\frac{E_{th}}{E}\right)^{1/2}$$

(See April 11 notes)

This gives

$$[\sigma v] = 1.67\times10^{9}\,\rho\,\sigma_m\left(\frac{E_{th}}{A}\right)^{1/2}\exp(-11.6\,E_{th}/T_9)$$

for σ_m in barns, E_{th} in MeV. I think it is better than any law involving $T^{1/2}$ etc or ΔE.

WF.

We must now finalize $[\sigma v]$ for neutron reactions by choosing forms for $Q>0$ and $Q<0$.

Fig. 9.8 One of Willy Fowler's (almost daily) memos.

Fig. 9.9 Outside the Kellogg Radiation Lab in January, 1967. I (recognizable?) am next to Willy; Don Clayton is next to Fred.

Caughlan, and Zimmerman 1967). Dave Arnett was developing a similar code to follow explosive nucleosynthesis in stars. The major problem was the stability of the numerical integration algorithm, which turned out (after months of exasperation and perspiration) to have a straightforward solution.

Our first results were presented at the annual meeting of the National Academy of Sciences in April, 1966 (Wagoner, Fowler, and Hoyle 1966). We noted that "In addition to D^2, He^3, and He^4, it is found that possibly significant quantities of some elements heavier than He^4 can be produced . . ." The amount of helium was close to that obtained by Peebles (1966a). Four months later, we submitted our paper (Wagoner, Fowler, and Hoyle 1967) to the *Astrophysical Journal.* Fred's continuing enthusiasm for supermassive objects as the elusive first generation of stars had infected Willy and me, so we allowed our major parameter (the number of baryons per photon) to also range over the larger values appropriate to their explosions (dubbed "little bangs"). If such explosions occurred from temperatures near 10^9 K, amounts of carbon and nitrogen or much heavier elements could be produced in amounts of order that observed in the oldest stars.

Our major conclusion was that deuterium, ^3He, ^4He, and (to a lesser extent) ^7Li could be produced in amounts in reasonable agreement with observed abundances if the present matter density was about 2×10^{-31} grams per cubic centimeter, as shown in fig. 9.10. This baryon density was within the range allowed by observations. However, the cosmic significance of this comparison was not yet clear, because the abundance data came mostly from the solar system and also because other sites for the synthesis of these nuclei had been proposed. In addition, uncertainties in some important nuclear reaction rates remained.

Deviations from the standard model were also considered. One was the effect of inhomogeneities in the matter density, particularly with respect to increasing the production of ^7Li. Another was the effect of a large lepton number, producing neutrino or antineutrino degeneracy which strongly affected the production of all nuclei, as shown in fig. 9.11. We also pointed out a "remarkable coincidence" which Fred had stressed: if the observed helium had instead been produced within stars, the resulting release of nuclear energy, *if thermalized*, would have a temperature of order the 3 K observed. Thus the need for a hot big bang might be obviated, if the other light elements could also be produced in other ways.

Although the agreement between the observed abundances and those produced in the standard hot big bang was encouraging, a cloud had appeared on the horizon. Sargent and Searle (1966) and Greenstein (1966) had discovered a few blue halo stars that appeared to contain about one hundred times less helium than the 25–35% mass fraction found in the younger Population I objects. This possibility of a primordial helium

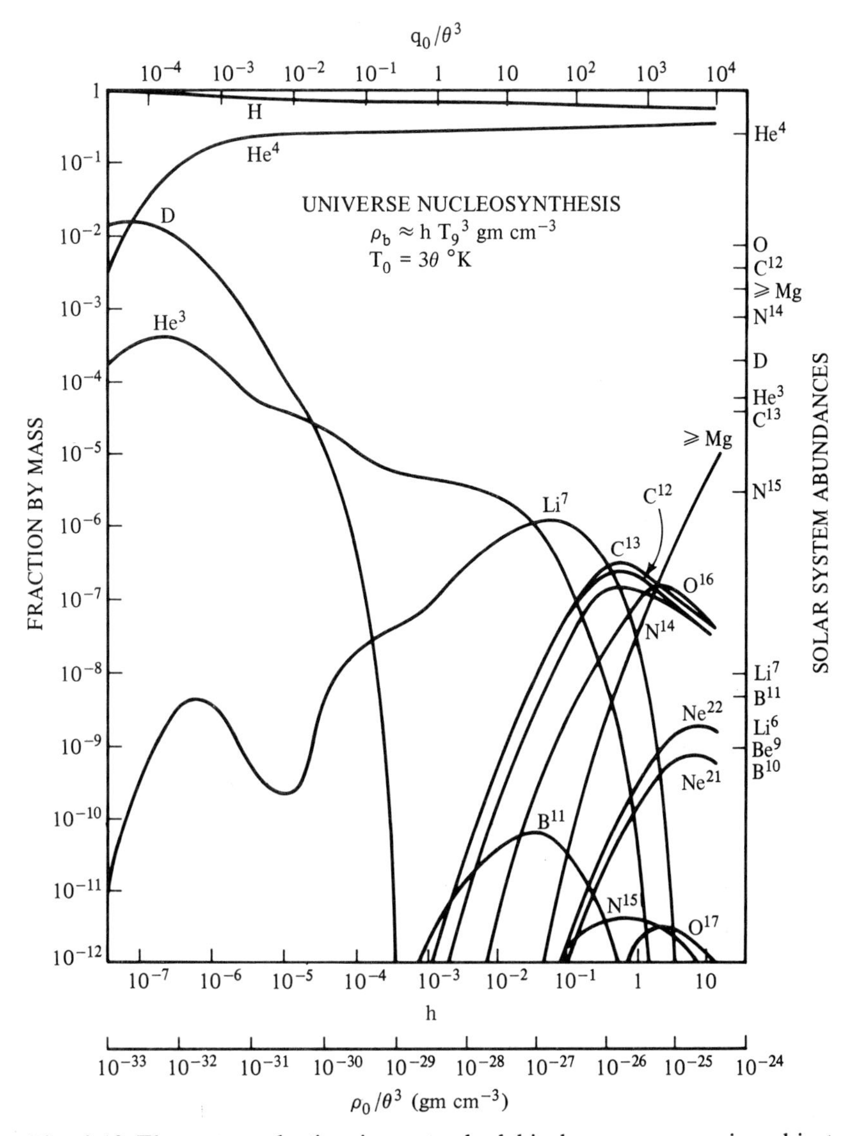

Fig. 9.10 Element production in a standard big bang or a massive object expanding from a temperature $T > 10^{10}$ K, according to Wagoner, Fowler and Hoyle (1967). The single parameter h related the baryon density to the temperature. (ϱ_0 is the present baryon density, q_0 is the deceleration parameter produced by it, and $T_9 = T/10^9$ K.)

abundance much lower than that produced in the standard model (20–30%) motivated me to investigate a more comprehensive class of other models of the hot big bang (Wagoner 1967). The guiding principle was the fact that the abundances produced in any hot early universe were sensitive *only* to the local values of the expansion rate, baryon number and lepton number per photon, and neutrino distribution function (if the expansion

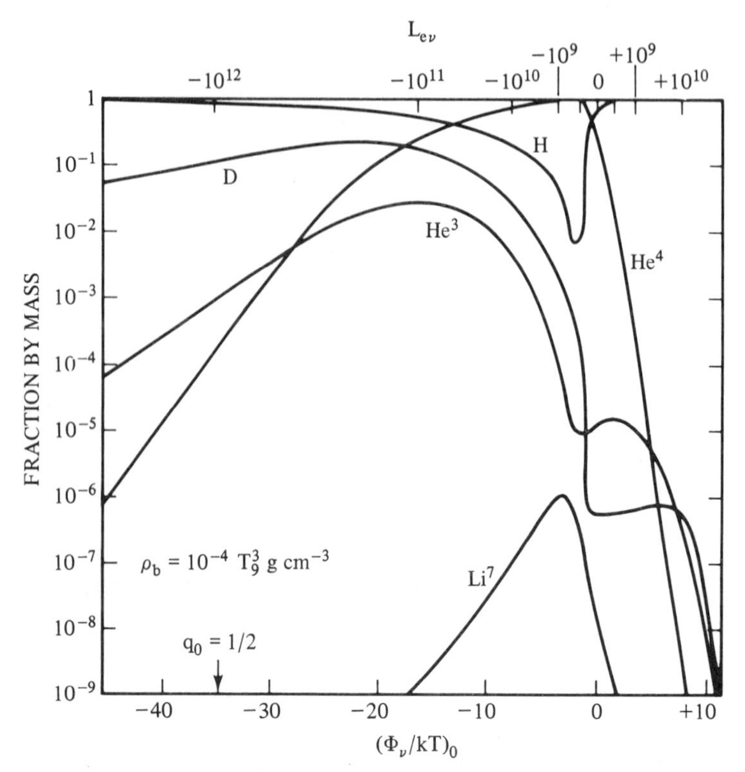

Fig. 9.11 Abundances produced in big bang models with a fixed density parameter $h = 10^{-4}$ but with various numbers of electron neutrinos (+) or antineutrinos (−) per baryon, $L_{e\nu}$. The present ratio of neutrino Fermi level to photon energy is also indicated, as well as its value necessary to produce a closed universe. [From Wagoner (1967).]

was anisotropic). The effects of new particles, other theories of gravitation (such as that of Brans and Dicke 1961), and anisotropy (Hawking and Tayler 1966; Thorne 1967) on the expansion rate were elucidated. As had been indicated previously, both slower or much more rapid expansions (Peebles 1966b; more extensive results are shown in fig. 9.12) as well as even mildy degenerate neutrinos (see fig. 9.11) could drastically reduce the helium production while keeping the other abundances below their observational bounds.

However, by 1972 the observational situation had clarified considerably. Baschek, Sargent, and Searle (1972) argued that the extreme helium deficiencies were confined to the surfaces of the blue halo stars. Even more important was their discovery of dwarf blue galaxies in which the abundance of helium was normal, yet oxygen and neon were significantly below their normal (Population I) values (Searle and Sargent 1972). The galaxies appeared to be young, and thus provided strong evidence for a large universal pregalactic source of helium. In addition, a flood of observations had established both the extreme isotropy and an accurate

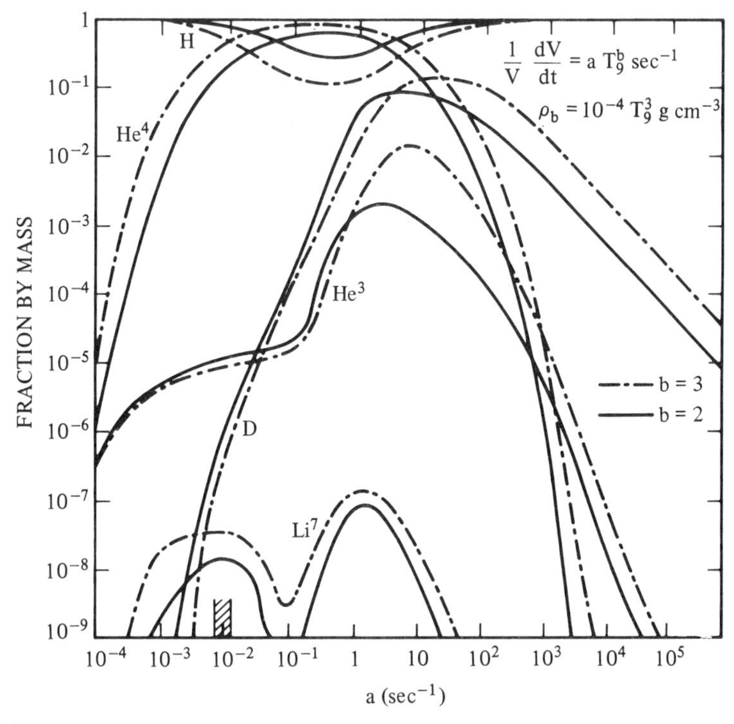

Fig. 9.12 Abundances produced in nondegenerate universes with the same density parameter as Figure 9.11 but with various expansion rates ($V^{-1}dV/dt$) during nucleosynthesis. [From Wagoner (1967).]

temperature (2.6–2.8 K) of the microwave background, as discussed by Wilson (this volume).

On the theoretical side, no process other than the hot big bang had been found to be capable of producing more than a mass fraction of 10^{-6} of deuterium (Reeves, Audouze, Fowler, and Schramm 1973). On the other hand, the spallation process involving collisions of the observed cosmic rays with the interstellar gas was found to be capable of producing the observed amounts of ^6Li, ^9Be, ^{10}B, and possibly ^{11}B (Reeves, Fowler, and Hoyle 1970; Meneguzzi, Audouze, and Reeves 1971). And evidence had accumulated since the work of Burbidge, Burbidge, Fowler, and Hoyle (1957) that stars could probably produce the bulk of the heavier elements ($A > 11$). The fact that the remaining nuclei (^2H, ^3He, ^4He, ^7Li) were precisely those that we had produced in our big bang calculation began to assume increasing significance, at least to me (although stars could still not be ruled out as sources of the ^3He and ^7Li).

Thus in 1972, four years after joining the Cornell Astronomy Department, I decided that it was time to attack the problem again. More information about many of the important reaction rates had continued to be supplied to me by the heroic efforts of Fowler, Caughlan, and

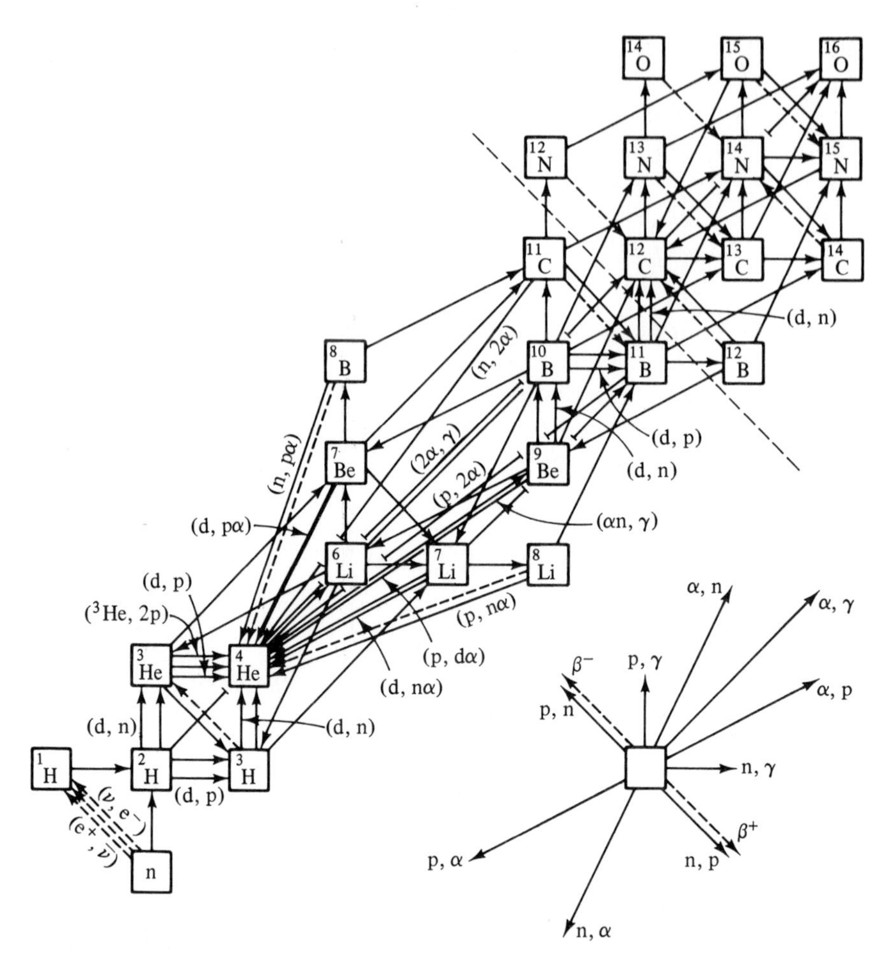

Fig. 9.13 The nuclear reaction network employed in the calculation of Wagoner (1973).

Zimmerman (1975) since my most recent compilation (Wagoner 1969). The reactions that were included at that time are indicated in fig. 9.13. In addition, the computer program had been significantly improved. In the resulting paper (Wagoner 1973), two major issues were addressed; one controlled mainly by the helium abundance and the other controlled mainly by the deuterium abundance.

The first issue was the limits that could be placed on deviations from the standard model. The expansion rate as well as the strength of the weak interaction (i.e., the neutron lifetime) were allowed to vary. When combined with previous results, the new observational bounds on the primordial helium abundance placed strong limits on most types of deviation.

The second issue, whose importance was beginning to be realized, was the limit that could be placed on the present universal baryon density. Allowing for an arbitrary amount of nonuniformity in the baryon density, a firm upper limit of 2×10^{-34} g cm^{-3} was placed on its product with the primordial deuterium abundance. However, estimating this abundance was complicated by the fact that the interstellar abundance of deuterium was still undetected, but known to be less than the standard terrestrial and meteoritic value (Weinreb 1962; Black 1972; Geiss and Reeves 1972), presumably due to chemical fractionation. On the other hand, any primordial deuterium abundance would have been reduced by stellar processing (Truran and Cameron 1971; Talbot and Arnett 1972), but this uncertainty would not affect the upper limit on the present baryon density. It was becoming clear that with reasonable limits on the amounts of fractionation and inhomogeneity, the universe could not be closed by baryons.

My paper was published in January, 1973, which turned out to be a very eventful year. As I was preparing my talk for the I.A.U. Symposium (No. 63) in Poland that summer (Wagoner 1974), rumors began to build that absorption by the Lyman lines of deuterium had been detected with the Copernicus satellite, giving an interstellar mass fraction of order 2×10^{-5}. This result meant that the present baryon density in the standard model was at most 6×10^{-31} g cm^{-3}, compared to the value of 5×10^{-30} $(H_0/50)^2$ required to close the universe (H_0 is the Hubble constant). The significance of this soon reported detection of interstellar deuterium (Rogerson and York 1973) was discussed in an article with David Schramm, in which it was argued that noncosmological sources of deuterium generically overproduce beryllium and boron (Epstein, Arnett, and Schramm 1974; Schramm and Wagoner 1974). The evidence was building that deuterium as well as helium was bringing us information from the early universe.

9.5 The power of this probe

Although we must end our story here, some general reflections on subsequent developments and what the future may hold cannot be resisted. An authoritative review of the observational and theoretical status of the subject has been provided by Boesgaard and Steigman (1985). Detailed references provided by them will not be repeated below.

In my opinion, the major advance during the past 15 years has been the increasing evidence that primordial lithium may also have been discovered. The Spites found that old (Population II) stars contain about 10 times less lithium, and the abundance does not vary with surface temperature, indicating that it has not been affected by mixing. On the other hand, new cross section data has led to an increase in the predicted primordial ^{7}Li abundance, as shown by Yang, Turner, Steigman, Schramm, and Olive (YTSSO) and Kawano, Schramm, and Steigman (1988). Both of these changes have produced a better agreement between the predicted and observed abundance for the choice of baryon density which matches the deuterium and helium (whose dominant primordial component has also been better determined). The fact that new astrophysical and nuclear data have produced this concordance must be considered as additional evidence for the validity of the standard big bang model.

Another advance was the discovery of the tau neutrino and the carriers of the weak interaction, in particular the Z° boson. Soon its decays will tell us directly how many types of light neutrinos there are. This will be a powerful test of the upper limit provided by their effect on the cosmological helium abundance, as emphasized by YTSSO and others.

Most recently, there have been attempts to reconcile the belief among many (especially those who favor a universe which inflated during its infancy) that the universe is spatially flat with the prediction of standard big bang nucleosynthesis that its baryon density is at least 6 times too low to achieve flatness, without invoking some form of nonbaryonic dark matter to make up the difference. Two modifications of the standard scenario with the higher required baryon density have recently been proposed. The first invokes a first-order phase separation during the quark-hadron transition which precedes the era of nucleosynthesis (Alcock, Fuller, and Mathews 1987; Applegate, Hogan, and Scherrer 1987; Malaney and Fowler 1988), leading to large inhomogeneities in the nucleon abundances. The main problem has been great difficulty in avoiding the overproduction of ^{7}Li. The second modification, introduced by Dimopoulos, Esmailzadeh, Hall, and Starkman (1988), involves late-decaying massive particles whose electromagnetic and hadronic showers strongly affect the products of the previous standard nucleosynthesis (with the exception of ^{4}He). Yet the abundances which emerge from this reprocessing can agree with those observed. The major difference is an excess of ^{6}Li relative to ^{7}Li, which future observations of primordial material may be able to test.

Indeed, the quest for primordial abundances is the key challenge for the future. Mapping the space and time (i.e., redshift) dependence of the amounts of these light nuclei (and their heavier brethren) via both ground-based observations with the largest telescopes and space observatories such as HST and FUSE will be required. Of special urgency is deuterium, since its abundance is only known in our neighborhood, where it appears to be nonuniform. The possibility of observing the ground-state abundances of HeI and HeII at large redshifts, as well as that of deuterium at smaller redshifts in the apparently heavy-element poor clouds seen in quasar spectra makes one's mouth water.

Why has nucleosynthesis been such an informative probe of the early universe? (For instance, we should also mention that it provides the best limit on the rate of change of the "gravitational constant" predicted by many alternate theories of gravitation.) The fundamental reason is that the physics of the probe is known (i.e., has been tested by experiments). In addition, since the working model of the system being probed is simpler than the stars, the cleanest predictions in the field of nucleosynthesis are primordial.

Can we reverse this process, and use the very early universe to learn about physics at energies not accessible to accelerators? This is a much more dangerous enterprise, since we have little experimental or observational information about the actual cosmological laboratory then. The most promising experimental probes appear to be direct detection of nonbaryonic dark matter candidates and experiments (like those looking for proton decay) which explore very high energies indirectly. And perhaps we should devote more attention to the highest energy cosmic rays. The most promising observations are those of large-scale structure, if they can be cleansed of their astrophysical complications enough to reveal their origin. But the most penetrating information will probably come from an unexpected source.

Finally, we must continually be wary of being channeled into a parochial view of the scope of possible universes. We must remember that the study of the early universe represents the greatest extrapolation in science. Are warning signs becoming visible? (For instance, is the microwave background too isotropic, when compared to the evidence of structure that we continue to discover as we explore larger scales?) As J. B. S. Haldane said, "the universe is not only queerer than we suppose, but queerer than we can suppose." I hope that he was right.

Acknowledgements

This paper was prepared with the support of the National Science Foundation (grant PHY 86-03273) and the National Aeronautics and Space Administration (grant NAGW-299). The hospitality of the organizers of this symposium was also very much appreciated.

References

Alcock, C., Fuller, G. M. and Mathews, G. J. (1987). The Quark-Hadron Phase Transition and Primordial Nucleosynthesis. *Astrophysical Journal*, **320**, 439–447.

Alpher, R. A. (1948). A Neutron-Capture Theory of the Formation and Relative Abundance of the Elements. *Physical Review*, **74**, 1577–1589.

Alpher, R. A. and Herman, R. C. (1948). On the Relative Abundance of the Elements. *Physical Review*, **74**, 1737–1742.

(1949). Remarks on the Evolution of the Expanding Universe. *Physical Review*, **75**, 1089–1095.

(1950). Theory of the Origin and Relative Abundance Distribution of the Elements. *Reviews of Modern Physics*, **22**, 153–212.

(1951). Neutron-Capture Theory of Element Formation in an Expanding Universe. *Physical Review*, **84**, 60–68.
(1953). The Origin and Abundance Distribution of the Elements. *Annual Review of Nuclear Science*, **2**, 1–40.
Alpher, R. A., Bethe, H. and Gamow, G. (1948). The Origin of Chemical Elements. *Physical Review*, **73**, 803–804.
Alpher, R. A., Follin, J. W. Jr. and Herman, R. C. (1953). Physical Conditions in the Initial Stages of the Expanding Universe. *Physical Review*, **92**, 1347–1361.
Applegate, J. H., Hogan, C. J. and Scherrer, R. J. (1987). Cosmological Baryon Diffusion and Nucleosynthesis. *Physical Review* **D35**, 1151–1160.
Baschek, B., Sargent, W. L. W. and Searle, L. (1972). The Chemical Composition of the B-Type Subdwarf HD 4539. *Astrophysical Journal*, **173**, 611–618.
Black, D. C. (1972). On the Origins of Trapped Helium, Neon and Argon Isotopic Variations in Meteorites – I. Gas-Rich Meteorites, Lunar Soil and Breccia. *Geochimica et Cosmochimica Acta*, **36**, 347–375.
Boesgaard, A. M. and Steigman, G. (1985). Big Bang Nucleosynthesis: Theories and Observations. *Annual Review of Astronomy and Astrophysics*, **23**, 319–378.
Brans, C. and Dicke, R. H. (1961). Mach's Principle and a Relativistic Theory of Gravitation. *Physical Review*, **124**, 925–935.
Brown, H. (1949). A Table of Relative Abundances of Nuclear Species. *Reviews of Modern Physics*, **21**, 625–634.
Burbidge, E. M., Burbidge, G. R., Fowler, W. A. and Holye, F. (1957). Synthesis of the Elements in Stars. *Reviews of Modern Physics*, **29**, 547–650.
Cameron, A. G. W. (1957). *Chalk River Report*, **CRL-41**.
Chandrasekhar, S. and Henrich, L. R. (1942). An Attempt to Interpret the Relative Abundances of the Elements and Their Isotopes. *Astrophysical Journal*, **95**, 288–298.
Dicke, R. H., Peebles, P. J. E., Roll, P. G. and Wilkinson, D. T. (1965). Cosmic Black-Body Radiation. *Astrophysical Journal*, **142**, 414–419.
Dimopoulous, S., Esmailzadeh, R., Hall, L. J. and Starkman, G. D. (1988). Is the Universe Closed by Baryons? Nucleosynthesis with a Late-Decaying Massive Particle. *Astrophysical Journal*, **330**, 545–568.
Epstein, R., Arnett, W. D. and Schramm, D. N. (1974). Can Supernovae Produce Deuterium? *Astrophysical Journal (Letters)*, **190**, L13–L16.
Field, G. B. and Hitchcock, J. L. (1966). Cosmic Black-Body Radiation at $\lambda = 2.6$ mm (From Observations of Interstellar CN). *Physical Review Letters*, **16**, 817–818.
Fowler, W. A., Caughlan, G. R. and Zimmerman, B. A. (1967). Thermonuclear Reaction Rates. *Annual Review of Astronomy and Astrophysics*, **5**, 525–570.
(1975). Thermonuclear Reaction Rates, II. *Annual Review of Astronomy and Astrophysics*, **13**, 69–112.
Friedmann, A. (1922). Über die Krümmung des Raumes. *Zeitschrift für Physik*, **10**, 377–386.
Gamow, G. (1946). Expanding Universe and the Origin of Elements. *Physical Review*, **70**, 572–573.
(1948). The Evolution of the Universe. *Nature*, **162**, 680–682.

Geiss, J. and Reeves, H. (1972). Cosmic and Solar System Abundances of Deuterium and Helium-3. *Astronomy and Astrophysics*, **18**, 126–132.
Goldschmidt, V. M. (1938). Geochemische Verteilungsgesetze der Elemente. IX. Die Mengenverhältnisse der Elemente und der Atom-arten. I. *Matematisk-Naturvidenskapelig klasse*, No. 4, 1937. Oslo.
Greenstein, J. L. (1966). The Nature of the Faint Blue Stars. *Astrophysical Journal*, **144**, 496–515.
Hawking, S. W. and Tayler, R. J. (1966). Helium Production in an Anisotropic Big-Bang Cosmology. *Nature*, **209**, 1278–1279.
Hayashi, C. (1950). Proton–Neutron Concentration Ratio in the Expanding Universe at the Stages Preceding the Formation of the Elements. *Progress of Theoretical Physics*, **5**, 224–235.
Hoyle, F. and Tayler, R. J. (1964). The Mystery of the Cosmic Helium Abundance. *Nature*, **203**, 1108–1110.
Kawano, L., Schramm, D. and Steigman, G. (1988). Primordial Lithium: New Reaction Rates, New Abundances, New Constraints. *Astrophysical Journal*, **327**, 750–754.
Klein, O., Beskow, G. and Treffenberg, L. (1946). On the Origin of the Abundance Distribution of Chemical Elements. *Arkiv för Matematik Astronomi och Fysik*, **33B**, No. 1.
Lemaître, G. (1931). Contributions to a British Association Discussion on the Evolution of the Universe. *Nature*, **128**, 704–706.
(1933). L'univers en Expansion. *Annales Société Scientifique Bruxelles I*, **53A**, 51–85.
(1950). *The Primeval Atom*. New York: D. Van Nostrand.
Malaney, R. A. and Fowler, W. A. (1988). Late-Time Neutron Diffusion and Nucleosynthesis in a Post-QCD Inhomogeneous $\Omega_b = 1$ Universe. *Astrophysical Journal*, **333**, 14–20.
Mayer, M. G. and Teller, E. (1949). On the Origin of Elements. *Physical Review*, **76**, 1226–1231.
Meneguzzi, M., Audouze, J. and Reeves, H. (1971). The Production of the Elements Li, Be, B by Galactic Cosmic Rays in Space and its Relation with Stellar Observations. *Astronomy and Astrophysics*, **15**, 337–359.
Peebles, P. J. E. (1966a). Primeval Helium Abundance and the Primeval Fireball. *Physical Review Letters*, **16**, 410–413.
(1966b). Primordial Helium Abundance and the Primordial Fireball. II. *Astrophysical Journal*, **146**, 542–552.
Penzias, A. A. and Wilson, R. W. (1965). A Measurement of Excess Antenna Temperature at 4080 Mc/s. *Astrophysical Journal*, **142**, 419–421.
Reeves, H., Audouze, J., Fowler, W. A. and Schramm, D. N. (1973). On the Origin of Light Elements. *Astrophysical Journal*, **179**, 909–930.
Reeves, H., Fowler, W. A. and Hoyle, F. (1970). Galactic Cosmic Ray Origin of Li, Be, and B in Stars. *Nature*, **226**, 727–729.
Reines, F. (ed.) (1972). *Cosmology, Fusion, and Other Matters*. Boulder: Colorado Associated University Press.
Rogerson, J. B. Jr. and York, D. G. (1973). Interstellar Deuterium Abundance in the Direction of Beta Centauri. *Astrophysical Journal (Letters)*, **186**, L95–L98.
Roll, P. G. and Wilkinson, D. T. (1966). Measurement of Cosmic Background

Radiation at 3.2 cm Wavelength – Support for Cosmic Black-body Radiation. *Physical Review Letters*, **16**, 405–407.
Sargent, W. L. W. and Searle, L. (1966). Spectroscopic Evidence on the Helium Abundance of Stars in the Galactic Halo. *Astrophysical Journal*, **145**, 652–654.
Schramm, D. N. and Wagoner, R. V. (1974). What can Deuterium Tell Us? *Physics Today*, **27**, 40–47.
Searle, L. and Sargent, W. L. W. (1972). Inferences from the Composition of Two Dwarf Blue Galaxies. *Astrophysical Journal*, **173**, 25–33.
Suzuki, S. (1928). On the Thermal Equilibrium of Dissociation of Atom-Nuclei. *Proceedings of the Physico-Mathematical Society of Japan*, **10**, 166–169.
Talbot, R. J. Jr. and Arnett, W. D. (1971). Helium Production by Pulsationally Unstable Massive Stars. *Nature Physical Science*, **229**, 150–151.
Thaddeus, P. and Clauser, J. F. (1966). Cosmic Microwave Radiation at 2.63 mm from Observations of Interstellar CN. *Physical Review Letters*, **16**, 819–822.
Thorne, K. S. (1967). Primordial Element Formation, Primordial Magnetic Fields, and the Isotropy of the Universe. *Astrophysical Journal*, **148**, 51–68.
Tolman, R. C. (1922). Thermodynamic Treatment of the Possible Formation of Helium from Hydrogen. *Journal of the American Chemical Society*, **44**, 1902–1908.
Truran, J. W. and Cameron, A. G. W. (1971). Evolutionary Models of Nucleosynthesis in the Galaxy. *Astrophysics and Space Science*, **14**, 179–222.
Urey, H. C. and Bradley, C. A. Jr. (1931). On the Relative Abundances of Isotopes. *Physical Review*, **38**, 718–724.
Wagoner, R. V. (1967). Cosmological Element Production. *Science*, **155**, 1369–1376.
(1969). Synthesis of the Elements within Objects Exploding from Very High Temperatures. *Astrophysical Journal Supplement*, No. 162, **18**, 247–295.
(1973). Big-Bang Nucleosynthesis Revisited. *Astrophysical Journal*, **179**, 343–360.
(1974). Cosmological Synthesis of the Elements. In *Confrontation of Cosmological Theories with Observational Data*, ed. M. S. Longair, pp. 195–210. Dordrecht: Reidel.
Wagoner, R. V., Fowler, W. A. and Hoyle, F. (1966). Nucleosynthesis in the Early Stages of an Expanding Universe. *Science*, **152**, 677.
(1967). On the Synthesis of Elements at Very High Temperatures. *Astrophysical Journal*, **148**, 3–49.
Weinreb, S. (1962). A New Upper Limit to the Galactic Deuterium-To-Hydrogen Ratio. *Nature*, **195**, 367–368.
Zel'dovich, Ya. B. (1963). The Theory of the Expanding Universe as Originated by A. A. Friedmann. *Uspekhi Fizicheskikh Nauk*, **80**, 357–390 (reprinted in *Soviet Physics Uspekhi*, **6**, 475–494).

Discussion

Hoyle:

I wonder if it worries you, as it does me, that the widely quoted spectroscopic determinations for ^{4}He, ^{7}Li, refer to small samples of material, e.g. H II regions in dwarf galaxies for ^{4}He. If we had values for a wide variety of samples it is quite possible that a wider spread in the density–temperature relation $\varrho = hT_9^3$ would show up. If there has been a spread in h, it is easy to see, for example, that particular samples will hit the bottom of the ^{7}Li curve. With the observers looking for the least ^{7}Li value it would then be more or less inevitable that they would hit the minimum, without it being general for all samples.

Wagoner:
I agree that the interpretation of and the strategy for obtaining abundances should allow for possible inhomogeneity in the entropy per baryon. Indeed, there is much current interest in such consequences of the quark-hadron phase transition which precedes the epoch of primordial nucleosynthesis. However, it is difficult to prevent overproduction of lithium due to regions of high baryon number per photon.

Sivaram:
(1) The lifetime of the neutron is known to have been reduced from 10.6 minutes in recent measurements. How sensitive is the abundance to this?
(2) The constraint on number of neutrino types applies to any relativistic particles at the decoupling time. Is that right?

Wagoner:
As shown in my 1973 paper, the production of helium depends logarithmically on both the lifetime of the neutron and the number of weakly-interacting particle species which were relativistic at that epoch (mass $\leqslant 1$ MeV).

IV
The great cosmological debates

10

The cosmological scene 1945–1952

Hermann Bondi

Here, some of us seem to be writing our own history and there's nothing against that, as far as I know. Let me add that I am really giving one and a half talks because Tommy Gold was originally asked to speak and said yes, but later found that he couldn't. He has asked me to stray from the area of my talk into the area originally intended for him. Therefore, I will start with my own talk which was intended to be on the background, the scene, in which the steady-state theory was born and then follow with some reminiscences on how it was born and its first reception.

The figure of Hubble was more than life sized in the 30s and 40s, I think quite rightly so. What he had done in establishing that one could talk sensibly about the spiral nebulae by showing that he could find stars in them, that indeed they were galaxies much like our own, in putting Slipher's work on the redshift on an enormously extended basis and thus firmly establishing the velocity-distance relation – all this was quite fundamental. He really did a wonderful amount. This enormous prestige, I think, more – and here I differ a little from McCrea – than Eddington's calculations, gave a great sanctity to his estimate of $1.8 \cdot 10^9$ years as the inverse Hubble constant. A name that should be mentioned here is Arthur Holmes of Newcastle who was the first person to establish the age of the rocks, finding $3\text{–}4 \times 10^9$ years for the oldest ones. This led to the great time-scale problem which was oppressing us all then. You must of course remember that those of us who, led by Einstein, rejected the cosmic constant had to deal not with $1.8 \cdot 10^9$ years but with a shorter time because, with gravitation slowing it down, expansion must have been faster in the past: with the simple Einstein–de Sitter model it would have been $1.2 \cdot 10^9$ years since the initial big bang. Thus the conflict was extreme.

At that time (30s and 40s) the work of Eddington, of Milne, and of Dirac showed that cosmology was ripe for imaginative ideas, ripe for something to bring order into it. I think it was Milne's pugnacious unwillingness to accept that people might swallow only part rather than the whole of his theory which makes us all give him less credit than he really deserves, first, for inventing the radar method of measuring distances years before radar became public property and two and a half decades before it became the method of establishing distances on the Earth and in the solar system. Although in cosmological terms the times are absurdly large, it was most important to establish in principle how to define distance without the concept of the rigid rod. I found this immensely valuable in clarifying and explaining the concept of special relativity. Secondly, Milne should be praised for his idea that there might be different time scales for different clocks. This still seems to me one of the most imaginative ideas of my time. True, we have not found a good application for it so far but the idea that different clocks of different construction might register different times is a profound idea. Even if it turns out not to be so, it is one that I think deserves more discussion and debate than it has had. The attitudes that influenced us all at the time owe also something to a man whose name has not been mentioned here, Otto Heckmann. He wrote a magnificent book on cosmology that because of the war and the difficulties of getting anything across afterwards has not received anything like the credit it deserves. He was a magnificent person and it was a very good book. When people here say kind things about my book I would very much want to say that I owe a great deal to him.

Let me now come to some of the other ideas. It was Lemaître who first put real physics into cosmology. The question then beginning insistently to be asked concerned the origin of the elements. Again it is interesting to observe how the different aspects of astrophysics come together. In the early days of astrophysics and at the time when Eddington was still writing his 'Internal Constitution of the Stars' the dominance of hydrogen and helium was not appreciated at all. If all elements were present in large amounts there would be much less incentive to try to trace everything back to hydrogen. Almost at the same time as Lemaître saw that cosmology could say something about the origin of the elements, the dominance of hydrogen in stars became appreciated through the work of Hoyle and others and especially of McCrea who developed the first model of a large hydrogen star. Very gradually this led to a realisation of a situation in which the heavy elements seemed to be a rare oddity calling for an explanation. As has been pointed out, Gamow as a nuclear physicist entered the subject at that time. Another item that became significant then was concern about general relativity. Though most physicists refused to say anything in public the great majority of them thought of general relativity as an obscure backwater which took an awful lof of learning to get into. If I put my strong personal bias into this question, I don't think the way Einstein or Eddington or Weyl or others presented it was sufficiently physical for the common or garden physicist to appreciate how logically and inevitably it followed from Galileo's principle that all bodies fall equally fast. Though general relativity was on the back burner, it was thought to have a unique ability to account for a universe that showed universal recession. Milne never liked the theory at all. He and McCrea then created the subject of Newtonian cosmology. To me, one of the most interesting features of this work was the question of why it hadn't been done fifty or a hundred years earlier. It was simply because people then had a preconception of the static universe firmly fixed in their minds. The mathematical and physical apparatus for Newtonian cosmology was available long before, but until the 1920s, certainly until de Sitter's paper, nobody took the idea of a universe in a state of motion, any state of motion, as worth spending five minutes on. So the work of Hubble, of Einstein, of Milne, of McCrea, of Eddington, of Lemaître, of Dirac, set the scene.

Having described the background I will now turn to the actual origin of the theory. It was Hoyle who brought Gold and me into astronomical questions, with a sceptical arguing attitude. It was his inspiration, first when we were all working for the Admiralty during the war and then when we were all back in Cambridge, that brought both of us into the subject. We discussed all sorts of astronomical questions from morning to night. The time-scale problem of cosmology caused us considerable worries. We looked at the solutions then on offer, especially the Lemaître–Eddington model for which we developed a considerable dislike. This was first because it seemed to be a bit of a fiddle. To imagine that the cosmological constant is just but only just above the critical value and that the model has a long, almost static, dwell time didn't look at all attractive. Next, it barely approached the question of the thermodynamics of the universe. So we searched for other solutions. As I think is historically well known, it was Gold's idea that there might be a process of continual creation that allowed a steady state to go on. Fred and I said: 'Ach, we will disprove this before dinner.' Dinner was a little late that night, and before very long we all saw that this was a perfectly possible solution to the question. Then Fred on the one hand and Tommy and I on the other went somewhat different ways. Hoyle followed the field-theoretical direction of putting in his creation field, while we went in a much more philosophical direction. Again it is worth mentioning the kind of trouble we had in those days because it shows how attitudes change. At that time people were not necessarily quite so shocked by the idea of continual creation as by the question of how the newly created matter would know with what velocity to move. It is my impression that nobody had seriously considered at that time that the universe was not Lorentz-invariant and that its isotropy defined a preferred velocity at every point, namely that from which the universe looked most isotropic. We had a lot of arguments about this and nobody liked our statement that in any case the universe was not Lorentz-invariant. Of course, with the observations of the day the preferred velocity was only rather approximately defined. Twenty years later, with the discovery of the microwave background, everyone knows that there is a preferred velocity. Even today, if I may boast, few people are quite as logical as I, who says that the principle of special relativity now can really only be properly phrased as: the universe provides a preferred velocity at every point. To date, nobody has devised a laboratory experiment to reveal this velocity.

Let me return to our philosophical attitude and the irony of the situation of the day. The idea of the steady-state universe certainly came first; the philosophy came afterwards. But the essential point of the philosophy was and is that if the universe was evolving and changing, then there is no reason to trust what we call the laws of physics, established by experiments performed here and now, to have permanent validity. This argument impressed us very strongly. Of course it was easy at the time to be greatly impressed. For the freedom to construct theories of the universe in which the 'constants' were changing had been employed to great effect by Dirac and by Milne (with his two time scales) in their cosmological theories of the day. But the validity of this criticism of any evolving model struck us then, and strikes us today, as being of the utmost importance. This argument does not demonstrate what the universe must be like, but shows clearly that an evolving model must handle the question, 'Why is it that this feature is regarded as conserved and permanent and that feature as time dependent?' This is what worries me about all the modern developments – the big-bang model, including the inflationary side, is an extrapolation of our knowledge into incredibly different conditions. The belief that our present physical concepts apply unchanged then strikes me as daring, to put it politely.

My talk is perhaps slightly more philosophical than I originally intended it to be but I will make some more remarks on such themes. You will remember that the phrase with which Lyell, the geologist, revolutionised geology 150 or 160 years ago was 'the present is the key to the past'. What he meant was that it is the processes we now see here and now that must be used to account for the formation of the rocks of the Earth. Now, the essential idea is always that we can tell something about the universe only because it is in interaction with us, interaction that conditions things here today. By Mach's principle, our inertia is due to the universe. By Olbers' argument, our thermodynamical situation is due to the universe. Whichever resolution to Olbers' paradox is taken, the universe has everything to do with what we observe here and now. The more modern observation of the microwave background – that we are in a heat bath of 2.8 degrees or so – is the kind of observation and the kind of background that seems to me a much better test of cosmology than anything else. So, to paraphrase Lyell, an understanding of how the universe shapes conditions here and now is the key to what it is like, more than observations taken at the limits of our instruments. If what goes on here and now is so dependent on the state of universe, then our ability to imagine conditions in a very different state of the universe must be precarious. Gold wanted me particularly to stress that on these grounds he still feels very critical of what has been done on evolving models because such work does not discuss these options of a changing physics, even where they are implicitly denied.

To return to the history of the day, when Fred was attempting to publish his paper, Gold and I were left a little high and dry. There was Gold's idea, there was our philosophy that was beginning to form, but you cannot publish a paper in physics/astronomy on an idea and a little bit of philosophy. The key to publication came when I noted that Hubble's number counts could very well be accounted for by the steady-state model. This is the peg on which our publications was initially hung. I wasn't as worried yet about observations at great distances as I later became but I still find it rather amusing that it wasn't the thermodynamics that led to publication, it wasn't any other item but the number counts which before long I came to mistrust so profoundly. Of course the number of people who found the theory attractive was relatively small. McCrea was perhaps the first to like it and I still recall how he encouraged me, at the Zurich IAU of 1948, to describe it to Lindblad, then its president, while we were on an excursion on a lake steamer. However, the number of people who took it quite seriously was rather large, though they generally didn't like it. Yet in the many discussions I had with people who preferred an evolving model, what I always threw at them was: if the universe has ever been in a very different state from what it is now, show me some fossil remains of what it was like a long time ago. The only fossil remains that anyone could possibly think of at that time were the heavy elements. A paper of the early fifties that has perhaps not been mentioned enough is one by Salpeter. He showed that, in fact, the Gamow–Lemaître model of the early fireball state of the universe would not produce any elements higher than helium (except for a little lithium).

Then came the great work that I regard as the full justification of the steady-state theory, when Hoyle, the Burbidges and Fowler gave us our modern theory of nucleosynthesis. The fact that the steady state inspired it is to me an important historical fact. However, there was already a fossil I was beginning to be worried about, and that was the amount of helium in the universe. I remember in particular discussing this question with Andrew McKellar who was then the astronomer most involved in looking for elements like helium to see whether one could or could not get to a good estimate of its abundance. If, in fact, there is a good deal more helium than corresponds to the helium synthesis responsible for starlight, this is just the kind of evidence that to me is most important in cosmology. What the abundance of helium is here and now seemed to me then, as it still does now, to be a key question in understanding whether the universe ever was very different from what it is like now. We are pretty sure that we don't owe the condensations to the big bang, we are certain we don't owe the heavy elements to the big bang, but we may owe the helium to the big bang according to the standard interpretation, and there is a great deal to be said for this view. While I had foreseen the helium problem I had not in any way foreseen the microwave background, which is so readily interpreted as a fossil from those days. So my challenge of whether fossils could be found has had an answer long after I posed it.

Let me finish by saying a word or two more about the time-scale problem and how it affected our thinking. With steady state the average age of any objects is only a third of the Hubble constant. So even in steady state we had to say that our galaxy was much older than the average, not at first sight a very attractive proposition. What is today often forgotten is that with the then accepted Hubble distance scale ours was an enormous galaxy. I remember somebody said at that time, that whereas the other galaxies were islands, ours was a continent. So it was very natural to suppose that our, supposedly unusually large, galaxy was also unusually old. This was not done just in order to fit the age of the rocks. All we had to assume is that galaxies grow in time. Then an exceptional size and an exceptional age fit well together.

I think the person from whom we had most help on the philosophical side was Popper. His analysis of science encouraged one to be imaginative, and encouraged one to go for something that was very rigid and therefore empirically disprovable. He always gave preference to the more testable theories. We saw certainly in those days and I think still to some extent today that an evolving universe model can always explain almost any results by ascribing them to the evolution in time of the galaxies or other features. The fact that this way out is not available to steady state makes it such a testable theory. Of course it is the natural fate of testable theories to be disproved, that is what science is all about, but it is much better to be testable than to be untestable. (I always tell my students that if my weather forecast for tomorrow is 'it will be wet or dry, warm or cold, cloudy or fair', they won't be able to disprove me but I haven't helped them a lot.) And so we were proud of the rigidity of our theory and the fact that it could be tested. Of course all this happened forty years ago. Let me finish with a remark that relates to McCrea's final point (and pat myself just a little on the back) when he regretted the great people, such as Milne (to whom I think he should have added, in a slightly different context, Schrödinger), who spent so many years of their lives digging away in a particular direction where nobody was following them at the time and nobody has followed them since, which is why he regards such work as fruitless. One shouldn't keep hacking away at things for too long. I can proudly say I haven't written a paper in cosmology for well over three decades.

Discussion

North:

I feel rather ashamed to make such a trivial observation after such a splendid lecture, but my point is simply that cosmology is not only created by the interaction between the cosmologist and the universe, but by the interaction between cosmologists and other cosmologists. This is really a warning to historians who might consider that because an idea was published, therefore it was somehow in the public domain, and so must have influenced people. I was reminded of this during the morning session, for many of the wonderful ideas then brought to light had not, in their day, been particularly influential. An example of an extremely good book that was slow to have any influence is Otto Heckmann's *Theorien*, first published during the war. It has since been reprinted, but nearly thirty years ago, when I was writing my history of cosmology, it was almost impossible to find a copy in England. I eventually found one in the library of the RAS. This is just one example of the danger, but I could produce many others.

Sciama:

I wonder if I might make a comment. I am slightly worried about the advice that both McCrea and Bondi have given you all not to hack away for years and years and years because I do have to remind you that in discovering general relativity Einstein had to wait at least for nine years without anyone supporting him. When you are successful it is all right but when you aren't it isn't.

Bondi:

I was not suggesting one should change direction every year. A decade I would regard as tolerable. When it becomes twenty years then it begins to be bad.

Sullivan:

Just a brief comment on Heckmann's book. It was actually that book that Van der Hulst read as a graduate student and then based his first suggestions for the use of radio astronomy in cosmology, in 1944, on which he had learnt from it.

McCrea:

I thought it was Einstein himself who brought the concept of radar distance measurements into astronomy.

Ehlers:

I remember that the mathematician Constantin Caratheodory in Göttingen made an axiomatics of special relativity where he used optical means solely and where he emphasised that the concept of distance should be considered as a derived concept, derived effect from radar reflections that was a little known.

Bondi:
I certainly learnt about radar distances from Milne, who did not give a reference to anybody else.

Wagoner:
I was just wondering if you were aware at the time of the Alpher–Hermann prediction of the microwave background temperature?

Bondi:
No. No.

11

Personal recollections: some lessons for the future

William McCrea

11.1 Introduction

Contributions to this book include one or two systematic accounts by professional historians of the history concerned, such as I am not qualified to present. It happens, however, that I knew personally several of the creators of modern cosmology. Sadly, there are few left who did. Therefore I feel an obligation to record something about them as persons and about the way in which their involvement came about. Also I have one such obligation in particular because the organizers have invited me to speak more especially about Georges Lemaître. It seems possible to meet such obligations in the space available only if I restrict personal recollections to some concerning, in about the years 1929 to 1935, developments leading up to what is now generally seen as Lemaître's major contribution and regarded as the start of 'physical cosmology'. After comments on the scientists involved (in order of date of birth), I mention three phases in those developments before enlarging somewhat upon Lemaître's ideas. I offer a few general reflections prompted by the narrative.

11.2 Persons

Sir Joseph Larmor 1857–1942 is the scientist with whom it seems natural to start such a narrative. In Cambridge from 1903 to 1932 he held the Lucasian Professorship of Mathematics – that held by Isaac Newton 1669–96 and at the present time since 1979 by Stephen Hawking. All the other British scientists who figure in my account were pupils, or pupils of pupils, of Larmor, and they continued to value his advice on scientific matters. Recalling Larmor at this point may therefore impart some coherence and perspective to this account.

Although Larmor made no claim to be any sort of astronomer, he did maintain an interest in current work in astrophysics, and he occasionally published interesting comments in the *Observatory* magazine. Up to his retirement he was a man of wide influence. In mathematical physics he had done much to prepare the way for the great developments that he saw occurring during his later years. But he found himself out of sympathy with them when they came. In Cambridge it used to be said that Larmor believed real science to have come to an end with Lord Kelvin. Unhappily, he felt that the new generation of practitioners had no use for him; he made himself a recluse, and he was disinclined to receive visits by these persons (but after he died I found that he had fairly recently nominated me to succeed him in a certain rather nominal office).

Willem de Sitter 1872–1934 was Professor of Astronomy in Leiden Observatory from 1908, and Director from 1918, until his death. He was an astronomer of remarkable versatility and rare genius. He was a deeply respected international figure, and he served as President of the International Astronomical Union 1925–28. De Sitter was a good personal friend of all the people I am about to mention in this section. He was a man of gentle graciousness. He carried a wonderful air of serenity, and yet he got through a prodigious amount of work, while being far from physically robust.

De Sitter had early been interested in any possible significance for astronomy of the theory of special relativity (SR). Then when in 1916, during World War I, Einstein published his main great paper on general relativity (GR), de Sitter was one of the first to grasp its importance. About that time in the neutral Netherlands, he and colleagues in Leiden were able to receive visits by Einstein. De Sitter was then the very first to make the work known to the outside world. This he did by sending a copy of Einstein's paper to Eddington, and by himself writing in 1916–17 three papers for *Monthly Notices* of the Royal Astronomical Society (de Sitter 1916a, 1916b, 1917). These were partly expositions of Einstein's work and partly his own investigations of its consequences for astronomy. In the latter regard de Sitter had an important part in stimulating the British Astronomer Royal, Sir Frank Dyson, to organize the eclipse expeditions in which in 1919 were made the first measurements of the gravitational 'bending' of light. They obtained results in good agreement with the predictions of GR.

In his third paper on the subject, de Sitter propounded the 'de Sitter universe', which historically became the reason for the existence of the present book. Later he discussed the classification of cosmological models, and in 1932 he discussed in particular the 'Einstein–de Sitter model' about which we shall have more to say.

Sir Edmund Whittaker 1873–1956 was Royal Astronomer of Ireland 1906–12 and Professor of Mathematics in Edinburgh University 1912–46. He was extraordinarily versatile. Nowadays he is perhaps most remembered for his great *History* (Whittaker 1910, 1951, 1953). Among many other things he had expounded much of the mathematics that turned out to be needed in the formulation and application of quantum mechanics and wave mechanics. He too had been a pioneering expositor of GR; we shall see that he went on to make a crucial advance required in the interpretation of the predictions of relativistic cosmology. As a young Cambridge 'don' he had lectured to Jeans and Eddington; some forty years later he edited for publication Eddington's posthumous work *Fundamental Theory* 1946.

Whittaker was a man of sparkling wit and charm with an immense capacity for getting acquainted with people and their work. It can be no exaggeration to say that he had some personal contact with a majority of the 600 or more 'authors cited' in the second volume of his *History*. In fact it was his own appreciation of the uniqueness of the scope of this acquaintanceship that, he felt, laid him under the obligation to write that book.

Sir James Jeans 1877–1946 was one of the most notable pioneers of modern astrophysics. Also he was 'Physical Secretary' of the Royal Society 1919–29 so that he was responsible for attending to the publication by the Society of much of the famous work of the period including that of Rutherford, Bohr, C. G. Darwin and particularly of Dirac. This was only one of Jeans' public-spirited services to British science. His manner gave the quite misleading impression of his being somewhat arrogant. In fact he was a man of much kindness and generosity.

Without professing to make any direct original contribution to cosmology, in the early days Jeans participated effectively in many discussions of the 'expanding universe', and his own work proved to be relevant in numerous ways. Indeed during the symposium on which the present book is based, it was impressive to note how many times there were explicit references to work done by Jeans.

Albert Einstein 1879–1955 was the only one of those listed in this section with whom I had no personal contact. Therefore I may not write about him as a person under the heading of 'personal recollections'. But it would be absurd not to include him in the narrative, since it was his paper of 1917 entitled 'Cosmological Considerations on the General Theory of Relativity' that initiated the whole of the history with which this book is concerned (Einstein 1917, 1923). His direct contribution will be mentioned later. Here it may be useful to try briefly to analyse the reasons for Einstein being concerned at all with cosmology, for he never evinced any strong interest in the actual cosmos as observed by astronomers.

Any problem in classical physics, or in special relativity, deals with some well-defined material system that is postulated to exist in a given space-time. This space-time is indifferent to the presence of the matter. In GR, on the other hand, any material exists only as a feature of the space-time.

Until we know the geometry of the whole of the space-time concerned, in GR we do not know completely what material system we are discussing. Thus *any* problem in GR is concerned with what, according to the theory, could exist on its own as a whole 'universe'. In 1916–17 this situation was all very novel and Einstein wondered whether it would present difficulties for his theory.

At the time Einstein did not explicitly describe the situation in quite those terms. He came to it, I believe, by worrying about the status of 'boundary conditions' in his new theory. In particular, it was not clear to him what in GR might correspond to boundary conditions 'at infinity' as invoked in classical theories. Einstein then saw that one way of avoiding any problem about boundary conditions was not to have a boundary. At any rate, so far as a spatial boundary is concerned he could avoid having a boundary by treating space as the 3-dimensional surface of a *cylinder* in 4 dimensions. This is spatially closed so that there is no spatial infinity. The time direction is taken to be that of the generators of the cylinder. So every spatial section is the same and so the system is *static*. This was thought to render the infinity in time unobjectionable.

According to GR as Einstein had at first formulated it, the model would, however, have zero material content. The theory could be retained, and yet the model could have non-zero content, if Einstein slightly modified his definition of the 'matter tensor' to include 'cosmical terms' having non-zero constant coefficient Λ, say. In due course it was appreciated that this implied, as regards this particular model, the postulation of a 'cosmical repulsion' to counterbalance the self-gravitation of the matter present.

Evidently Einstein considered that the model was qualitatively sufficiently like the actual cosmos for him to claim that his theory admitted one physically acceptable model that encountered no difficulty occasioned by infinities and boundary conditions. So long as there was one, he could be satisfied that no such difficulty was inescapably inherent in his new theory. That was really all he wanted to know. At that stage, Einstein was interested in what the actual universe had thus been able to tell him about his theory; he was not greatly interested in what his theory might tell all of us about the universe. So in 1917 he explicitly did not discuss any quantitative comparisons with astronomical measurements.

A few comments: At the end of the 1917 paper Einstein remarked that $\Lambda \neq 0$ was required only in order to admit a static cosmological model. It is well known that he himself later declared the introduction of Λ to be the worst mistake he had ever made. After 1917 he contributed no further considerable advance in cosmology, although in 1932 he joined de Sitter in commending the 'Einstein–de Sitter cosmological model' as a useful norm for discussing cosmological theory (Einstein and de Sitter 1932). Also in editions from 1946 onwards of his small book *The Meaning of Relativity* Einstein had a well-informed critical summary of the GR theory of the expanding universe. Finally, although Einstein had earlier been impressed by the writings of E. Mach on the meaning and origin of inertia, there seems to be no reason to believe that they had much immediate influence upon Einstein in the formulation of GR or in the derivation of a cosmological model. It used to be supposed that Einstein believed his cosmological model to be in some way a realization of a 'Mach principle'; evidence seems lacking.

Sir Arthur Eddington 1882–1944, in consequence of the action by de Sitter mentioned above, became the most effective of all expositors of GR. In 1919 he successfully led one of the two British eclipse expeditions that first observed the gravitational 'bending' of light and found its amount to be in good accord with Einstein's prediction. In 1923 he published his classic *Mathematical Theory of Relativity*. Therein he described the cosmological models of Einstein and of de Sitter; he discussed de Sitter's prediction of a recession of the galaxies, and the modest measure of empirical support then available for its occurrence in the actual universe. At the time he was hard at work also on stellar structure; his great *Internal Constitution of the Stars* appeared in 1926. Eddington did usually have two lines of research going at the same time, as well as some book of general exposition. About 1928 he was thinking more seriously than hitherto about 'fundamental theory'; it seems to have been in this context that, apparently sometime in 1929, he asked his graduate student G. C. McVittie to study the stability of the Einstein static universe. Shortly afterwards an embarrassed Eddington had to confess to forgetting having had from G. Lemaître a reprint that demonstrated implicitly the instability of the Einstein model. Thereupon Eddington published a discussion of the astronomical significance of this property. Also he arranged for Lemaître's paper to be republished in translation by the Royal Astronomical Society (Lemaître 1931a).

As part of the history of cosmology we shall later resume this account. At the moment, in regard to Eddington himself, a paradox to be noted is that, while he was one of the few professional astronomers figuring in the early days of this history, his personal interest was not primarily in the astronomy of the subject, but in its physics. As indicated above, he was developing a consuming interest in the foundations of physics. Here the meaning and values of the constants of physics are a basic topic. According to Eddington's view of GR the cosmical constant Λ is a very important one of these. Its expected value was such that it could have direct observable consequences only on the cosmic scale. Eddington's interest in the expansion of the universe was therefore in the possibility it offered of determining that value. Thus another paradox was that whereas Eddington looked to the expansion to enable him to evaluate Λ, Einstein hailed it as allowing him to discard Λ, since he had introduced it only in order to admit a universe that was not expanding!

Personally Eddington gave the impression of being agonizingly shy with people he did not know well. 'Shy' was not, however, an accurate description. His upbringing simply left him at a loss to know how to get going with such people. If he faced an audience, large or small, without having prepared every word of what he had to say, he was hesitant and floundering. But if he had prepared adequately he was apparently at ease in front of an audience of any size, and he seemed to enjoy entertaining them with rather crafty witticisms. He was unassuming and un-self-seeking. At the same time he behaved as though he had special revelation of the secrets of Nature. He certainly placed much trust in his own scientific intuition. But his intuition did not cause him to avoid mathematics; it could lead him to the mathematics he needed. In his fundamental theory, however, there were gaps in the arguments that he apparently bridged by intuition, but that made it impossible for others to follow him all the way. All who had worked with Eddington retained profound respect for him and always spoke of his personal kindliness.

Edwin Hubble 1889–1953 was one of the great observational astronomers; he played the key part in establishing the empirical basis of the history under review. It was predominantly his work that led to the discovery that any remote galaxy appears to be receding from us with speed that, to a good approximation is a constant (Hubble constant H_0) multiple of its distance, where H_0 is independent of the distance and direction to the galaxy. This is the phenomenon of the apparent expansion of the universe. Hubble began to publish his work on the subject in 1929; in much of it he had the collaboration of M. L. Humason.

Astronomers determine distances of galaxies by measuring the apparent brightness (or some other feature) of objects in them which they identify as 'standard candles' of various sorts, or they may seek to treat some galaxies as standard candles themselves. Almost by definition, Hubble was dealing with the most difficult observations possible, since he was observing the most remote objects he could find in the universe. When he measured a feature of some object no one seemed to dispute his value, and apparently he was never shown to have been seriously mistaken. When Hubble identified the object as a particular standard candle, nobody disputed that either – until the astronomers W. Baade and A. D. Thackeray independently did so about two decades later. Then it soon became clear that Hubble's own value of H_0 had been of order 10 times too large. It has to be said, however, that now sixty years after Hubble's original determination, astronomers are still uncertain about the value of H_0 to within a factor 2. Also from, for instance, the closing pages of Hubble's classic *The Realm of the Nebulae* (1936, reprinted 1982) we see that he was far from dogmatic about his results – unfortunately he did not himself recognize which features were most vulnerable.

When Hubble visited Britain – and presumably other countries – he showed himself to be a friendly and attractive individual; rightly or wrongly, on his own ground in the USA he gained a reputation for being somewhat domineering. Anyhow, he was the first man to observe what appears to be a concerted behaviour of the whole universe – cosmologists are still profoundly unsure about how the concert came about.

Monsignor Georges Lemaître 1894–1966, Belgian cleric and mathematician, was the greatest pioneer of modern cosmology as a study of the physics of the actual universe in the large. As we have seen the relevant interests of Einstein and of Eddington were primarily in the basic physics, not primarily in the observed astronomical universe. It was only after de Sitter knew about Lemaître's work that he showed interest in the astronomical applications of developments from his own work of 1917. When in 1927 Lemaître published his first fundamental paper in the field, he was quite unaware of the papers of 1922, 1924 in which A. A. Friedmann (1888–1925) had presented almost the same mathematics. But Friedmann, a meteorologist in Petrograd, had not been in an environment to suggest possible astronomical developments; tragically he had died before he could know the great significance of his contribution. Lemaître, on the other hand, after having had an early interest in relativity theory, had had the advantage of spending the academical year 1923–24 with Eddington in Cambridge. After that he had travelled widely in Canada and USA where he had discussions with theorists like L. Silberstein and with many leading astronomers. But the outcome of all this in Lemaître's work was the fruit of his own genius. Indeed after his 1927 paper was published he encountered positive discouragement from Einstein and de Sitter and only belated – though effective – recognition from Eddington. The sequel will be discussed in a later section.

Lemaître had an intellect of great power and versatility. From 1911 his lifelong academic attachment was to the University of Louvain. He graduated in engineering in 1913 and in mathematics in 1919 and gained a doctorate in mathematics in 1920. From 1925 he held a succession of Professorships in Louvain. After about 1934 he made few attempts to contribute further to cosmology although he published many reviews and discussions; he hoped for some observational validation of his concept of a big-bang origin of the universe. In 1965 his successor in Louvain, O. Godart, was able to tell him about the discovery of microwave background radiation which appeared to be evidence of the kind he had awaited – until he was on his deathbed.

Lemaître probably regarded his work in cosmology as his most significant contribution to science. But he devoted far more time and labour to other valuable mathematical endeavours. He did much good work in celestial mechanics and on the motion of charged particles in the geomagnetic field and in galactic magnetic fields. Many of the problems he and his pupils encountered could be solved only by numerical methods. He had an enduring interest in such methods. They became the dominant interest of his later years when he did pioneering work in the exploitation of electronic computers and in setting up a computer laboratory in Louvain.

Until his last illness, Georges Lemaître was a man of robust vigour. He appeared to be the complete extrovert; he had a stentorian laugh which was readily provoked. To some extent, however, I think all that appearance was a protection for a sensitive personality. He was a man of courage. He was in the Belgian army throughout World War I and was decorated for his services. In World War II he strove to maintain some vestige of academic and scientific activity in Louvain even when food rations had fallen to starvation-level. After the war, he was awarded the highest national honours in recognition of his courageous leadership. In Louvain he is still remembered with the deepest respect and affection.

Edward Arthur Milne 1869–1950, a Cambridge mathematician who became the first Rouse Ball Professor of Mathematics in Oxford, was another of the pioneers of modern astrophysics. He is mentioned here mainly as a representative of those who in the early days of talk about the expanding universe, explicitly or implicitly, considered it necessary to pose a certain question. When Hubble observed the apparent expansion of the universe, taking note of the predictions of de Sitter, of Friedmann and of Lemaître, people said in effect: 'hitherto GR had predicted only minute emendations of classical gravitation theory that seemed scarcely worth all the fuss, but now we see that this tremendous concerted behaviour of the entire cosmos was predicted by GR, and by GR alone, and surely this is the overwhelming vindication of GR'. The question was, is it? If so, what is in GR that enables it uniquely to treat the phenomenon? If not, how else can the phenomenon be treated?

Milne started from a very simple idea which amounted to this: if somebody's string of beads breaks and spills in the middle of the dance-floor, the beads start rolling in all directions; one second say, after the start, a bead that is *n* metres from where it fell has been travelling at speed n metres per second. So the beads that have been receding from where they fell have been doing so according to a 'Hubble law' of speed proportional to distance. Not only so, but if the speeds are about uniform any one bead would 'see' all the rest to be receding from itself according to that same law. So we should have a model of the expanding universe of galaxies. Milne was therefore inclined to say that the expansion of the universe was basically nothing more than this exceedingly elementary kinematic phenomenon. But of course there are several important features that must be put right before we could say that we have in principle a satisfactory model: a) We should use 'relativistic' kinematics. b) We should want every 'fundamental' observer in the system to see the same history of the rest of the system. c) We should take account of the mutual gravitational attraction of the parts of the system. d) We should want to know how the system got its kinetic energy.

Almost all cosmologists said that when they took account of all these requirements in the best ways they knew how, they ended up with precisely the Friedmann–Lemaître range of models.

Milne did not accept this. He believed that by developing his model so as to satisfy certain almost self-evident 'cosmological principles' he could demonstrate that there had to exist in it, for instance, the phenomenon of gravitation satisfying certain 'laws', and so forth. Thus he believed he was on the road to showing that the actual Universe is the only universe that can exist logically. He did construct an ingenious system of *Kinematic Relativity* (1948). Unfortunately, but inevitably, it contained unacknowledged assumptions. Ultimately it achieved nothing in the general advancement of cosmology. But it did involve searching discussions of particular topics such as time-keeping. Also Milne discussed one possibility that has attraction for some other physicists: if $\tau = t_o \ln(t/t_o) + t_o$, where t_o is a disposable constant, and if τ is the time-scale in which statements about Newtonian mechanics and gravitation take their standard forms then, Milne asserted, t is the time-scale in which statements about electromagnetism and photons take their usual forms. Ideas of this sort do form part of 'cosmology in retrospect'; so far they have not proved fruitful.

Milne's contributions to astrophysics proved to be of great value and importance. He was a man of brilliant intellect, and it is distressing to have to make adverse remarks about his work on relativity and cosmology, which he regarded as so much more significant. In common with others of my generation I owe a great deal to Milne's inspiration and advice and to his personal kindness. All the individuals discussed in this section were concerned to find 'meaning' in the universe that they were trying to understand; Milne was perhaps the most concerned of all.

11.3 History

Most of the developments up to about 1935 have been mentioned in the course of the preceding section in connection with the workers chiefly concerned in producing them. We now summarize them in chronological order; in the present short-term treatment three phases may be discerned.

11.3.1 *Theory phase*

In 1917, after the introduction of 'cosmical terms' containing the cosmical constant Λ into the formulation of GR, it yielded the Einstein static cosmological model. In the interval 1917 to 1929 de Sitter, Friedmann, Lemaître separately produced reasonable non-static models. Eddington and others discussed de Sitter's model – which had originally purported deceptively to be another static model – and its possible relevance to the radial velocities of spiral nebulae observed by V. M. Slipher. But until about 1930 these workers ignored the work of Friedmann and of Lemaître. Einstein was aware of that work; he became convinced of its mathematical validity, but not yet of its possible significance for cosmology.

Early in 1930 Eddington showed – or showed that Lemaître showed in 1927 – that the Einstein static universe is unstable. According to GR, therefore, the universe cannot remain in a static state; it must be expanding (or contracting, or conceivably be in some more complicated non-static condition).

11.3.2 *Observation phase*

The important pioneering observations by Slipher around 1920 could not themselves indicate any systematic behaviour of spiral nebulae because at the time there were no observational determinations of distances of these objects. In fact Slipher's observed redshifts, interpreted as velocity-shifts were just beginning to be taken as one of the indications that such objects are certainly not permanently attached to our own Galaxy.

In the mid-1920s Hubble himself was the first to measure distances of spiral nebulae – as he preferred to call them rather than 'galaxies' – even to within about what are now accepted as the correct orders of magnitude. Then by 1929 he found he could relate the measured redshifts in the spectra of the nebulae to his distances. So that year he was able to start announcing results in support of his velocity-distance 'law' of expansion.

Hubble and his colleagues were able to claim further that observation did not invalidate the supposition that, so far as they could observe it, the universe in the large is homogeneous and isotropic.

Quantitatively, Hubble's observations gave the new cosmic parameter, the Hubble constant H_0. He and his colleagues found quite persistently that their measurements led them to the approximate value $H_0 \approx 550\,\mathrm{km\,s^{-1}\,Mpc^{-1}}$.

As has been indicated, astronomers at the time were profoundly impressed by the discovery of such concerted behaviour of the universe in the large. Because of the theoretical developments outlined above, and because of Slipher's earlier observations seen in the light of those developments, astronomers could not be said to have been taken by surprise. But they did appreciate that they were participating in a second 'Copernican revolution' – the acceptance of the demonstration that our own Galaxy has no special status in the universe of galaxies.

I do not recall that there was a rush to try to fit a particular Friedmann–Lemaître model to the observations. The immediate interest was in the indication from Hubble's results that the universe had been in a very congested state apparently no more than about 2 billion (2×10^9) years ago.

Evidence from radioactivity had been indicating ages of the oldest rocks on Earth of about 4 billion years. Both these times were clearly subject to revision; what impressed astronomers and geologists was that they were of the same order. So these workers were ready to suppose that a few billion years ago the universe, composed of much the same constituents as at present, went through a congested state. This would have been favourable for the occurrence of the collisons between stars that were then deemed the most likely requirement for the formation of planetary systems.

At the time astronomers seemed to believe that the Friedmann–Lemaître models were too simplified for their application to be pressed very far. Were the inference about a congested state valid, it was probably asking too much of the models to seek to learn from them how the universe, or maybe just our part of the universe, got into that state. Certainly, nobody at the time seemed to expect the models to reveal anything about the *creation* of the universe or its earliest moments.

Here it ought to be recalled that reportedly it was not until 1930 when Einstein visited California, and heard from Hubble himself about his observational results, that Einstein was ready to accept the idea of the universe being not in an effectively static state. His conversion seems to have been swift and complete so that he became perhaps too eager to reject once and for all his cosmical constant Λ.

11.3.3 *Exploratory phase*

Only a comparatively very small number of scientists were directly involved in the history here briefly recounted. To any one of them developments followed one after the other in a manner that was certainly exhilarating but that seemed natural and even inevitable. By contrast, to an interested but not involved scientist of say 1915 who watched these developments of just the ensuing fifteen years from the sidelines, they must have come as a string of ever more startling astonishments. In 1915 he would know that to astronomers the Galaxy was the universe. He would begin to hear about a new theory of space-time and gravitation that contemplated this universe as filling finite but unbounded space, and as obeying a law of gravitation that could be expressed only in a complicated mathematical form, but evidently making predictions differing from Newton's by only minute amounts. He might be aware that one of these predictions was supported by the evidence in the only known case in which the difference from Newton was then detectable – the advance of the perihelion of the planet Mercury. In 1919 he would share in the general excitement about British astronomers at the solar eclipse of that year checking another of Einstein's predictions of a small 'bending' of light in the gravitational field of the Sun. But our scientist would gather that what ought most to impress him at that stage was that Einstein had produced a theory of gravitation that in application was so very close to Newton's and that nevertheless was fully 'relativistic'. This was apparently far more significant than any new observable consequences.

Our friend would quite soon become aware that astronomers were being convinced that spiral nebulae are indeed systems of the same species as our Galaxy. This meant that in the course of a few years the astronomers' universe had expanded from one galaxy to an innumerable multitude of galaxies. Then about 1929 he would learn that this whole system is observed to be apparently literally expanding – every galaxy receding from every other remote galaxy. Then he would be reminded that this stupendous phenomenon had been predicted by general relativity theory (GR). Thus the theory that a few years earlier had seemed to be in practice no more than an infinitesimal advance upon Newton's was suddenly seen to predict the behaviour of the entire Cosmos – something that had always been regarded as quite beyond the scope of Newtonian treatment.

If about 1930 the onlooker asked the astronomers what it all implied then, as we have seen, all they could tell him was that the whole of what they could observe of the universe apparently must have been in some very congested state a few billion years ago. At the time they did not encourage speculation about the nature of that state.

If the onlooker wanted to know what next in the way of observation the answer could have been so disturbing that it claims special notice here.

11.3.4 *Interlude: big science*

What the world had witnessed was the start of big science. This was science depending inescapably upon the employment of unique large equipment that could not be equalled or excelled within normal funding. Hubble's awesome discoveries depended upon the Mt Wilson 100-inch telescope being there for his use, and he was using it to the limit of its capability. These discoveries were the greatest possible vindication of its construction.

Hubble's work is also, however, the classic demonstration of the drawbacks of such a situation. His small team had the monopoly in the field. His work was superbly conceived and executed and he deserves the greatest credit for his wonderful discoveries. But he was making mistakes of a sort that had crucial effects upon the development of astronomy and cosmology throughout almost the whole of the next four decades. He was not observing the sort of standard candles he thought he was.

It has to be said for one thing that other astronomers were far too ready to accept Hubble's numerical results without query. For another thing the mistakes were actually discovered only after two other unique telescopes had come into operation. One was the 200-inch Hale telescope at Palomar as used by W. Baade; the other was the 74-inch Radcliffe telescope in Pretoria (the largest at the time in the southern hemisphere) as used by A. D. Thackeray. Their results became known about 1952.

Our onlooker of 1930 knew nothing of all this. Astronomical friends would merely assure him that Hubble and Humason were doing their utmost to extend the scope of their observations. In those days he would entertain no suspicion that his friends might be wrong to accept everything that Hubble told them.

11.3.5 ***Exploratory phase*** (resumed)

After the initial excitement around 1930 the situation became generally frustrating. For the reasons just indicated, the general body of observers could do nothing. Without more observational information there was nothing further for theorists to work on in the main development of cosmology. Consequently somewhat auxiliary issues came to be explored.

As was mentioned in connection with Milne's development of kinematic relativity, people wanted to know if, and why, GR seemed to be so essential. Perhaps the best description of the GR model is that it is the exceedingly simple kinematic model from which Milne started when this is made relativistic and when gravitation has been taken into account. GR as such is needed because it supplies the only way we know in which to satisfy these two requirements consistently.

The introduction of the cosmical terms having the constant Λ as a factor into GR is equivalent to postulating the existence of 'cosmical repulsion'. An inertial observer in the space-time would regard any test-particle in his vicinity as being subject to a repulsive force per unit mass away from himself proportional to Λ times its distance, in addition to any other force acting upon it. This phenomenon can be automatically taken into account in the Friedmann–Lemaître treatment. That being noted, it could be said that it soon came to be recognized that in predicting the expansion of the universe GR had employed only physical concepts that were already familiar. It should be noted too that GR could be said to predict expansion whether or not cosmical repulsion was operative, i.e. whether or not Λ is zero.

This situation was confirmed in another way. In 1934 Milne and McCrea showed that Newtonian mechanics and gravitation, with only the addition of a simple convention for the treatment of an unbounded system, led to precisely the Friedmann–Lemaître equations for a pressure-free cosmological model, with or without cosmical repulsion. That is to say every locality of a model universe would behave in the same way according to classical mechanics as it would do according to GR. The ways in which the localities are pieced together to make a universe in the large are different in the two treatments since classical space-time is given independently of its contents while the geometry of space-time in GR is determined by its particular contents. So the authors concerned had no intention of abandoning GR! All they suggested was that quite a lot of cosmology may be quite well understood in classical terms. In particular, we saw how Einstein had had difficulty when he tried to find a static cosmological model in conformity with his original presentation of GR. We now see that this is matched by classical theory; any 'natural' classical model must be non-static.

We can, of course, reverse the argument and say that by trying our best to do cosmology by classical methods we learn to appreciate the need for the GR type of treatment.

Just one example of this will have particular relevance a little further on. Most GR cosmological models of interest, taken literally, start from a singularity. This can be looked upon as the creation of the (model) universe. All the matter in the model, with the energy needed for the expansion, emerges from that singularity. It is impressive that GR produces this feature in such a simple way. This is something that is not well matched in a Newtonian treatment. We shall return to it later. But one reason for mentioning it here is that I believe it was Milne who, having come across a particular example of the feature, was probably the first to venture to suggest that physics might have to take note of creation! It might have been given more attention at the time had not Hubble's value of the Hubble constant implied an unacceptably recent date for such creation.

In the early days of GR its predictions were compared with observation simply by looking for what appeared to be the best match with the formalism of the classical treatments of what could be regarded as the same physical phenomena. Rather more precisely, the GR results might be transformed to a coordinate system locally like that of special relativity (SR), and then the predictions could be interpreted as in SR. Such procedures failed in dealing with regions of space-time over which 'curvature' had significant effects, as was generally required in interpreting predictions of cosmological models.

In 1931 Whittaker was, I thought, the first explicitly to point out in quite general terms the need, in GR models of physical situations (not necessarily with any special reference to cosmology), to express observable quantities as *invariants*, i.e. quantities independent of any particular coordinate system. For example, astronomers measure the distance of a remote radiation-source by observing its apparent brightness, size or motion. Followed by H. S. Ruse, A. G. Walker and others, Whittaker derived invariant expressions for such observed quantities. After I had recalled this, colleagues have been kind enough to point out that in 1928 G. Y. Rainich derived an invariant expression for an example of gravitational redshift, and V. Fréedericksz and A. Schechter derived expressions for aberration and parallax in Friedmann cosmological models. I happen to have heard Whittaker announce his first work on the topic and I am convinced that he was unaware of any earlier work, and that his mathematical approach was altogether more general.

Such expressions are not themselves susceptible to observational test. Only predicted relations between any two or more such can be tested. About 1934, mainly for Friedmann–Lemaître cosmological models, R. C. Tolman, McCrea and others derived *observable relations*, e.g. the relation between apparent magnitude and spectral redshift, that between apparent magnitude and number-density, of galaxies in the models, and so forth. At the time there were insufficient observations for detailed testing of the predicted relations.

Eddington published in 1933 his small book *The Expanding Universe*. This was widely treated as a popular account of the astronomical discoveries. This was not at all Eddington's intention. Actually he was giving a serious preview of what became his *Fundamental Theory* (1946). At the time he declared his only interest in the expansion of the universe to be his expectation of its supplying him with the empirical value of Λ. So this work did not directly advance the study of cosmology.

However, it had one potentially very interesting connection with cosmology. This was in regard to a topic that began to attract attention at the time and has done so intermittently ever since – that of 'Large Numbers'. Professor J. Ehlers tells me that H. Weyl discussed it in 1930 in an address to students in Göttingen, which may have been its first mention in modern times. But Eddington was beginning to write of it about the same time. His suggestion was that all the dimensionless constants of physics are either of order not very different from unity or else some such number multiplied by a simple power of a number N, say, equal to about 10^{79}. Eddington believed that the constants other than N would have some fairly simple mathematical significance in a fundamental theory of physics. He thought N would have to be in some way characteristic of the cosmos as a whole, actually the total number of particles in the universe or the number that is logically found to result from any exercise designed by physicists to 'count' those particles.

A little later Dirac (1937, 1938) had much the same general ideas. But he had one big difference in taking $N^{1/2}$ to be the Hubble time T_0 expressed in 'natural' (i.e. 'atomic') units. Thus N was dependent upon the cosmic epoch, not a constant of physics. Since powers of N appeared in others of the dimensionless 'constants' of physics, Dirac's hypothesis required at least one other 'constant' to be also dependent upon cosmic epoch. Dirac obtained consistent results by simply taking the gravitation constant γ to be inversely proportional to the 'age of the universe'. The pursuit of these ideas led Dirac also to propose the use of two time-scales. But his scales were effectively the Milne scales interchanged! This circumstance doubtless caused either proposal to be taken a great deal less seriously than it would have been had it been supported by the other.

In the times we speak of, these notions were naturally highly exciting because they proposed intimate connections between the new study of the cosmos on the largest possible scale and the laboratory study of matter on the smallest known scale. But it has come to be a matter of history that they have hitherto had negligible influence upon the development of cosmology or of laboratory physics. Nevertheless, they seem to have had a lasting effect upon physicists' attitude to the foundations of their science. Most physicists probably now believe that something along the general lines of Eddington's thinking must be possible even though no one has yet had any definite success in achieving it. Also there has been in recent years another line of thought in this whole field, that arising from the presentation of various *anthropic principles*. This can be traced back to the ideas of Eddington and Dirac, which is one reason for recalling those ideas here. If these principles are as significant as they appear to be, it is not clear how they affect the prospect of achieving something like what Eddington set out to do. This seems to depend upon what we are to understand by the uniqueness of the experienced universe.

There is one rather unhappy detail in the history that has to be mentioned because it probably had a considerable effect at the time. In the course of his book *The Expanding Universe*, Eddington gave the impression of predicting on purely physical grounds the value of H_o and recovering almost exactly that which Hubble claimed for his astronomical determination. Undoubtedly this helped to delay the discovery of Hubble's mistakes. Eddington's reasoning was not convincing, but even so the irony of the situation as later noticed by Slater (1954, 1957) was that had Eddington carried out his calculation in what Slater regarded as a more self-consistent manner he would have obtained a value of H_o smaller by a factor 4/9. This would have given him a value of H_o close to Baade's observational value of about 1952. This had been the first determination to cast doubt upon Hubble's value; it is well known that the reduction needed is actually between 1/5 and 1/10. But had it not been for Eddington's own inconsistency, instead of winning credit for confirming one wrong value, he could have had credit for predicting another still wrong (but less wrong) value!

11.4 Physical cosmology: Lemaître around 1933

Lemaître was a scientist of superbly robust common sense. All of us who knew him must ever since have wished we had paid better attention to his ideas. In order to explain why, the following digression is needed.

In GR, Einstein sought a field theory that would be a unification of concepts of space-time, matter and gravitation. It was to be given by a single tensor-field g_{pq} – symmetric, and having appropriate signature.

Space-time was to be the geometry having g_{pq} as metric-tensor. *Matter* was to be described by some derived tensor-field G_{pq}. This was to have zero tensor divergence, because the equations expressing this property were the sort needed as equations of motion of matter possessing density and stress. Moreover, these equations included terms that were interpreted as showing matter so described necessarily to exhibit the property of *self-gravitation*.

Einstein identified such a tensor G_{pq} which naturally is called the 'Einstein tensor'. Then a tensor

$$aG_{pq} + bg_{pq},$$

where a, b are any numerical constants, also has the same property of zero divergence.

When interpreting the metric tensor in physical terms a constant c makes its appearance, and takes the rôle of a fundamental speed. This is naturally interpreted as light-speed in free space. We can treat it as the natural unit of speed, and so we put $c = 1$.

Then in writing out what we are to regard as the equations of motion we find that a plays the part of a simple multiple of γ^{-1}, where γ is the Newtonian gravitation constant. We can put $\gamma = 1$, which then serves to determine a natural unit of mass. So the constant a has been fixed.

If for the moment we forget about the constant b we now have a theory of space-time and gravitation that involves *no physical constants*. For the fixing of c, γ as stated implied no physical assumptions whatever, simply algebraic convenience. The theory is in fact standard GR.

Manifestly this was the way Einstein wanted GR to be. Although he did not say so himself, it acquired a status similar to that of pure thermodynamics. This is another theory that involves no physical constants. Both theories are claimed to be ones to which any physically acceptable matter must conform; they can forbid certain properties; they cannot say what properties actual matter must have.

Digressing briefly, we remark that, although Lemaître also did not make this particular comparison with thermodynamics, he did make another that is very relevant here. Lemaître (1931) argued that

> in atomic processes, the notions of space and time are no more than statistical notions, they fade out when applied to individual phenomena involving but a small number of quanta. If the world has begun with a single quantum, the notions of space and time would altogether fail to have any meaning at the beginning; they would only begin to have a sensible meaning when the original quantum had been divided into a sufficient number of quanta.

This was the beginning of Lemaître's suggestion to regard the start of the universe as the decay of a 'primeval atom' – this terminology being merely a convenient way of expressing his concept.

The view of thermodynamics as a theory of statistical behaviour of particulate systems is very old. The corresponding notion of space-time and gravitation being treated as statistical phenomena has long been the subject of rather casual speculation. Lemaître may be regarded as the first to give specific significance to it. Nowadays his line of thought survives in the widely held concept that 'quantum gravity' must be significant for the very early universe. There follows the corollary that, if macroscopic gravity is a statistical property of quantum gravity, then quantum gravity is expected to 'look' very different from macro-gravity. (End of digression.)

Returning to physical constants, in contrast to Einstein, Eddington thought that GR must incorporate one universal constant that was somehow needed as a basis for measurement. This he took to be in effect the above constant b, which is proportional to the cosmical constant Λ as usually defined. But Eddington's arguments for this were somewhat occult! He certainly did insist upon a very special status for Λ.

Lemaître took an attitude independent of those of Einstein and Eddington. He said, in effect: 'the constant b has got there quite naturally; keep it and compare the consequences with experience; it may be found that agreement is best with $\Lambda = 0$, but we must wait and see'.

More philosophically, he might have said, once any pair of values of a, b has been assigned we have a theory with the same status as that for any other pair; there is nothing sacrosanct about a pair with $b = 0$. He could have pointed out that the whole formulation of GR is arbitrary and tentative – why choose four dimensions? why take g_{pq} to be symmetric? why restrict G_{pq} to contain derivatives of g_{pq} up to only second order? . . . why choose $b = 0$?

Lemaître then considered the range of solutions for $\Lambda > 0$ of the Friedmann–Lemaître equations for the cosmical expansion-factor $A(t)$. This range included the particular case $A = A_E$, a constant, with, say, $\Lambda = \Lambda_E$, which gives the original Einstein static model universe. Then for Λ greater than Λ_E by a relatively small amount he found the solution to be qualitatively as shown in the figure.

The Lemaître cosmological model with this expansion factor has the features:

- It is a closed, homogeneous isotropic universe.
- This universe has the *initial singularity* demanded by GR, which as Lemaître showed, is not peculiar to a model having the symmetries of this particular model.
- *Inflation against gravity* results from the singularity.
- The era of *near stagnation* ensues, in which gravitation and cosmic repulsion nearly balance, resulting in very slow expan-

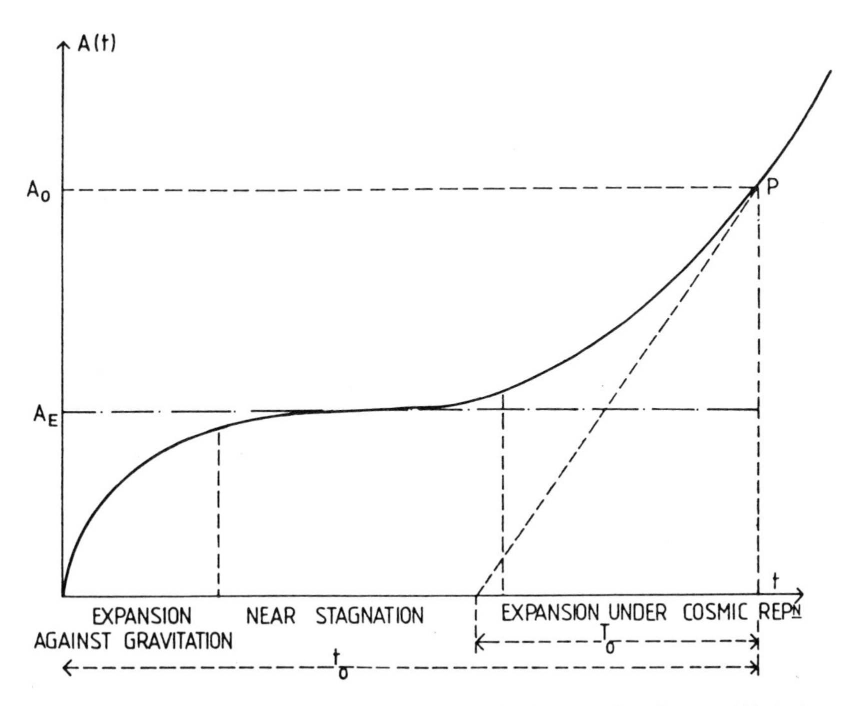

Fig. 11.1 Lemaître cosmological model: graph of expansion factor $A(t)$ (schematic only). Cosmical constant Λ exceeds Λ_E by relatively small amount; $A = A_E$ is Einstein static model having $\Lambda = \Lambda_E$; P is present epoch having $t_0 =$ age of Universe, $T_0 =$ Hubble time. (Courtesy of the Editors of the Quarterly Journal of the Royal Astronomical Society)

sion. The universe is 'coasting' close to the Einstein static state. Lemaître considered that the contents evolve during this era with the formation of galaxies, etc.

- The era of *accelerated expansion* follows in which expansion proceeds under the predominant effect of cosmic repulsion.
- The Hubble time T_o is shown for an observer existing at age of universe $t = t_o$. If conditions are as implied by the figure then $t_o \gg T_o$. The universe is then much older than the Hubble time. This was Lemaître's resolution of the 'age problem' presented by the shortness of the Hubble time given by the work of Hubble himself.

In order to reap the advantage of a long stagnation era the amount by which Λ exceeds the critical value Λ_E must be very small compared with Λ_E. It is interesting that for some cosmological models favoured at the present time the density-parameter Ω must differ from the critical value Ω_c for the so-called Einstein–de Sitter model by an exceedingly small amount compared with Ω_c. Thus Lemaître's concept of a critical parameter

Table 11.1. *Lemaître*'s ideas: their enduring character

Lemaître's concepts; mainly from about 1933	Some counterpart concepts current about 1988
Initial singularity: big bang ('fireworks')	Hot big bang
Retension of Λ	Inflation implies effective (short-lived) Λ
Retension of pressure terms: possible importance of radiation	'Radiation era'
Initial unique quantum unit: 'primeval atom' spontaneous decay gives initial inflation	Quantum gravity deemed necessary in very early universe
Relics of decay produce cosmic-ray background	Relics of energetic photons produce microwave background
Particles from initial decay go to produce stellar matter	'Helium synthesis' produces raw material for first stars
Rôle for random density-fluctuations	Random fluctuations postulated
Galaxy-formation inferred during near-stagnation: fluctuations produce regions where gravitational attraction overcomes cosmic repulsion	Galaxy formation following onset of 'matter era', discussed in several ways
Computer program proposed for problem of clusters of galaxies	Clustering of galaxies discussed
Age problem (see text) using $\Lambda > \Lambda_E$	Age problem using $\Lambda = 0$

Problems discussed by Lemaître, and still under discussion, include: the topology of space-time, neighbourhoods of space-time that may be unobservable to us, initial uncertainties and their consequences, rebounds of a 'Phoenix universe', star-formation and the possible rôle of shocked material, the possible rôles for cosmic magnetic effects.

continues to play a rôle, although the particular parameter concerned happens to be different.

Lemaître indeed identified most of what are still regarded as the basic problems in the field, and he recognized many of the concepts still considered as likely to be required for their solution.

In order, as concisely as possible, to display this situation table 11.1 shows some of the concepts advanced by Lemaître, mostly around 1933, side by side with some counterparts in recent or current work around 1988. In addition, the lower part shows a number of topics which he was one of the first to discuss and which others have been discussing at intervals ever since.

Most of the details of Lemaître's own work have not endured; also in some respects his treatment was at best somewhat sketchy. But the speculative element was mostly quite restrained compared with much that is prevalent at the present time!

Lemaître's great achievement in the field was to make it plausible for scientists to expect to understand the *physical history of the universe* in the large. This means, by known or discoverable physical theory, to account for the observed state of the actual universe in terms of inferred earlier states back to something truly rudimentary. This is not to claim that Lemaître's, or our own, picture has much validity. More than anyone else, however, he convinced us that we should go on seeking, and he pointed the way. He himself might have achieved more had his contemporaries grasped better the significance of his endeavours, and responded more constructively.

Some of Lemaître's own accounts of his work are Lemaître 1927 *et seq* in the references. 'Reflections on Early Work on 'Big Bang' Cosmology' (Alpher and Herman 1988), referring explicitly to Lemaître as its originator, appeared after the writing of this paper. The 'Proceedings of a Symposium in honour of G. Lemaître fifty years after his initiation of Big-Bang Cosmology' (Berger 1984) includes biographical material and an updated bibliography.

11.5 Possible lessons from this recital

Heed the right prophets. Einstein, Eddington and Milne may have been greater scientists than Lemaître, and more famous in their day. But on the subject of cosmology and its importance for astronomy Lemaître had more to impart. He talked better sense.

Recognize no monopoly. Hubble was in possession of the field of observational cosmology for about two decades – and a fine pioneer he was. Because of his misidentification of standard candles, however, he did delay much possible progress for about that length of time. It is true, of course, that he used the 100-inch Mt Wilson telescope, and at the time there existed no other fully comparable instrument. But surely for observations of such fundamental importance some other observer could have found a means of making some check. It is difficult, too, to understand why no theorist was prepared to query Hubble's conclusions.

Let cosmologists honour their Humasons. As a professional scientist, Humason's standing was modest. He showed genius, however, in identifying remote galaxies on which Hubble could profitably exercise his observing skills, and in assisting Hubble in doing so. To outsiders, he seemed in his lifetime to win scant attention. He and his like should be assured of a worthy place in the history of man's exploration of the universe.

Avoid getting bogged down. To a considerable degree Einstein, Eddington and Milne may be judged to have misspent much of the time of their later years working on splendid ideas that were getting them almost nowhere! I hope I have done them no injustice, but it does appear that, had they broken away to devote themselves for a while to different scientific pursuits, and maybe returned later to these ideas, they would have done even greater service to their science.

Let students know what their teachers have achieved. I went up to Cambridge only four years after the British work at the 1919 eclipse on Einstein's 'bending of light' in the gravitational field of the Sun. I frequently saw Eddington going about his daily avocations. Nobody told me how he had played the chief part in that historic occurrence. Had somebody done so I suppose I should have experienced a thrill every time I saw him – and I might have wanted to learn some relativity much earlier than I did. It is to be hoped that today's students are better served in such matters.

References

Alpher, R. A. and Herman, R. (1988). Reflections on Early Work on 'Big Bang' Cosmology. *Physics Today*, **41**(8), 24–34.

Berger, A. (1984). *The Big Bang and Georges Lemaître*, ed. A. Berger. Reidel: Dordrecht.

Dirac, P. A. M. (1937). The Cosmological Constants. *Nature*, **139**, 323.

(1938). A New Basis for Cosmology. *Proceedings of the Royal Society*, **A165**, 199–208.

Eddington, A. S. (1923). *Mathematical Theory of Relativity*. Cambridge: Cambridge University Press.

(1926). *Internal Constitution of the Stars*. Cambridge: Cambridge University Press.

(1930). On the Instability of Einstein's Spherical World. *Monthly Notices of the Royal Astronomical Society*, **90**, 668–678.

Eddington, Sir Arthur (1933). *The Expanding Universe*. Cambridge: Cambridge University Press. Reissued (1987) with Foreword by Sir William McCrea.

(1946). *Fundamental Theory*, ed. Sir Edmund Whittaker. Cambridge: Cambridge University Press.

Einstein, A. (1917). Kosmologische Betrachtungen zur allgemeinen Relativitätstheorie. *Sitzungsberichte der Preussischen Akademie der Wissenschaften*, 141–152. Translation (1923) in *The Principle of Relativity*, ed. G. B. Jeffery and W. Perrett, pp. 175–188. London: Methuen.

Einstein, A. and de Sitter, W. (1932). On the Relation between the Expansion and the Mean Density of the Universe. *Proceedings of the National Academy of Science Washington*, **18**, 213.

Friedmann, A. A. (1922). Über die Krümmung des Raumes. *Zeitschrift für Physik*, **10**, 377–386.

(1924). Über die Möglichkeit einer Welt mit konstanter negativer Krümmung des Raumes. *Zeitschrift für Physik*, **21**, 326–332.

Hubble, Edwin (1936). *The Realm of the Nebulae*. New Haven and London: Yale University Press. Reissued (1982) with Foreword by James E. Gunn.

Lemaître, G. (1927). Un univers homogène de masse constante et de rayon croissant rendant compte de la vitesse radiale des nébuleuses extragalactiques. *Annales de la Société scientifique de Bruxelles*, **47A**, 49–59.

(1931a). A Homogeneous Universe of Constant Mass and Increasing Radius Accounting for the Radial Velocity of Extra-Galactic Nebulae. *Monthly Notices of the Royal Astronomical Society*, **91**, 483–490.
(1931b). The Expanding Universe. *Monthly Notices of the Royal Astronomical Society*, **91**, 490–501, erratum 703.
(1931c). The Beginning of the World from the Point of View of Quantum Theory, *Nature*, **127**, 706.
(1934). Evolution of the Expanding Universe. *Proceedings of the National Academy of Science*, **20**, 12–17.
(1946). *L'hypothèse de l'atome primitif.* Neuchatel: Editions du Griffon. English trans. 1950. *The Primeval Atom*. New York: Van Nostrand. Rev. edn. 1972. *L'hypothèse de l'atome primitif.* Bruxelles: Edition culture et civilisation.
(1958). The Primeval Atom Hypothesis and the Problem of the Clusters of Galaxies. In *La structure et l'évolution de l'univers*, 11th Solvay Conference, pp. 1–31. Brussels: Stoops.
McCrea W. H. (1935). Observable Relations in Relativistic Cosmology. *Zeitschrift für Astrophysik*, **9**, 290–314.
(1988). Cosmology: Background to Developments. *Quarterly Journal of the Royal Astronomical Society*, **29**, 51–60.
Milne, E. A. (1935). *Relativity Gravitation and World-Structure*. Oxford: Clarendon Press.
(1948). *Kinematic Relativity*. Oxford: Clarendon Press.
Sitter, W. de (1916a). On Einstein's Theory of Gravitation and its Astronomical Consequences. *Monthly Notices of the Royal Astronomical Society*, **76**, 699–728.
(1916b). *Monthly Notices of the Royal Astronomical Society*, **77**, 155–184 (481 erratum).
(1917). *Monthly Notices of the Royal Astronomical Society*, **78**, 3–28.
Slater, N. B. (1954). Recession of the Galaxies in Eddington's Theory. *Nature*, **174**, 321.
(1957). *Eddington's 'Fundamental Theory'*. Cambridge: Cambridge University Press, p. 18.
Tolman, R. C. (1934). *Relativity Thermodynamics and Cosmology*. Oxford: Clarendon Press. Part IV.
Whittaker, E. T. (1910). *A History of Theories of Aether and Electricity: The Classical Theories*. London: Nelson.
(1931). On the Definition of Distance in Curved Space, and the Displacement of the Spectral Lines of Distant Sources. *Proceedings of the Royal Society*, **A133**, 93–105.
Whittaker, Sir Edmund (1951). *A History of Theories of Aether and Electricity: The Classical Theories*. Revised and enlarged edition. London: Nelson.
(1953). *A History of Theories of Aether and Electricity: The Modern Theories 1900–1926*. London: Nelson.

12

An assessment of the evidence against the steady-state theory

F. Hoyle

In a sense, the steady-state theory may be said to have begun on the night that Bondi, Gold and I patronised one of the cinemas in Cambridge. The picture, if I remember rightly, was called *The Dead of Night*. It was a sequence of four ghost stories, seemingly disconnected as told by the several characters in the film, but with the interesting property that the end of the fourth story connected unexpectedly with the beginning of the first, thereby setting-up the potential for a never-ending cycle. When the three of us returned that evening to Bondi's rooms in Trinity College Gold suddenly said: 'What if the Universe is like that?' These were the days, late 1946 or early 1947, when anything that anybody suggested was avidly discussed, before everybody became bowed down with committees and other forms of unproductive administrative work. Within a short time we saw that a continuous universe would demand the creation of matter, which Bondi and I found unpalatable, Gold being less impressed by what seemed to the two of us to be an insuperable difficulty.

After that nothing further happened for about a year, until early 1948, by which time the Bondi–Gold partnership was working on problems different from those in which I had become interested, mostly to do with the synthesis of elements in stars. From February to May 1948 our paths converged to a considerable extent, however, in what became the published versions of the steady-state theory. Bondi describes elsewhere in this volume the approach followed by Gold and himself. What I did was to couple a scalar field to gravity, a procedure that leads inevitably to what is nowadays termed an inflationary solution of the Einstein field equations, a solution with the de Sitter metric

$$ds^2 = dt^2 - \exp(2Ht)[dr^2 + r^2(d\theta^2 + \sin^2\theta\, d\phi^2)],$$

with H a constant determined by the coupling of the scalar field. Since de Sitter had originally obtained this metric for an empty universe, but with the cosmical constant term Λ included in the field equations, it followed that the coupling of the scalar field to gravity played the same role as the cosmical constant in de Sitter's solution. But with the term interpreted physically, so that the universe was no longer empty.

The situation differed from modern inflationary cosmology in that the 'steady' condition as evidenced by the exp(2Ht) factor in the metric was considered to persist at all times, so that the universe had neither a beginning nor an end. In contrast, modern inflationary theory is like a sandwich, the inflationary condition being the filling of the sandwich, with a white slice of bread on one side and a brown slice on the other. The white slice is the big bang and the brown slice is the modern astrophysical era of the universe. This difference provides a causal directivity to the sandwich leading to evolutionary properties in the universe.

The steady-state theory on the other hand has no evolution on a very large scale, although there is much evolution on local scales. There was no disagreement between Bondi, Gold and myself about what local scales were involved, because observational cosmology showed spatial variability to be on the scale of large clusters of galaxies, with masses extending upwards to the range 10^{15} to 10^{16} times the Sun. Large as this seemed, it still meant there were 10^6 to 10^7 clusters within the observable horizon, sufficient for the spatial irregularities to be thought of as cosmologically small.

The field equations implied a creation of matter but they did not specify the form in which creation took place. For this we had to speculate. We thought about creation of matter in stars and rejected it. We thought about smooth creation throughout spacetime, and there not being any immediate astrophysical objection to this concept, we accepted it as a working hypothesis. Where all three of us were curiously blind was in not realising that the units of creation must be so-called little big bangs giving rise to whole clusters of galaxies, with temporal irregularities thus connected directly to the spatial irregularities, and with the physical processes occurring at the genesis of each such unit being closely the same as those which are taken in modern cosmology to apply to the whole universe. The physicist loses nothing in such an interpretation of the steady-state theory, although the mathematician loses the advantage of the simplicity of ordinary differential equations, since the equations governing the expansion of a little big bang involves a radial coordinate as well as the time and are therefore of a partial character. Partial, non-linear, and therefore awkward, but not too difficult to visualise in their properties. One has to think of the creation of an object leading to a new cluster of galaxies, not as an object projected into a prescribed spacetime structure, but as an object whose contents are powerful enough to modify the previously existing spacetime structure. Each new object makes room for itself among the previously existing units, forcing the previously existing units to move apart from each other, and so providing a physical *raison d'être* for the expansion of the universe. Instead of thinking of matter as being created in preordained expanding spacetime, as one tends to do in cosmology, it is better to invert the logic. Think of the creation as being driven by ascertainable physical processes, and of the inexorable introduction of new units of creation as forcing the others apart, much as the introduction of new guests into a cocktail party forces earlier guests to move outwards from the initial gathering point, although as always in cosmology this concept has to be formulated without reference to any particular spatial centre. Looking at things from this point of view, with each new unit having to struggle for *lebensraum*, you can see why each cluster of galaxies creates just enough room for itself, enough but not an excess of room as in a mathematically parabolic situation.

These perceptions did not begin to emerge until the early 1960s, and then only rather distantly. Their fate was to be chopped-off by the discovery of the microwave background. Even before this discovery, the theory had run into trouble on several counts, which happened in the following way. Big bang cosmology can explain pretty well anything at all by claiming that what we observe now was caused by an earlier state of the universe, and so on progressively back to a beginning which is not itself explained, a procedure which Tommy Gold pithily refers to as: 'The Universe is what it is because it was what it was.' In effect, nothing is ultimately explained. In complete contrast, the steady-state theory has the immense task of explaining everything in terms of ongoing processes, with the difficulty that gaps in our knowledge appear as inadequacies in the theory, and the still more unfortunate difficulty that errors of observation can be interpreted as contradicting the theory. I will give examples of these difficulties at a later stage.

The theory requires that all observed physical properties must be capable of regenerating themselves, otherwise a property would already have died out over the indefinitely large number of previous generations of the universe, defining H^{-1} as the generation length. Particularly, each generation of clusters of galaxies must be capable of regenerating itself, as in biology each generation of plants scatter seeds some fraction of which germinate and reproduce the next generation of plants, a process which in the cosmological case is connected with the partial differential equations I mentioned earlier. It is a consequence of such a point of view that aggregates or clusters of galaxies grow from generation to generation by a kind of dendritic accretion, while a void remains a void, giving a rather good correspondence with the generally hierarchical picture of the distribution of baryonic matter that has been revealed by modern observations.

The reproduction of physical properties from generation to generation is what I have myself always meant by the description 'steady'. This is not equivalent in meaning to constant or uniform. A car can move along the road at a constant speed without all the properties of the engine being uniform in time. Some properties such as the oil pressure are indeed nearly uniform, but other properties such as the fuel cycle are variable over the cycle time. Thus some properties may be considerably variable in the steady-state theory over time-scales of order H^{-1}, while other properties may be smooth on such a scale. Unlike big bang cosmologies, in which galaxies are required all to have closely the same age, there are wide variations of age in the steady-state theory, from about $\frac{1}{3}H^{-1}$ on the low side to about $5H^{-1}$ on the high side. Young objects can lie more or less side-by-side with old objects, just as in a wood young trees may grow side-by-side with old trees. So long as one is discussing the properties of young objects there is an approximation to smoothness, but for older objects, as for instance arrangements in superclusters, the situation can be markedly irregular, more irregular than in the big bang cosmologies. The smoother a property the less difficult it is to test by observation, so that it is sensible to focus on such properties, at any rate initially. Unfortunately, however, observers in the 1950–65 period treated every property as if it were smooth, without giving much thought to whether this should be so or not.

So let me come now to the observations that were claimed to disprove the theory, extending the confrontation with observation to 1970, but allowing myself to give updates on the various situations that arose over the two decades which followed the inception of the theory in 1948. There were five main confrontations:

1. The existence of the microwave background.
2. The need for a primordial synthesis of D, ^{4}He and ^{7}Li.
3. A test of angular diameters versus redshifts.
4. The value of q_o.
5. The counting of radio sources.

There has never been a difficulty in the steady-state theory over the energy-density of the cosmic microwave background. The average cosmic baryon density is about $3 \times 10^{-31}\,\mathrm{g\,cm^{-3}}$, i.e. an energy density $\sim 3 \times 10^{-10}\,\mathrm{erg\,cm^{-3}}$. So only a small fractional conversion of baryonic energy into electromagnetic energy inevitably yields the correct order of magnitude for the energy of the microwave background, as in the following example:

Taking 1/4 of the baryonic material to be ^{4}He, i.e. $\sim 7.5 \times 10^{-32}\,\mathrm{g\,cm^{-3}}$ of helium cosmologically, and taking the ^{4}He to have been synthesised from hydrogen, with an energy release of 6×10^{18} erg per gram of ^{4}He produced, gives a cosmic electromagnetic energy density of $(6 \times 10^{18}) \cdot (7.5 \times 10^{-32}) = 4.5 \times 10^{-13}\,\mathrm{erg\,cm^{-3}}$. Were this electromagnetic energy to be thermalised the resulting temperature T would be given by $aT^4 = 4.5 \times 10^{-13}$, i.e. $T = 2.78$ Kelvin.

Why then was the discovery of the microwave background seen as a devastating blow to the steady-state theory? Because of an (unsubstantiated) assertion that electromagnetic energy produced by astrophysical objects could not be thermalised, an assertion which amounted to denying Nature the ability to convert radiation to its thermodynamic state. In all branches of science of which I am aware, Nature turns out to be far more efficient at degrading energy than we expect beforehand, an amazing example being that of chemoautotrophic bacteria.

An unexpected device available to Nature in this instance comes from a study of particles condensing from metallic vapours. Instead of being more or less spherical blobs, the condensates tend to lengthen dramatically when they reach a size of about 10^{-2} μm, the lengths attained being rather uniformly of order 1 mm, giving a length-to-diamter ratio of about 10^5. Such long rods or 'whiskers' absorb and scatter radiation like infinite cylinders, for which exact calculations can be done given the optical constants of a metal, provided the electromagnetic frequency ν satisfies an inequality of the form

$$\nu \text{ in Hz greater than } 3\sigma \frac{a^2}{\ell^2} \ln\left(\frac{\ell}{a}\right)$$

where σ is the conductivity in $(\text{seconds})^{-1}$. Using the experimental value $a/\ell \simeq 10^{-5}$ and putting $\sigma = 10^{18}\,\text{s}^{-1}$ for the low-temperature, low-frequency conductivity of iron (astrophysically the most interesting metal) gives $\nu > {\sim}3 \times 10^9$ Hz, i.e. the wavelength less than about 10 cm. At wavelengths in the main microwave region and in the far infrared, such a conductivity implies the immensely high mass absorption coefficient of ${\sim}10^7\,\text{cm}^2\,\text{g}^{-1}$. If only a few tenths of a percent of the average baryonic density were in the form of extragalactic iron whiskers the optical depth over a distance cH^{-1} would be of order 10 in the microwave region. A similar absorptivity does *not* apply in the visual region of the spectrum, because the optical constants of metals are such that σ falls by about two orders of magnitude at wavelengths less than 10 μm. Thus an optical depth of 10 in the microwave range would be accompanied by only an optical depth of 0.1 in the visual range. A universe that was quite black for microwaves would be essentially translucent in the visual range.

On this view the primary source of radiation feeding the microwaves would come largely from the infrared, say over the range of wavelengths from 30 μm to 300 μm, in which range even 'normal' galaxies like our own can be very powerful emitters.

There are many astrophysical arguments one can make for the existence of iron whiskers. An interesting example is afforded by the pulsed radiation from the Crab Nebula. With increasing ν going from the radio region the intensity falls steeply to become undetectable at $\nu > {\sim}3 \times 10^9$ Hz, i.e. at just about the value given by the above inequality. Over the range ${\sim}3 \times 10^9 < \nu < {\sim}3 \times 10^{13}$ Hz the pulsed radiation from the inner pulsar is missing, requiring that either the inner pulsar radiates for some mysterious reason only very weakly over this range, or that the envelope of the Crab surrounding the pulsar contains particles that absorb very strongly over this range. Iron whiskers condensed from the supernova ejecta would do so, the low frequency end of the range at $\sim 3 \times 10^9$ Hz arising from the above inequality, and the high frequency end of the range at 3×10^{13} Hz arising from the fall of σ occasioned by the optical constants of iron.

A 'little big bang' expanding parabolically rather than hyperbolically satisfies the following relation between the total mass density ϱ and the temperature T_9 measured in units of 10^9 Kelvin,

$$\varrho \simeq 10^5 \left(\frac{\odot}{M}\right)^{1/2} T_9^3 \ \mathrm{g\,cm^{-3}}.$$

Putting $M/\odot = 10^{16}$ for our little big bang therefore gives

$$\varrho \simeq 10^{-3} T_9^3 \ \mathrm{g\,cm^{-3}},$$

whence the baryonic density is given by

$$\varrho_{\mathrm{baryons}} \simeq 10^{-3} f T_9^3 \ \mathrm{g\,cm^{-3}},$$

where f is the fraction of the total mass density that is baryons. The best modern value for the average baryonic density in the universe is $\sim 3 \times 10^{-31}$ g cm^{-3}, not much different from the value estimated by Oort more than two decades ago. Thus the modern cosmological trend towards a total mass density at its so-called closure value of $\sim 10^{-29}$ g cm^{-3} implies that baryons form about 3 percent of the mass, giving $f \simeq 1/30$, and hence $\varrho_{\mathrm{baryons}} \simeq 3 \times 10^{-5} T_9^3$ g cm^{-3} in little big bangs. This is an average relationship, with individual samples of material varying from the mean perhaps by a factor 3. That is to say, if we put $\varrho_{\mathrm{baryons}} = h T_9^3$ g cm^{-3}, the value of h varies in little big bangs over the range 10^{-4} to 10^{-5}, with 10^{-5} a good number for yielding the least observed abundances of ^{4}He and ^{7}Li, the least value of ^{4}He being from H II regions in dwarf galaxies and the least ^{7}Li value being from extreme Population II stars. The larger value $h = 10^{-4}$ gives ^{4}He, ^{7}Li values agreeing very well with abundances found in Population I stars.

In preparation for the Paris Symposium on Radioastronomy, Rudolf Minkowski asked me to review the various forms of observational tests of cosmological models, with particular reference to any new ones I could think of. It was in preparing my contribution that I found remarkable differences for the redshift dependence of a standard length held at right angles to the line of sight – this was according to the spacetime metric one assumed. The Minkowski metric gave a simple $1/z$ dependence, the de Sitter metric of the steady-state theory gave a $(1+z)/z$ dependence, and a Friedmann metric at closure ($q_o = +\frac{1}{2}$ for $k = 0$, $\Lambda = 0$) gave $(1+z)^{3/2}/[(1+z)^{1/2} - 1]$, which has the remarkable property of possessing a minimum apparent diameter at $z = 1.25$. In 1958 this test was somewhat visionary, as it lay beyond the available techniques of the day, both in visual astronomy and in radioastronomy. By now, however, a test had been made. To everyone's surprise, I would think, the test has yielded the Minkowski dependence, $1/z$. Although the steady-state theory gives the next best fit to the data, for some reason I do not understand it has been concluded that the data disprove the steady-state theory, a supposed disproof which arises from confusing 'steady' with 'constant', the situation to which I drew attention at an earlier stage. What the data do *not* show is the minimum predicted for uniform sources by the Friedmann cosmologies. This appears to be the situation *both* for visual data and for radio data. Of course it is always possible for the Friedmann cosmologies to escape from this, or any other contradiction, by making arbitrary assumptions that the past was different to the present, and if one calls the difference 'evolution' the situation is semantically changed from bad to good. Arbitrary *ad hoc* assumptions sound bad, but 'evolution' sounds good. Realistically, however, it boils down once again to Tommy Gold's precept: 'The universe is what it is because it was what it was.'

In the early months of 1953 I had the privilege of walking most Sunday mornings with Edwin Hubble. Hubble returned repeatedly to a point which appeared to intrigue him in a high degree, namely that the apparent photographic luminosity of selected cluster galaxies appeared to vary as the inverse square of their redshifts, which relationship he said held within only small margins of deviation. A similar result was subsequently obtained by Sandage and his collaborators, holding to larger and larger values of the redshift as time went on. Assuming the selected galaxies to be intrinsically the same, the result provides evidence for the closed Friedmann model with $q_o = +1$. The steady-state theory with $q_o = -1$ has the dependence $z^{-2}(1+z)^{-2}$ instead of z^{-2}, and so differs in the apparent luminosity which it predicts from what is claimed to be the observational result by the factor $(1+z)^{-2}$. For a galaxy at $z = 1$ this is a difference of 1.5 mag, which would be strong evidence, if it were entirely unequivocal, against the steady-state theory. The problem comes in ensuring that like is being compared with like. First, there is the technical difficulty of comparing the same wavebands for galaxies observed at widely different redshifts, a difficulty that requires extensive corrections to be applied to the actual data, corrections that are as large, or larger, than the cosmological effects under test. Such situations always make me uneasy, and I think I can fairly say I would be just as uneasy if I were at the observing end of the argument instead of at the theoretical end. The second question is, how can we be sure the selected galaxies are intrinsically similar to each other? The short answer is we cannot. Unquestionably it would be possible to select galaxies at various redshifts for which the apparent luminosities fitted the expectations of the steady-state theory, so that the issue boils down to which of two sets of galaxies is uniform in their intrinsic properties. The usual claim rests on the similarity of clusters of galaxies to each other, but if age-correlations are present this will not be the case. I know of no example of a situation where the sampling of a few individuals reveals the full range of properties possessed by a large number of individuals. A morning spent in assessing the heights attained by humans might well lead to the assertion that the maximum height attained by the human male is about 2 metres, and ten mornings spent in the streets of ten cities might well lead to the same conclusion, give or take a few centimetres. Yet the maximum height of the human male is actually about 2.5 metres, a circumstance that one would be unlikely to notice except from a very large sample. The problem is therefore one of selection of material, with the possibility of systematic bias arising as more and more galaxies become available for observation with increasing redshift. Lest this be thought unfair let me point out that the dependence of apparent luminosity on redshift for the Friedmann cosmology with $q_o = 0$ is quite close to the steady-state behaviour. A few years back, when the consensus view among astronomers favoured this $k = -1, q_o \simeq 0$ model, nobody to my knowledge rejected it on the grounds of a wrong redshift-magnitude for galaxies.

Once again I might add that all this is subject to the restrictive assumption that 'steady' be interpreted as 'uniform'. If the creation of little big bangs, while being steady on the average, has cyclic variations like the fuel in a car, then the factor exp Ht in the metric becomes replaced by a function of the time, F(t) say with the property that $\dot{F}(t)$ is larger at times of rapid creation than the average H. In such a situation $q = -\ddot{F}/(\dot{F})^2$ fluctuates about the value -1. What one can prove is that, if over some interval creation falls to a sufficiently low ebb for it to have effectively ceased, then in that interval $q = +1$.

The story of how radiosource counts were used to disprove the steady-state theory has often been told, wrongly as it turns out. From about 1955 onwards Ryle had the idea that by counting radiosources as a function of their fluxes he could disprove the steady-state theory. His programme, which he pursued relentlessly over the years, does not seem to have been directed towards any other end. There was no question of establishing the correct cosmology, but only of disproving the views of a colleague in the same university, a situation which I have never felt to have deserved the plaudits which the scientific world showered upon Ryle.

If radiosources were intrinsically similar to each other, and if their density per unit proper volume was homogeneous at all times past, then the function N(S) of sources with fluxes at a particular wavelength greater than S should take a form dependent on the radio spectrum of the sources and on the cosmological metric. For sources with a spectrum $d\nu/\nu$ in Euclidean space, one has $N(S) \propto S^{-3/2}$, while for sources in non-Euclidean cosmological spaces one obtains $N(S) \propto S^{-\alpha(S)}$, where the function $\alpha(S)$ depends on the intrinsic power of the sources and on their redshifts. In general, however, for intrinsically similar, homogeneously distributed sources, $|\alpha| < 1.5$. Ryle's first claim, made around 1955, was $|\alpha| = 3.5$. By the time of the Paris Symposium in 1958 this first claim had been reduced to $|\alpha| \simeq 2.5$. This was for sources counted down to a lower limit S_o in what became known as the 3C survey. Eventually the value of $|\alpha|$ became reduced to 1.7 to 1.8 for the sources of the 3CR survey. Throughout these reductions, Ryle claimed that although he had been wrong before, he was right now. Any value of $|\alpha|$ above the Euclidean value of 1.5, no matter how small, would establish that the sources could not be intrinsically similar, and so the steady-state theory was disproved. But this was to interpret 'steady' as meaning 'uniform'. We saw at an earlier stage that physical properties must be reproducible on time-scales $\sim H^{-1}$. Fluctuations on time-scales shorter than this are permitted by the theory, just as the movements of inlet and outlet valves encompass the entire cycle of an ordinary car. In 1961, Narlikar and I showed that fluctuations on a time-scale $\sim 0.1\,H^{-1}$, only a tenth of the permitted value, were sufficient to give values of $|\alpha|$ for sources at reasonable distances up to 2, thereby exceeding the observational value of 1.7–1.8.

At the Paris Symposium of 1958, Hanbury-Brown argued for caution until the nature of the sources became known. To many people, including myself, it was unclear how conclusions of great cosmological significance could be distilled from ignorance, ignorance of the nature and properties of the sources and ignorance of their distances. And so it has proved. About 60 percent of the sources have turned out to be radiogalaxies, most of the rest being quasars. The radiogalaxies have $|\alpha| < 1.5$, while the quasars have $|\alpha| > 2$ thereby giving a slope greater than 1.5 for the combined distribution. Nobody who reads without prejudice the recent book *Quasars, Redshifts and Controversies* by Halton Arp, should doubt that there are uncertainties over the nature and distances of quasars. Uncertainty is therefore maintained by continuing to include them alongside the radiogalaxies, which by themselves should reveal the effect sought for by Ryle, if it exists.

The data now give the number of radiogalaxies with fluxes at a particular wavelength greater than S_o and with redshifts less than z. In recent work, Das Gupta, Narlikar and Burbidge have shown that subject to a choice of metric, and subject to the strict 'uniform' hypothesis of zero evolution, it is possible to divide the data with respect to flux and with respect to redshift. From the redshift data *alone* the radioluminosity function of the sources can be determined. Then the flux data can be examined for consistency with the hypothesis of zero evolution. This is done as follows. Plot all the sources in the sample, about 160, in the S, z plane. For each such point count the number of sources with fluxes greater than S and redshifts less than z. Also from the determined luminosity function, suitably normalised with respect to the total number of sources in the sample, the expected number of sources with fluxes >S and redshifts <z can be obtained. Because the expected numbers so determined vary smoothly with S and z, whereas the actual counts involve discrete numbers, there are inevitable statistical fluctuations when expected numbers and actual numbers are compared. To decide whether or not the fluctuations are significant, artificial counts were generated by a standard Monte Carlo technique, which could of course be done very many times. The Kolmogoroff-Smirnov measure of statistical deviation was used in comparing the simulated counts with the actual counts. An ideal result was for the Kolmogoroff-Smirnov statistic to exceed the actual observed value in about 50 percent of Monte Carlo cases, which is just what was found. In about 60 percent of Monte Carlo cases the statistical deviation was greater than for the observed sources. A more usual way of describing this result would be to say that the observed sources agree with the hypothesis of zero evolution to within about half a standard deviation. Subject to the metric being de Sitter, the radiogalaxies show no evolution. A similar result has also recently been derived for the visual and near infrared properties of galaxies.

This total inversion of what was originally claimed shows the unsoundness of attempting to disprove theories with messy data. Messy data can be used to give a modest measure of positive weight to a theory, but only accurate results of excellent quality should be used in attempting to negate a theory. This point is so evident in logic that I am left wondering why anybody should have believed Ryle, especially after his repeated changes in the value of $|\alpha|$. The only reasonable explanation I can offer is that Ryle's claims were widely considered to be culturally desirable. If the claims had been undesirable, as the steady-state theory has always been judged undesirable, then I think the entire argument would have been laughed out of court. People would simply have said that the repeated adjustments in the value of $|\alpha|$ made the entire procedure doubtful. And it would surely have been emphasised that Hubble had been obliged to abandon a similar method applied to visual galaxies, in spite of the visual nature of galaxies being known and in spite of redshift determinations being available for many of them.

The only philosophical objection to the steady-state theory with which I have felt much sympathy is that it dumps the universe in a never-ending rut, although why this was considered any less attractive than a universe that evaporated into nothingness I have also never understood. To deny 'evolution' in the sense in which big bang cosmologists have considered 'evolution' is no loss at all, for the issue of whether galaxies have grown a little brighter or a little fainter, or a little larger or a little smaller, over the past 10^{10} years is singularly uninteresting. The only evolutionary processes of real subtlety in astronomy are those which relate to stars, and strikingly enough stellar evolution proceeds essentially without reference to cosmology.

The proper philosophical point of view, I believe, for thinking about evolution cosmologically involves issues that are superastronomical, as one inevitably gets as soon as one attempts to understand the origin of biological order. Faced with problems of a superastronomical order of complexity, biologists have resorted to fairy tales. This is shown by a consideration of the order of the amino acids in any one of hundreds of enzymes. To offer vague handwaving explanations of the genesis of such amazing structures is simply a counsel of desperation. History shows that whenever humans encounter problems of vastly greater dimensions than they can cope with successfully the normal pattern is to retreat into absurdity, and so it seems to be here. To have any hope of solving the problem of biological origins in a rational way a universe with an essentially unlimited canvas is required, a universe in which the entropy per unit mass does not increase inexorably, as it does in big bang cosmologies. It is to provide just such an unlimited canvas that the steady-state theory is required, or so it seems to me.

Discussion

Ehlers:
Which data were you referring to when you discussed the log (diameter)/log (z)-data?

Hoyle:
The data obtained by radio astronomers using the radiotelescope at Ooty in S. India.

McCrea:
What would you say is the best observational test of your model? What do you predict about the uniformity of the microwave background.

Hoyle:
The most immediate objective would be to demonstrate by observation the existence of the particles capable of thermalising the microwave background. While this would not prove the model in itself, it would provide the basis for seeing it as a viable alternative to so-called standard cosmology. At centimetre wavelengths very little in the way of fluctuations in the background is to be expected, just as is observed. It is conceivable, however, that fluctuations might be greater at the centre of the microwave distribution, viz. 1 to 2 mm. This is for fine-scale fluctuations.

13

Steady-state cosmology, the arrow of time, and Hoyle and Narlikar's theories

José M. Sánchez-Ron

13.1 Introduction

Although it has been long discarded by practically all physicists, astronomers, and cosmologists, the steady-state approach to cosmology, in its different versions, played a rather influential role during the late 40s, the 50s, and even in the early 60s, although by then its influence had decayed enormously, especially outside Great Britain.

There exist, however, two different periods in the history of the steady-state cosmology. The first was dominated by the 1948 contributions of Hermann Bondi and Thomas Gold, on one side, and of Fred Hoyle, on the other: more philosophical the former, more physical the latter. As for the second period, it was the result of a mixture of different components and interests: Wheeler and Feynman's time-symmetric action-at-a-distance electrodynamics (1945 and 1949), Hogarth's elaboration (1953 and 1962) of this electrodynamics in a cosmological scene, and the way in which he connected such elaboration with the question of the arrow of time, a connection that would be emphasized by such people as Bondi, Gold, and Sciama. With all these elements, Hoyle and Narlikar developed, in the 60s and 70s, a series of theories, with which they tried to keep alive the steady-state cosmology, relying in part, naturally, on Hoyle's previous formulations.

In the present article I am going to study, from a historical point of view, this second period that, by contrast with the first one, has not yet been considered by historians of science. I will focus mainly, though by no means uniquely, on the years in which the transition from one period to the other took place.

13.2 The cosmological scenario during the 50s and early 60s

I am going to deal with developments and/or theories connected, directly or indirectly, with the steady-state cosmology, an approach that the present community of physicists, with the exception, perhaps, of a few of its members, now considers superseded, refuted. Although it is not my intention to discuss whether or not such opinion is justified, it is important to point out that at the time when the first developments I am going to refer to took place – that is, during the 50s and even during the early 60s – it was by no means clear that the steady-state cosmology had run into serious trouble. Martin Ryle's radio stars counts, with its "log N – log I curves,"[1] on the basis of which he saw "no way in which the observations can be explained in terms of a Steady-State theory" (Ryle 1955) were received with caution. Speaking at an important conference, held at the University of North Carolina, Chapel Hill, in January 1957, A. E. Lilley (1957) could say:

> [Ryle and Scheuer's observations] will suggest departures from an isotropic and uniform universe and the results, if valid, are not consistent with a steady-state universe. However, the interpretation of this [log N – log I] curve has been discussed by Bolton, who has suggested that when one has observational errors which increase with decreasing intensity, even an isotropic distribution can produce a curve of the form [given by Ryle and Scheuer].

Six years later, the conflict had not yet been settled – although here I have to say that it seems to me that the reluctance to completely abandon the steady-state point of view came basically from British astronomers – and W. H. McCrea (1963) could still point out, during his Presidential Address at the Royal Astronomical Society that

> M. Ryle and his colleagues showed almost beyond doubt that their radio surveys reach much further into the universe than any present optical surveys can do. The first of Ryle's surveys conflicted greatly with the steady-state model. His recent surveys conflict less violently but still very definitely. Recent optical surveys appear on the whole to conflict but the conflict is probably not yet outside the limits of uncertainty in the observations. Indeed, the general picture is that the observational results that have from time to time been in gross conflict with the steady-state model have mostly had to be revised and the revision usually seems to reduce the conflict; for the rest the situation taken as a whole is inconclusive.[2]

13.3 Cosmology and the arrow of time: Hogarth's elaboration of Wheeler and Feynman's electrodynamics

It may seem surprising that a key fact in the history of the steady-state cosmology was, as we shall see, the publication, in 1945 and 1949, of two papers in which John Wheeler and Richard Feynman presented an action-at-a-distance time-symmetric version of classical electrodynamics.[3]

[1] N = number of radio stars per unit solid angle having intensity greater than I.

[2] Hoyle and Narlikar (1961) were also among those who reacted to the claims made by Ryle and his co-workers.

[3] Wheeler and Feynman (1945, 1949). I have considered several aspects of the history of Wheeler and Feynman's electrodynamics in Sánchez-Ron (1983).

I have said, "it may seem surprising," because what Wheeler and Feynman wanted was to face some problems in the physics of the *microcosmos*. Wheeler, as he said many years later (Wheeler 1979), was struggling with the idea of whether the electron would be after all the basic particle whose theoretical description would enable him to set up a satisfactorily nuclear theory. As for Feynman, one reads in his Nobel lecture (Feynman 1966), that his reason for focusing precisely on the action-at-a-distance concept was due to the idea – common to numerous physicists throughout our century – that if many of the problems which appear in quantum theory, and specially in relativistic quantum theory, are recognized as due to the infinite degrees of freedom introduced by fields, why not to dispense with them in the most straightforward manner?

Wheeler and Feynman's theory, which – they claimed – permitted an explanation of how to introduce on a time-symmetric basis electromagnetic radiation, initially received considerable attention. Afterwards, well into the 50s, and with the successes of quantum field electrodynamics, interest in Wheeler and Feynman's theory diminished considerably in the physics of the microcosmos.

Stimulated by Wheeler and Feynman's papers, a Canadian, J. E. Hogarth, prepared a Ph.D. thesis, supervised by McCrea, that was submitted to the University of London in 1953.[4] In his thesis, entitled *Particles, Fields, and Rigid Bodies in the Formulation of Relativity Theories*, Hogarth (1953) studied Wheeler and Feynman's electrodynamics as well as Alfred Whitehead's action-at-a-distance special relativistic theory of gravitation (Whitehead 1922), a theory that, as it is well known, predicts the same values as general relativity for the three classical tests. Hogarth somewhat extended Wheeler and Feynman's definition of distance, introducing also covariant potentials in the de Sitter space, generalizing in this way Wheeler and Feynman's and Whitehead's theories. "It is shown," he wrote, "that in the flat-space theory of Wheeler and Feynman radiation damping is indeterminate, but that in de Sitter space and in conjunction with the steady-state cosmology the irreversibility of radiation is closely related to the phenomenon of the creation of matter" (Hogarth 1953). What Hogarth was introducing in the discussion, still in a rather primitive and incomplete manner, was a very important idea, namely, that the existence of a definite electromagnetic arrow of time (only retarded solutions of Maxwell equations seem to be detected) could be used to discriminate between different cosmological models, or, more dramatically, that, as he put it later, "the electrodynamical arrow-of-time is determined by the cosmological properties of the universe" (Hogarth 1962).

Apparently, Hogarth's thesis was noticed, although it is now difficult to

[4] During his stay in England Hogarth was attached to the Department of Mathematics, Royal Holloway College, Surrey.

know to what extent; for instance, he gave some seminars (including one at Cambridge, arranged by McCrea). There were certainly people in England who did not know of Hogarth's researches, but it is probably significant that around 1954 a then young Peter Scheuer, who was working with Ryle at Cambridge, knew of Hogarth's work.[5] However, not much effort was dedicated during the next few years after the completion of the thesis, to improve its results (Hogarth, for instance, left England after obtaining his degree, returning to Canada). But to one thing his work did indeed contribute: the increase in the popularity of the problem of the arrow of time among relativists. In 1958, during the 11th Solvay Congress, dedicated to the structure and the evolution of the universe, Thomas Gold argued quite generally that in a universe in which all detailed physical theories were time-symmetrical, the arrow of time must ultimately be associated with the large-scale properties of the universe.[6] Soon afterwards, Sciama converted the idea that irreversible local processes are connected to the irreversible expansion of the universe, in the second law of his "Three laws of cosmology."[7] (As a matter of fact, in his discussion Sciama paid considerable attention to the ideas that Hogarth had put forward in his thesis.)

It was in the early 60s when things began to speed up. Hogarth, by then at the Queen's University, Kingston, Ontario, reassumed his researches. Moreover, he came back to England for a couple of months, staying this time with Bondi's group at King's.[8] As a result, he published, in 1962, a paper in the *Proceedings of the Royal Society of London*, "Cosmological Considerations of the Absorber Theory of Radiation" (Hogarth 1962). This paper was communicated to the Royal Society by Bondi, who, as we shall see later, came to play a key role in the developments which followed.

To understand Hogarth's ideas, as expressed in his 1962 paper, it is necessary to consider in more detail the Wheeler and Feynman articles mentioned before. Thus, it is important to point out that although it is true that Wheeler and Feynman's action-at-a-distance theory solved the self-action infinities problems, the same could not be said with respect to the *ad hoc* choice of retarded solutions, i.e., with the explanation of the *electromagnetic* arrow of time. In fact, the model of the universe assumed by Wheeler and Feynman was static and hence time-symmetric; i.e., a change of t to $-t$ left their arguments invariant, the effect being a radiative reaction with opposite sign (that is, an accelerated charge would gain energy instead of losing it, and only advanced effects would be locally

[5] Information obtained through personnal communications to the author by Profs. McCrea and Scheuer.

[6] Gold (1958). See also Gold's Richtmyer Memorial Lecture, delivered on January 25, 1962 (Gold 1962).

[7] Sciama (1961). This paper is the text of a lecture that Sciama gave at the Institut Henri Poincaré on May 4, 1959.

[8] Personal communications to the author by Profs. McCrea and Bondi.

observable). Wheeler and Feynman avoided this possibility by rejecting the advanced potential solution on causal statistical grounds; that is, they sought to relate the electrodynamical arrow of time to asymmetry of initial conditions, or what is the same, to the *thermodynamical* arrow of time. However, this was not a satisfactory answer to the problem inasmuch as such argument could also be invoked to reject the retarded solution of Maxwell electrodynamics. In fact, Wheeler and Feynman themselves rightly conjectured the solution to this question: they knew that in their theory the universe acts as a perfect absorber; that is, it absorbs all electromagnetic disturbances emanating from a typical charge on it. Wheeler and Feynman considered this feature of their theory "an idealization," pointing out that "present observations suggest that the absorption of radiation is far from complete ... a complete description of the mechanism of radiation would require us to take into account not only the curvature of space but also the phenomena summarized under the term 'expanding universe'" (Wheeler and Feynman 1945).

These were the kind of ideas that Hogarth elaborated. What he did was to examine Wheeler and Feynman's theory in various models of the expanding universe, taking into account that for retarded (advanced) solutions to be consistent, the universe must act as a perfect absorber on the future (past) light cone. A static Euclidean universe is a perfect absorber in the past as well as in the future, therefore there is an ambiguous situation. "In the author's view," Hogarth (1962) wrote, "the theory of Wheeler & Feynman is deficient in its explanation of the arrow-of-time. This deficiency is related to the assumed static nature of their universe." On the contrary, in a non-static universe the absorber, as viewed along the future branch of a null cone, may be fundamentally different from that as viewed along the past branch of the same null cone. The interaction with the future absorber, and the resultant force of radiative reaction need no longer be a symmetrical or indeterminate function of time. In other words, the expansion of the universe brings a continuous modification in the absorber so that it is different in the past from what it is in the future.

In his paper Hogarth considered two classes of universes: the "stationary density" model (the proper density of matter, and of charged particles, is assumed to be constant; continual creation of matter is required to compensate for the reduction in density which would otherwise result from the expansion), and the "no continual creation of matter" model (the expansion of the proper density of matter is a decreasing function of time), arriving to the conclusion that "in most (but not all) cases, continual creation of matter is necessary for compatibility with observation. According to our analysis the steady-state cosmology predicts the observation of retarded fields, while the Einstein–de Sitter universe [i.e., Friedmann model with $k = 0$] predicts advanced fields."

These were the sort of results that attracted the interest of Hoyle and Narlikar.

13.4 From Bondi's comments to Hoyle and Narlikar's theories

During the summer of 1961 a course dedicated to "Evidence of gravitational theories" took place in Varenna (Italy); it was one of the prestigious courses of the Enrico Fermi International School of Physics. Bondi, Hoyle, and Narlikar, then a young Cambridge graduate (he did the Mathematical Tripos in 1960) and Hoyle's research student, were among the participants. According to both Hoyle and Narlikar,[9] their interest in Hogarth-type of approach to cosmology was aroused by Bondi, who at Varenna gave a talk on the work of Hogarth (such a talk does not, however, appear in the published version of the proceedings of the school).[10] On that occasion, Bondi told them that some years earlier Hogarth had demonstrated the need for a steady-state cosmological model, rather than a big-bang model, if the Wheeler and Feynman theory was to be successfully set in an expanding universe. (Most probably, Bondi's comments would have referred essentially to Hogarth's more elaborate researches as they were going to appear in his *Proceedings of the Royal Society* paper.[11])

In Varenna, Bondi apparently mentioned that it would be interesting to investigate whether a neutrino-absorber theory would yield a similar result to the photon-absorber theory of Wheeler–Feynman–Hogarth.[12] As he later put it,[13] Bondi's problem was the following:

> If we suppose that there are some models of the Universe in which the disequilibrium between past and future depends on the actual cross-section for absorption, then we are led to consider whether this is large enough in all cases, and naturally our minds turn to neutrinos. As you know, their interaction cross-section is very small. Could it be that with neutrinos the disequilibrium between past and future is not sufficient to produce a sense of time?

Let me say a few things before I proceed with this history: Although, as we are currently seeing, Bondi played an important role in the process that would lead to Hoyle and Narlikar's re-elaboration of Hoyle's previous formulation of the steady-state theory, it was to a large extent an accidental circumstance: throughout that period Bondi's interests in

[9] J. Narlikar to J. M. Sánchez-Ron, April 4, 1979; F. Hoyle to J. M. Sánchez-Ron, May 5, 1979.

[10] Møller (ed.) (1962). The Varenna course took place from June 19 to July 1, 1961.

[11] It was Bondi who communicated Hogarth's paper to the Royal Society. (The paper was received by the Royal Society on September 1, 1961, that is, after the Varenna course.)

[12] This is Narlikar's version of it (cf. note 9). However, in the paper published by Narlikar (1962) he stated: "The author is grateful to Professor Hoyle for suggesting the problem."

[13] Bondi (1962). This article was the Halley Lecture for 1962, that Bondi delivered at Oxford on May 16.

cosmology had waned; however, he was still very much interested in the link between thermodynamics and cosmology. At Varenna, Bondi was less interested in supporting the steady-state theory than in opposing the then popular notion of an oscillatory universe (Friedmann's model with $k = +1$), which to him was, as he put it, "a thermodynamical horror."[14] As for the question of the arrow of time, it was a problem that, according also to Bondi, only became apparent to him, in a way he now no longer remembers (due to his contacts with Hogarth, perhaps?), during the 50s.

On his return to Cambridge from Varenna, Narlikar followed Bondi's remark, showing (Narlikar 1962) that in some cosmological models neutrinos travelling both into the past and the future exist, whereas in other models only those going into the future can exist; the Einstein–de Sitter and the steady-state models being the respective examples of the two cases.

At the same time Narlikar was carrying out this research, he and Hoyle had joined forces to exploit Hogarth insights, a route that would reach its climax with the publication, in 1974, of their book *Action at a Distance in Physics and Cosmology* (Hoyle and Narlikar 1974). They discovered several loose ends in Hogarth's contributions, which Feynman had criticized, first at the Gravitation and Relativity Conference at Warsaw, 1962,[15] and later on at a meeting, dedicated to "The nature of time," that took place at Cornell University from May 30 to June 1, 1963 (Feynman is the Mr. X who appears in the published report of the meeting,[16] at which – besides Hogarth and Feynman – Hoyle, Narlikar, Bondi, Gold, Sciama, and Wheeler were present). Feynman's main point was Hogarth's use of the refractive index, which was based on collisional damping; he argued that if Hogarth was trying to link time asymmetries in electrodynamics and cosmology, he should not insert into his discussion a time asymmetry based on thermodynamics, especially since Hogarth was arguing that it was not necessary to invoke thermodynamic asymmetry in the whole set-up. Hoyle and Narlikar (1964a) were able to eliminate the collisional damping term from the argument and to replace it by the radiation reaction term which is of purely electrodynamic origin. In the same paper where they solved this problem, Hoyle and Narlikar generalized Hogarth's contributions in an important way: they showed how the Fokker action, which plays a central role in Wheeler and Feynman's electrodynamics, could be written in a Riemannian space-time, a result that helped them to apply the theory to Robertson–Walker–Friedmann spaces.

[14] Personal communication to the author by Prof. Bondi. The "thermodynamical horror" referred to by Bondi is connected with the Olbers' paradox, a theme which always has attracted Bondi.

[15] J. V. Narlikar to the author, April 4, 1979.

[16] Gold (ed.) (1967). This volume also contains an interesting contribution by Hoyle and Narlikar (1967).

13.5 Steady-state cosmology last stage: the works of Hoyle and Narlikar

It is important to point out that during the years I had been referring to, and generally during a great part of the 60s, Hoyle and Narlikar were working in problems related to general relativity and cosmology from different points of view; in other words, that there are several Hoyle and Narlikar theories. Indeed, it can be said that during the early 60s Hoyle and Narlikar were not really or properly concerned with providing a dynamical theory for the steady-state cosmology, or with developing an action-at-a-distance version of it. The steady-state cosmology only provided a background space-time in which to apply the Wheeler–Feynman electrodynamics. In fact, at that time (i.e., during 1961–63) they were using a field theory – the so called C-field of M. H. L. Pryce – to describe the continuous creation of matter within the framework of general relativity.[17] By using this theory, Hoyle and Narlikar arrived to conclusions rather similar to the ones previously obtained by Hogarth.

Although, as I just pointed out, their C-field theory is not particularly concerned with actions at a distance, in their efforts to develop that theory Hoyle and Narlikar arrived at results that enhanced the status of the action-at-a-distance concept; they showed, for instance, that the C-field could be represented as a direct particle "field" (Hoyle and Narlikar 1964b). Such sort of results gradually led Hoyle and Narlikar to stress the importance and role played by actions at a distance in their steady-state approach to cosmology. Thus, we arrive to their "conformal" theory of gravitation, first proposed in 1964 (Hoyle and Narlikar 1964d). In the paper in which they presented their new theory, Hoyle and Narlikar stated clearly that after realizing that with Hogarth's approach the empirical choice of the retarded solutions of Maxwell's equations could be avoided, "we became convinced that the further consequences of interparticle action should be explored. Evidently, it is not reasonable to dispense with the independent degrees of freedom of the electromagnetic field and yet to retain such degrees of freedom for other fields." In the years to come they would try to do precisely that for the gravitational field. During that process, however, they not only developed the 1964 theory,[18] but also proposed new ones, such as the theories with variable G, the first one developed in 1972 to try to understand the Dirac large number hypothesis (Hoyle and Narlikar 1972a, 1972b), and the second one included in their 1974 book (Hoyle and Narlikar 1974), a work which constitutes the culmination of their efforts in this field.

An aspect of Hoyle and Narlikar's post-1964 theories that is worth pointing out is that they allow for a Machian interpretation of inertia. This feature of their theories, which derives both from their conformal-

[17] See Hoyle and Narlikar (1964a, 1964b, 1964c).
[18] See, for instance, Hoyle and Narlikar (1966).

invariance and their non-local (action-at-a-distance) character, was appreciated by Hoyle and Narlikar at least in 1966, being more fully developed in 1972 and 1974 (Hoyle and Narlikar 1966, 1972a, 1972b, 1974). Since then, it has been one of the characteristics of their theory (as Hoyle and Narlikar seldom forget to point out).

Before putting an end to this schematic description of Hoyle and Narlikar's researches, I feel it is necessary to add a couple of final comments: First, that most of Hoyle and Narlikar's efforts during the 60s and 70s did not receive much attention, at least when one measures the somewhat elusive concept of "degree of attention," through published works or statements of other scientists. As a matter of fact, as far as I know only Roe (1969)[19] did some work in the line of Hoyle and Narlikar, while McCrea, Hawking, and Pirani and Deser criticized some points of their theories, criticisms that were answered by Hoyle and Narlikar (1966). My second, and final, point is that at the same time they were pursuing their search of what they considered a satisfactorily gravitational theory, Hoyle Hoyle and Narlikar were trying to solve two related problems: to extend their theory to quantum theory, and to set up an action-at-a-distance quantum electrodynamics which would free standard quantum electrodynamics of some of its fundamental problems. It was, and still is, an ambitious program.

References

Bondi, H. (1962). Physics and Cosmology. *The Observatory*, **82**, 133–143.

Feynman, R. P. (1966). The Development of the Space-Time View of Quantum Electrodynamics. *Physics Today*, August, **19**, 31–44.

Gold, T. (1958). The Arrow of Time. In *11th Solvay Conference "Structure and Evolution of the Universe,"* part 1, pp. 81–95. Brussels: Stoops.

(1962). The Arrow of Time. *American Journal of Physics*, **30**, 403–410.

(ed.) (1967). *The Nature of Time*. Ithaca: Cornell University Press.

Hogarth, J. E. (1953). *Particles, Fields, and Rigid Bodies in the Formulation of Relativity Theories*, Ph.D. thesis (University of London).

(1962). Cosmological Considerations of the Absorber Theory of Radiation. *Proceedings of the Royal Society*, **A267**, 365–383.

Hoyle, F. and Narlikar, J. V. (1961). On the Counting of Radio Sources in the Steady-State Cosmology. *Monthly Notices of the Royal Astronomical Society*, **123**, 133–142.

[19] Hogarth and Hoyle and Narlikar had examined only Robertson–Walker models with zero space curvature ($k = 0$); Roe studied the corresponding models with positive and negative space-curvature ($k = +1, -1$). His conclusion was that $k = -1$ models behave in the same way as the Einstein–de Sitter model in giving self-consistent advanced (and not retarded) solutions of Maxwell's equations, while the universes with $k = +1$ give both the advanced and retarded solutions as self-consistent. "It thus appears" – he stated – "that a direct connection between the cosmological and electrodynamic arrows of time cannot be established in any Friedmann model. Continuous creation appears to be both necessary and sufficient to establish such a connection." Of course, Roe's results and conclusions have been used by Hoyle and Narlikar.

(1964a). Time Symmetric Electrodynamics and the Arrow of Time in Cosmology. *Proceedings of the Royal Society*, **A277**, 1–23.
(1964b). The C-Field as a Direct Particle Field. *Proceedings of the Royal Society*, **A282**, 178–183.
(1964c). On the Gravitational Influence of Direct Particle Fields. *Proceedings of the Royal Society*, **A282**, 184–190.
(1964d). A New Theory of Gravitation. *Proceedings of the Royal Society*, **A282**, 191–207.
(1966). A Conformal Theory of Gravitation. *Proceedings of the Royal Society*, **A294**, 138–148.
(1967). Time-Symmetric Electrodynamics and Cosmology. In *The Nature of Time*, ed. T. Gold, pp. 25–41. Ithaca: Cornell University Press.
(1972a). Cosmological Models in a Conformally Invariant Gravitational Theory, I. *Monthly Notices of the Royal Astronomical Society*, **155**, 305–321.
(1972b). Cosmological Models in a Conformally Invariant Gravitational Theory, II. *Monthly Notices of the Royal Astronomical Society*, **155**, 323–335.
(1974). *Action at a Distance in Physics and Cosmology*. San Francisco: W. H. Freeman.
Lilley, A. E. (1957). Radio Astronomical Measurements of Interest to Cosmology. In Witt (ed.) (1957), pp. 55–59.
McCrea, W. H. (1963). Cosmology – a Brief Review. *Quarterly Journal of the Royal Astronomical Society*, **4**, 185–201.
Møller, C. (ed.) (1962). *Evidence for Gravitational Theories.* New York: Academic Press.
Narlikar, J. V. (1962). Neutrinos and the Arrow of Time in Cosmology. *Proceedings of the Royal Society*, **A270**, 553–561.
Roe, P. E. (1969). Time Symmetric Electrodynamics in Friedmann Universes. *Monthly Notices of the Royal Astronomical Society*, **144**, 219–230.
Ryle, M. (1955). Radio Stars and their Cosmological Significance. *The Observatory*, **75**, 137–147.
Sánchez-Ron, J. M. (1983). Quantum *vs.* Classical Physics: Some Historical Considerations on the Role Played by the "Principle of Correspondence" in the Development of Classical Physics. *Fundamenta Scientiae*, **4**, 77–86.
Sciama, D. W. (1961). Les Trois Lois de la Cosmologie. *Annales de l'Institut H. Poincaré*, **17**, 15–24.
Wheeler, J. A. (1979). Some Men and Moments in the History of Nuclear Physics: The Interplay of Colleagues and Motivations. In *Nuclear Physics in Retrospect*, ed. R. H. Stuewer, pp. 217–306. Minneapolis: University of Minnesota Press.
Wheeler, J. A. and Feynman, R. P. (1945). Interaction with the Absorber as the Mechanism of Radiation. *Review of Modern Physics*, **17**, 157–181.
(1949). Classical Electrodynamics in Terms of Direct Interparticle Action. *Reviews of Modern Physics*, **21**, 425–433.
Whitehead, A. N. (1922). *The Principle of Relativity.* Cambridge: Cambridge University Press.
Witt, C. M. (ed.) (1957). *Conference on the Role of Gravitation in Physics*, WADC Technical Report 57-216, ASTIA Document No. AD 118180. Ohio: Wright Air Development Center.

Discussion

Bondi:

Hogarth was a frequent attender at my seminar at King's College, London, and I got to know him and I took great interest in his work. Then and throughout this period my interest in cosmology had waned, but I was still very much interested in the link between thermodynamics and cosmology.

I was then (e.g., at Varenna) less interested in supporting the steady-state model than in opposing the then popular notion of an oscillating universe, which to me is thermodynamically a horror.

V

Cosmological observations and discoveries

14

The observational approach to cosmology: U.S. observatories pre-World War II

Donald E. Osterbrock

American physics in the days of Joseph Henry, Henry A. Rowland, A. A. Michelson, and Robert A. Millikan was strongly experimentally oriented, particularly toward precise measurements of well-defined physical quantities. Likewise American astronomy, except for its celestial mechanics component, was largely observational in character, with very little theoretical development. Thus in the years before World War II, the main contributions of American astronomy to cosmology came from observations. They were not planned to solve any cosmological problem, but rather to investigate the nature of specific objects. They gradually revealed the existence of highly luminous galaxies, making it possible to study and measure objects in the distant reaches of the universe. Some of these early observational studies are well known to students of cosmology; others are less so. In this paper an attempt is made to collect these fragments of history, particularly those which took place at Lick, Lowell, and Mount Wilson Observatories.

One of the first American observational cosmological studies, however, was based on theoretical ideas. Simon Newcomb, the long-time head of the Nautical Almanac Office, was primarily a great systematizer of solar-system orbital calculations and reductions of positional measurements (fig. 14.1). Generally considered the outstanding American astronomer of his time, he had wide-ranging interests and was a prolific writer. Like most astronomers then, he thought of our Milky Way Galaxy as "the universe." He knew of what we call Olbers' paradox (Harrison 1987), and realized that a measurement of "the total light of all the stars" would provide information on the size of our Galaxy (Newcomb 1901a, 1902, 1904). He made a "rude attempt" to measure this quantity himself while on his summer vacation in New Hampshire in 1901, by visual photometry (Newcomb 1901b), and suggested to W. W. Campbell, who had recently been appointed director of Lick Observatory, that the study be continued there (fig. 14.2). Only a few months before, Newcomb had strongly recommended Campbell for the first Nobel Prize in Physics (he did not get

Fig. 14.1 Simon Newcomb the "grim dean of American astronomers." (Mary Lea Shane Archives of Lick Observatory Photograph)

it; Wilhelm Roentgen did) and the young director gladly followed the advice of the "grim dean of American astronomy" who had provided counsel, solicited and unsolicited, to Lick Observatory astronomers from its earliest days. Campbell put Sidney D. Townley, a young Berkeley instructor, to work on the problem. He made his first measurements at Mount Hamilton in the summer of 1902, using a photographic method similar to Newcomb's visual method (Townley 1903). Neither of them realized the existence of the airglow in the Earth's upper atmosphere, and therefore did not eliminate it from their measurements.

Fig. 14.2 W. W. Campbell in 1901, soon after he became director of Lick Observatory. (Mary Lea Shane Archives of Lick Observatory Photograph)

Townley's preliminary results agreed fairly well with Newcomb's. Campbell wanted Townley to continue the work the following summer, but before then the young man got a better job at the International Latitude Observatory in Ukiah, California. Campbell arranged to have more photometric plates for measurements of the light of the sky taken for him at Mount Hamilton, and Townley returned once for a few nights' observing himself, but he was far more interested in visual photometry of variable stars, which he considered his specialty. He never published any more results on the light of the sky, and allowed this program to die.

Townley was too conventional to jump into a new field so far from his previous interests, even though the "greatest astronomer of his time" and the director of the outstanding Western observatory wanted him to do so.[1]

The main line of American observational research on what turned out to be cosmology began in 1898 with James E. Keeler's photographic survey with the 36-inch Crossley reflector at Lick Observatory. This telescope, the first "large" reflector at a professional observatory in the United States, came to the University of California as the gift of an English amateur astronomer. Edward S. Holden, the first director of Lick, considered obtaining this telescope the triumph of his administration, but he could not get any of his staff members to put it into operation and use it for research. It became the final episode in the staff rebellion which brought about his departure from Mount Hamilton in 1897, and his resignation.[2]

Keeler, who had been the first astronomer on the Lick Observatory staff, but who had left to become director of Allegheny Observatory in 1891, returned to Mount Hamilton as Holden's successor in 1898 (fig. 14.3). He announced that he intended to use the Crossley reflector himself for spectroscopy of faint stars and nebulae, leaving the 36-inch "great refractor" for the radial-velocity program that Campbell had begun after Keeler's earlier departure. The new director, an excellent observational astronomer, soon had the Crossley reflector in working condition, and began using it for direct photography of nebulae, while planning an efficient spectrograph to be built for it. Keeler marveled at the faint detail he could see on his long exposures of spiral "nebulae," taken at this superb clear-sky, good-seeing site. He could also see, on the same plates, smaller and smaller spirals, all similar in general form, down to the limit of resolution of his telescope. Keeler recognized them as flattened, rotating objects, seen in various orientations, of the same general size but distributed in space out to the most distant observable reaches of the universe. He estimated the total number that could be photographed with the Crossley reflector as conservatively 120,000 (Keeler 1899, 1900).[3]

However, Keeler, like most of his contemporaries, considered the spirals as "nebulae," rotating masses of gas or dust. He was familiar with the Chamberlin–Moulton hypothesis, that the solar system originated in a rotating nebula, drawn out of the Sun as the result of a tidal interaction with a passing star, and that spiral nebulae might be more recent examples

[1] This section is based on many letters between Campbell and Newcomb (1901–1903) and between Campbell and Townley (1902–1904) in the Mary Lea Shane Archives of Lick Observatory, University of California, Santa Cruz (afterward referred to as SLO).

[2] The complete story, with many references to original letters, is in Osterbrock (1984a).

[3] These are Keeler's first and final papers on the subject; references to his many other papers on gaseous and spiral "nebulae," as well as to hundreds of letters he wrote or received are given in Osterbrock (1984b). Keeler's work on spiral "nebulae" is fully described, especially on pp. 297–326 of this book.

Fig. 14.3 James E. Keeler, pioneer observational research worker on spiral "nebulae." (Mary Lea Shane Archives of Lick Observatory Photograph)

of this same process. Keeler corresponded with Thomas C. Chamberlin on this subject, and sent him a large set of copies of his photographs of spiral "nebulae"; he also continued to emphasize the very large number of spirals, which made this explanation unlikely.

Keeler had intended from the beginning to use the Crossley reflector to obtain spectra of faint nebulae. He was well aware that planetary and diffuse nebulae have emission-line spectra, indicative of a hot, low-density gas (much of his own earlier research was concerned with the spectra of these objects), but that the spirals showed no emission lines and appeared to have only faint continuous spectra. His direct, immediate interest in this problem was reawakened when he read the paper of Julius Scheiner (1899) who, with a 12-inch F/3 reflector at Potsdam, obtained a spectrum of M 31, the large spiral Andromeda Nebula. He reported the presence in it of two absorption features, the Fraunhofer H and G lines, characteristic of the Sun and other cool stars. On the basis of this spectrum Scheiner identified the Andromeda Nebula, and by analogy other spiral "nebulae," as consisting of stars like our Milky Way, and our Milky Way in turn as a spiral star system. In fact he had the answer! But Keeler and other astronomers of the time were skeptical. They knew how faint M 31 is, and how difficult it was to obtain a satisfactory spectrum of it. They knew Scheiner's result partly contradicted an earlier paper by William Huggins, who had reported the presence of emission lines also in the spectrum of M 31. In any case it was only one spiral.

Keeler decided he must get a low-dispersion, fast spectrograph for the Crossley, to check Scheiner's report and to obtain spectra of other spirals. He completed his design, ordered the necessary optics, and put his technician to work building the instrument. However, the spectrograph was not completed until the very day Keeler left Mount Hamilton for the last time, ill and destined to die unexpectedly a few months later on August 12, 1900, at the age of forty-two.[4]

After Keeler's death, Campbell became director of Lick Observatory. He shifted its program increasingly to the measurement of radial velocities, the study of stars which dominated astronomical research in the first half of this century. However, Campbell assigned Charles D. Perrine to continue direct photography with the Crossley reflector. Perrine, who had come to Lick Observatory as its secretary and had no training in science, had converted himself to an astronomer by dogged observational work. He had no creative new insights, but obtained better photographs of many of the nebulae Keeler had first taken, and also of fields Keeler never had time to get. Many of the photographs of spirals reproduced in the Keeler memorial publication (Keeler 1908) were actually taken by Perrine. From the plates which Keeler and he had taken, Perrine (1904) estimated the total number of "nebulae" within reach of the Crossley as 500,000. His result was much higher than Keeler's chiefly because the latter, the first researcher in the field, had consciously adopted a very conservative underestimate of the average number of nebulae he discovered per exposure.

[4] Keeler's ideas on spectroscopy of spiral "nebulae" are described more fully, with complete references, in Osterbrock (1984b: 351–355).

Edward A. Fath did pioneering work on the nature of the spiral nebulae in his Ph.D. thesis at Lick Observatory, completed in 1909, but it received very little recognition at the time (fig. 14.4). Born in Germany of American parents, he was educated at tiny Wilton College, Iowa, and at Carleton College, Minnesota, then taught mathematics, physics, and chemistry for three years before going to the University of Illinois for his first year of graduate work in astronomy. Joel Stebbins, Fath's professor there, described him to Campbell as not brilliant in mathematics but skillful in observing, a pretty fair mechanic, and a neat and accurate computer. On the basis of this recommendation Fath moved on to Mount Hamilton in 1906, with several years' more practical experience than the typical beginning students. He learned the technical aspects of observational research as Perrine's assistant during his first year at Lick.

Fath's thesis was devoted to a spectroscopic study of the spectra of spiral nebulae (Fath 1909). He designed and constructed a special fast nebular spectrograph for this problem himself. He recognized that its speed would depend entirely on the focal ratio of its camera, and used an F/3 lens, giving a dispersion of approximately 400 Å/mm, just enough to see absorption lines if present in a spectrum. He made the box of the spectrograph of wood, shellacked and screwed together. It was very fast, simple, inexpensive, quickly built, and incapable of being used for precision radial-velocity measurements, and was thus the exact antithesis of the other Lick spectrographs of the time. Fath estimated that to get a satisfactory spectrogram of M 31 would have required an exposure of 450 hours with the standard Mills spectrograph on the Lick 36-inch refractor. With his special nebular spectrograph on the Crossley reflector, Fath did obtain a good spectrogram of M 31 with an 18-hour exposure spread over three nights at the end of September 1908. (It followed an $8\frac{3}{4}$-hour exposure he had taken the previous month, which was not long enough.) On this spectrogram Fath could see a definite absorption-line spectrum, extending through the nucleus out to a distance of $2'.5$ on either side of it. He could identify fourteen absorption lines or features in the spectrum, and classified it as "of the solar type." There were no emission lines such as Huggins had earlier reported.

Fath (1909) also obtained spectrograms of six other spiral "nebulae," with exposures ranging from five to fifteen hours. One was too faint to show anything; the other five all were weaker than the spectrogram of M 31 but showed some of the same "solar-type" absorption lines. His spectrogram of NGC 1068 was confusing, for it showed the characteristic emission lines of a gaseous nebula as well as the solar-type absorption-line spectrum; today we know it as the brightest Seyfert galaxy or active galactic nucleus. In one other spiral, NGC 4736, Fath thought he saw an unidentified broad emission line, but in fact it was a bright region in the continuum between absorption features, as we know from better spectrograms today.

Fig. 14.4 Edward A. Fath, whose spectra showed the spirals are galaxies of stars. (Mary Lea Shane Archives of Lick Observatory Photograph)

Fath concluded that no spiral "nebula" has a pure continuous spectrum; they all had absorption lines of "solar type." As the center of the M 31 is an extended object it is clearly not a single star; yet its spectrum is continuous with absorption lines, and stars are the only known objects that emit such spectra. The conclusion was inescapable, he said; the center of the Andromeda Nebula must be a "star cluster" composed of many stars, individually too faint to be seen but numerous and close enough together to appear as a "nebula." He worried whether it was "reasonable" for all the stars in the "cluster" to have similar solar-type spectra. To check, Fath took spectrograms of several globular clusters, which were known to be composed of many individual stars. Putting the telescope slightly out of focus to smear the images of the stars, he obtained spectrograms of M 2, M 13, and M 15, and they were indeed continuous, with absorption lines of solar-type stars, fairly similar to the spectra of the spiral "nebulae." Fath had brilliantly discovered and proved the nature of the spirals. However, at the end of his paper he weakened his case by stating that his "hypothesis" must stand or fall on a determination of the parallax (that is, a direct distance measurement) of M 31. Very probably the conservative Campbell, who did not want anyone from his observatory ever to publish and later have to admit a mistake, had insisted that Fath include this disclaimer. It was the path of least resistance which most of his contemporaries followed, for Fath's result, so clear to us today, had little impact at the time, and the nature of the spirals remained an open question for at least another decade.

After he got his Ph.D. in 1909, Fath went south to join the staff of the Mount Wilson Observatory. He had met its young director, George Ellery Hale, on Mount Hamilton in 1909 and had asked him for a job; no doubt Campbell had recommended Fath strongly for it. The new 60-inch reflector, the largest telescope in the world, had just gone into operation, and Hale was glad to hire an experienced observer like Fath.

The Mount Wilson spectrograph was slower than the instrument Fath had used at Lick. It was mounted parallel to the tube, at the Newtonian focus, requiring two extra reflections, and the camera lenses were thick and absorbed much of the light. Nevertheless, Fath got good spectrograms of two spiral "nebulae," one of them of NGC 4736 which was much better than his earlier spectrogram of this same object taken at Mount Hamilton. Both spirals showed solar-type absorption spectra, further confirming his ideas. However, one other object he observed, NGC 650/1, actually a planetary nebula but with a somewhat spiral form, confused him. It had a purely emission-line nebular spectrum, but it appeared to belong to the class of spiral "nebulae." To explain it and NGC 6543, another planetary nebulae with a spiral form and an emission-line spectrum, Fath (1911) suggested a sequence of nebular forms associated with spectra. It ranged from the most irregular nebulae such as the Orion Nebula, with the strongest nebular emission-line spectra (H II regions in modern terminology), through "barely spiral nebulae" such as NGC 650/1 and "better spiral nebulae" such as NGC 6543, with nebular emission lines (planetary nebulae), to spiral "nebulae" with nebular emission lines plus solar-type absorption-line spectra like NGC 1068 (Seyfert galaxies), and ended in the spiral "nebulae" with pure solar-type spectra, like the M 31 (normal galaxies in modern terminology).

At Mount Wilson Fath also obtained integrated spectra of eight globular clusters, six of them well exposed enough to be studied. He described the globular-cluster spectra as "almost identical spectra, approximately F type," that is, somewhat earlier than the "solar-type" or G-type spectra of the spirals. From his description of the spectra of the clusters, it is clear that he recognized them as composite (though he did not use that word), containing features of stars of a range with spectral types.

Fath obtained spectrograms of several more spirals, with exposures lasting three, four, or five nights, ranging up to nearly 40 hours in total length. In his final summary of this work he reported that he had investigated the spectra of eleven nebulae, and that all of them but NGC 650/1 were spirals, indicating that he had recognized the latter as a planetary. Most of the spirals had G- or K-type absorption-line spectra. NGC 1068 was peculiar in showing nebular emission lines also, and NGC 4736 as having the unknown apparent emission feature. He also had spectra of four globular clusters, and they were composite, containing stars of F and G types. However, in this paper Fath (1913a) did not repeat that the spiral "nebulae," like the globular clusters, must be composed of stars; evidently he was concerned because the spectra of the spirals did not appear clearly composite to him.

At Mount Wilson Fath (1912) also obtained spectra of three of the brightest "star-cloud" regions in the Milky Way. For this program he borrowed the fast nebular spectrograph he had built at Lick Observatory. The exposure times were very long, ranging from 65 to 74 hours, each extending over a month or two of the parts of nights when the Moon was down. All three regions showed absorption-line solar-type spectra like the spiral "nebulae." Fath (1913b) explicitly compared "our entire stellar system, which we call the Milky Way" with M 31. He said that he had expected the Milky Way would have a mainly A-type spectrum, because the Harvard spectral survey, then recently published, showed that most of the individual brighter stars have A-type spectra, and that their fraction increased to fainter magnitudes. However, Fath (1912, 1913b) stated, his spectra proved that at still fainter magnitudes the situation must change, and that an increasing number must be G- and K-type stars. He obtained photometric evidence, based on plates taken with blue and red filters, which confirmed that many of the faint Milky Way stars are indeed red. All of Fath's results have held up; the earlier-type stars are in the nearby subsystem we call Gould's Belt, while the more distant regions of our Galaxy are dominated by the light of late G- and K-type giant stars.

With the 60-inch reflector, Fath (1914) obtained one-hour-long direct photographs of all 139 of J. C. Kapteyn's Selected Areas from the north celestial pole to declination $-15°$. Each plate covered an area of 1.88 square degrees. Fath examined these plates carefully and counted the number of "nebulae" on each. He confirmed Keeler's result that many of them are spirals, down to the limit of resolution of the telescope, and he found that the total number in the whole sky, estimated from the number observed in the Selected Areas, was 160,000. Comparing a few specific fields he had in common with fields photographed with the Crossley reflector at Lick, Fath found that both his survey and Keeler and Perrine's earlier survey reached essentially the same magnitude limit. However, Fath's resulting total number of spirals was only about one-third as large as the number Perrine had derived. Fath thought the difference was due to the fact that his survey, taken at points equally spaced in the sky (as the Selected Areas were laid out), correctly took into account the concentration of the spirals to the poles of the galactic plane, while Perrine's estimate did not. Keeler's program had been concentrated on known nebulae, especially on spirals, and thus his selection of fields was biased toward regions where there are spirals (the galactic poles), and away from regions where there are none (near the galactic equator). Thus Perrine's estimate, made under the assumption that his fields provided a fair sample of the entire sky (as he had stated), was somewhat too small. However, as Perrine (1916) later reported, the main reason for the discrepancy was that Fath used much larger fields at the 60-inch, which were badly affected by coma in their outer parts, causing him to miss many faint "nebulae." Still later Frederick H. Seares (1925) at Mount Wilson rediscussed Fath's data, confirmed that the correction for coma was important, and derived the total number of "nebulae" down to the limiting magnitude of these plates as 300,000. Fath (1914) also noted that the distribution of spirals was not a function of galactic latitude alone, and in particular was more concentrated to a clustering around the north galactic pole than to the south.

Campbell, who followed Fath's research carefully, was struck by the "most remarkable" distribution of the nebulae in the sky "in that the gaseous ones seem to prefer the Milky Way and the continuous spectrum ones seem to abhor the Milky Way." He wrote to Fath in 1914 to ask if his data gave any indication that the (observed) spirals are intrinsically brighter the closer they are to the galactic equator, which would indicate that the apparent scarcity of spirals near the Milky Way might be due to absorption or obscuration of their light close to the galactic plane. When he wrote this letter, however, Campbell was about to depart on an eclipse expedition to Russia, while Fath, who had left Mount Wilson Observatory for a faculty position at Beloit College in 1912, had moved on from there to become president of tiny Redfield College, South Dakota. Thus neither one of them could follow up the question then, and neither ever found time to return to it, but in fact Campbell was right, as other astronomers demonstrated a few years later (see below). Fath returned to astronomy in a faculty position at Carleton College in 1922, but he never went back to the research on spiral "nebulae" in which he had accomplished so much in a few short years.[5]

Fath was present at one of the most important astronomical meetings held in the first quarter of this century, the meeting of the International Solar Union at Pasadena and Mount Wilson in 1910. It was the occasion on which the new 60-inch telescope was first shown off to the astronomers of the world, and many of them were there to see it. The meeting in fact was devoted as much to stellar astronomy as to solar, though no papers were presented on research on spiral "nebulae." At this meeting Fath met Vesto M. Slipher, an astronomer from Lowell Observatory who was just five years older than himself (fig. 14.5). Slipher had received a bachelor's degree from Indiana University in 1901, and then had been hired almost immediately by Percival Lowell, to work as an assistant in his private observatory in Flagstaff, Arizona. Slipher's job was to get into operation and use for research a new spectrograph Lowell had just had built for the 24-inch refractor that was the major telescope of the observatory.

Lowell, the scion of an old and wealthy Boston family, had built his observatory to search for life on Mars. He confined himself to visual observing on his visits to Flagstaff, timed to coincide with the periods when Mars was near opposition, closest to the Earth, but he assigned Slipher to learning spectroscopy and applying it to planetary research. Lowell wanted Slipher to use the spectrograph to measure the rotational velocities of the planets (Hoyt 1980), and to detect and measure the rotation of M 31, which he thought might be an embryonic solar system in formation (Hall 1970).

Slipher, though he had practically no training in observational astronomy, was a slow, careful, analytical worker, and he essentially taught himself spectroscopy. He did take excellent spectrograms of the planets, measured their rotational velocities accurately, and found many new molecular absorption bands in the infrared spectra of the outer planets Jupiter, Saturn, Uranus, and Neptune. At his own research, in the time left over from his assigned planetary work, he made excellent radial-velocity measurements of several spectroscopic binaries. In the course of this work he discovered independently and correctly identified the sharp,

[5] This section is based on letters between Fath and Campbell (1906–1922), SLO, and letters about Fath from W. W. Payne and J. Stebbins to Campbell (1906), SLO; on letters between Fath and Hale or W. S. Adams (1903–1913) in the Hale Microfilm Papers; and also on the Annual Reports of the Director of Mount Wilson Solar Observatory (1909–1912) published in the *Carnegie Institution of Washington Year Books*.

Fig. 14.5 V. M. Slipher, first observer of the large radial velocities of the spiral "nebulae." (Lowell Observatory Photograph)

stationary absorption lines of Ca II as resulting from interstellar gas (Hall 1970; Hoyt 1980).

Thus in his early years at Lowell Observatory, Slipher had become an expert in planetary and stellar spectroscopy, with many important results to his credit. In 1909 Lowell reminded him that he wanted information on the spectra of the nebulae, too. At the meeting of the International Solar Union, at the end of August 1910, the young assistant had several long conversations with Fath about nebular spectroscopy. Soon after his return to Flagstaff, Slipher began getting into the field himself. On his own he worked out that for an extended nebula the speed of a well-designed spectrograph does not depend at all on the aperture of the telescope, but only on the focal ratio of the camera and the angular dispersion of the prism. Unfamiliar with the astronomical literature, he did not realize that this result had been mentioned in papers published by Keeler nearly twenty years before.

Slipher had a highly dispersive prism, and was able to get a fast F/2.5 Voigtlander camera lens to use with it for obtaining the spectra of nebulae. Its resolution was not perfect, but was adequate for his purposes. The only spectroscopist at Lowell, and with practically complete use of the telescope for his own research whenever the planets were not accessible for observation, Slipher was able within a few years to find many important new results.

First he obtained a good spectrum of the Pleiades Nebula. From its appearance it would be expected to be a gaseous nebulae, but its spectrum had proved too faint for previous observers to obtain. Slipher (1913a) took a 21-hour exposure with his fast spectrograph in December 1912. When developed, it clearly showed no trace of emission lines, but a continuous spectrum with exactly the same hydrogen and helium absorption lines that occur in the spectra of the bright stars of the Pleiades cluster. Slipher carefully tested that this spectrum was not due to light from the stars scattered in his telescope or spectrograph, but was the true spectrum of the nebula. It could not arise by emission in a gaseous nebula. It could not arise from a vast collection of faint stars; it was too uniform and too clearly associated with the Pleiades, a cluster of very bright stars. Slipher (1913a) correctly concluded that his observations could only be explained on the basis of "the assumption that the nebula is disintegrated matter similar to what we know in the solar system, in the rings of Saturn, comets, etc., and that it shines by reflected star light." He had discovered reflection nebulae. He even did a simple calculation which demonstrated, that Merope, the bright Pleiades star directly involved in the nebula, could easily provide the observed amount of light scattered by the "dust," as we now call it.

Soon thereafter Slipher (1913b) obtained a good, long-exposure spectrogram of the Andromeda Nebula. He confirmed Fath's result, that its spectrum is continuous, with solar-type absorption lines. Unfortunately, by analogy with his earlier result, he believed that the central nucleus of M 31 was a single star, and that the extended spiral was another reflection nebula, similar to but more regular than the Pleiades Nebula (Slipher 1913a, 1913b). Before long Slipher had a good exposure of the spectrum of another spiral, M 81, and it was very similar to that of the M 31. In letters to Campbell, Slipher argued that these spectra were each so "pure" (like the spectrum of a single star), M 31 of early K type, and M 81 of late G type, while the spectra of the globular clusters he had by now taken were clearly "composite in that their hydrogen lines and solar bands are mutually too intense to confirm to the established types of stellar spectra" that it was "hardly reconcilable with the theory that the spirals are galaxies of stars in that they are star clusters at irresolvable distances." He thought a spiral nebula might be "a sort of planetary system (larger than our solar system) composed of matter from dust-clouds to suns in size and development."

Slipher, with his rigid, metal spectrograph, was able to measure accurate wavelengths and thus determine the radial velocities of spiral "nebulae" from their spectra, as Fath had not been able to do with his wooden instrument. For M 31 he found the "extraordinary" velocity of −300 km/sec (approach), much larger than had been measured for any star. Campbell found it hard to believe, and advised Slipher to consider possible sources of errors carefully, and take several more spectrograms to check his result. Slipher had already taken three more, as he quickly advised Campbell. He also sent the Lick director photographic copies of his best spectrograms of the Pleiades and Andromeda Nebulae. Campbell could see that they were excellent, far better than any previous spectra of nebulae with continuous spectra that he had seen. His whole attitude toward Slipher's work changed from barely skeptical tolerance to real respect, and he gave him the ultimate Lick accolade "Your apparatus appears to be very efficient in collecting light and in conserving the collected light . . . [The spectrograms] are fine pieces of work, and I cordially congratulate you on them." Furthermore, Campbell assigned W. H. Wright to check Slipher's result for the velocity of M 31 at Mount Hamilton. Wright did so with an 18-hour exposure, spread over two nights, using one of the regular Lick spectrographs on the 12-inch refractor. He measured sixteen absorption lines; and his final result, −305 ± 10 km/sec agreed practically perfectly with Slipher's. From then on, Campbell never doubted Slipher's measurements, though he did sometimes question his interpretations.

By the end of 1913 Slipher had obtained the spectra of three more spirals, NGC 1068, 4565, and 4594. They had velocities even larger than that of M 31, each of order 1000 km/sec, all three positive. In a paper read for him at the American Astronomical Society meeting at Atlanta, Slipher (1914) reported these unheard of velocities. Henry Norris Russell, the brilliant young astrophysicist who presented his first version of what we now call the Hertzsprung–Russell diagram at this meeting, was frankly incredulous.

Within another six months Slipher had measured the spectra of several more spiral nebulae. He found that although M 31 was nearly unique in its negative velocity, almost all the fourteen velocities he had measured were numerically large, ranging from −300 km/sec for M 31 to the then fantastically high value +1100 km/sec for M 104. The average velocity without regard to sign was 400 km/sec, roughly 25 times as large as for stars, and strongly suggested that the "spiral nebulae" were "a long way from along the evolutionary chain" if what Slipher called the "Campbell–Kapteyn" increase in velocity with "advance in spectral type" were accepted (Slipher 1915). He presented this paper himself at the August 1914 meeting of the American Astronomical Society at Northwestern University, Evanston, Illinois. At its conclusion Slipher also reported that one spiral "nebula," NGC 4594, which he had observed with the slit parallel to its long axis, showed inclined spectral lines indicating a measurable rotational velocity. As he finished presenting this paper, Slipher received an unprecedented standing ovation (Hall 1970). To Campbell, Slipher's "results compose one of the greatest surprises which astronomers have encountered in recent times. The fact that there is a wide range in observed velocities – some of approach and some of recession – lends strong support to the view that the phenomena are real; nevertheless, the fact that so large a proportion of the velocities are absurdly high should lead us to hold in mind, I think, the search for an explanation other than the Doppler–Fizeau effect."

At Mount Wilson, Francis G. Pease, an optician and engineer whom Hale and Walter S. Adams put to work as an observer, also began measuring radial velocities of nebulae, including spirals, with a small, fairly fast spectrograph on the 60-inch. He essentially confirmed Slipher's results for M 31 and M 33, and found a somewhat different but still large velocity for NGC 1068 (Pease 1915a, 1915b, 1915c, 1916). It had become almost impossible to doubt Slipher's results. Astronomers who were familiar with analyzing radial-velocity measurements of groups of stars used the same methods to determine the motion of the Sun with respect to the spiral nebulae. O. H. Truman (1916) of the University of Iowa, who had worked with Slipher one summer at Flagstaff, and Reynold K. Young and W. E. Harper (1916) of the Dominion Observatory in Ottawa, using published data, mostly Slipher's, independently found that the Sun's velocity with respect to the system of spirals was very large, of order 600 km/sec, comparable with the velocities of these objects with respect to one another. As they stated, this provided strong evidence that the spiral "nebulae" are not objects similar to stars in our Milky Way system, but it would agree well with the idea that the spirals are independent stellar systems like the Milky Way itself. The Indiana amateur astronomer and writer Russell Sullivan (1916) reached this same conclusion simply by a qualitative comparison of the average radial velocities of the various types of nebulae and stars. Even earlier, Lowell himself had made the same deduction, and though he did not publish it, he stated it in a lecture he gave in Boston in 1915.[6] In his masterful summary of his work on the spectra of

[6] Lowell's statement that the spiral nebulae are not solar systems in formation but "something larger, and quite different, other galaxies of stars," is in the text of his lecture on "Nebular Motion" to the Melrose Club, Nov. 23, 1915. The manuscript is in the Lowell Observatory Archives. It was discovered there by the late William G. Hoyt, who mentioned it in his unpublished paper, "The Universe Began Expanding in Arizona," which he presented orally at the annual convention of the Arizona Historical Society in Flagstaff, June 11, 1976.

the spirals, Slipher (1917) repeated the quantitative analysis with the additional data he had by then obtained and concluded that it supported "the so-called 'island-universe' theory, ... which regards our stellar system and the Milky Way as a great spiral system" and "that the spiral nebulae are stellar systems seen at great distances."

Although M 31 and a few of the other spiral nebulae first measured have negative radial velocities, by 1916 it was clear that most of them were positive. The first paper to take this fact into account quantitatively was by George F. Paddock, a long-time Lick Observatory assistant. After completing his undergraduate work at the University of Virginia, he had worked for five years at the Lick southern station in Chile, then returned to the U.S., received his Ph.D. at Virginia, and continued at Mount Hamilton as Campbell's assistant. In his paper on the radial velocities of the spiral "nebulae," Paddock (1916) followed the standard method used for analyzing the motions of early-type stars, representing each object's velocity as the sum of its own random velocity plus the expansion of the entire system about the Sun. Solving from the available published data he found the mean expansion or "K-term" to be in the range from +300 km/sec to +340 km/sec (depending on the details of his analysis), several times larger than its probable error. However, Paddock believed for no apparent reason except his own conservative prejudices that "the apparent systematic quantity K" would "probably diminish with increasing numbers of observed velocities," that is, that it was a chance fluctuation in the data. In fact, in an unpublished, internal Lick Observatory report Paddock stated that in this paper "the high velocity and apex of the Sun with respect to certain 14 spiral nebulae are shown to be of a doubtful character and dependent on too great errors of observation."[7] He had found an important result, but he could not believe it.

The German astronomer Carl W. Wirtz, who had made a career of trying to measure proper motions and parallaxes of nebulae, found the same result as Paddock a few years later when he also discussed essentially the same published radial velocities of the spirals. Wirtz (1918), in a paper written while he was in military service in World War I, found a somewhat larger mean expansion velocity, K = 650 km/sec, and he did not cast doubt on his result, though he emphasized that more data were necessary, the eternal cry of the observational astronomer.

Meanwhile, in parallel with the spectroscopic observations, direct photographic studies of the spiral nebulae by Heber D. Curtis were underway at Lick Observatory (fig. 14.6). He had been trained in classical languages at the University of Michigan, and had begun teaching at Napa College and then the University of the Pacific, but shifted to astronomy, and after a few summers at Lick and Ann Arbor, earned his Ph.D. at

[7] G. F. Paddock, [~July 15, 1916], "Work for the year July 1, 1915 to July 1, 1916," SLO.

Fig. 14.6 Heber D. Curtis at the Crossley reflector, with which he proved the spirals are "island universes" composed of stars, nebulae, and "occulting matter" (dust). (Mary Lea Shane Archives of Lick Observatory Photograph)

Virginia in 1902. He joined the Lick staff and worked at Mount Hamilton for four years, then spent three years in charge of the Chile station (Paddock worked under him there) before he returned to the United States in 1909. From then on Curtis worked with the Crossley reflector, trying to understand the nature of the nebulae. He made important contributions to the study of planetary nebulae, and even more important contributions to our knowledge of the spirals (McMath 1942; Aitken 1943).

From his hundreds of direct photographs of nebulae Curtis (1918a) could see that the spirals all belong to one family, and he could recognize the most nearby edge-on members of the class, such as NGC 891 and NGC 4594. The dark lines along their equators, he said, must be due to "bands of occulting matter," similar to the "dark nebulae" and "coal sacks" in our own Galaxy, studied so thoroughly by E. E. Barnard in his early wide-field photographs of Milky Way fields (Curtis 1917a, 1918b). Curtis had long believed that the spirals were in fact composed of stars, individually too faint to be photographed separately or "resolved" with the Crossley, but the first direct proof came when George W. Ritchey, using the 60-inch Mount Wilson reflector he had built, discovered a nova, or "new star" in the spiral "nebula" NGC 6946. It was clearly a star of a type known in our own Galaxy. And it was not unique; Ritchey went back over all the plates he had taken of spirals with the 60-inch since 1908, and found other novae that had occurred in M 81, NGC 2403, M 101, and several in M 31 (Ritchey 1917a, 1917b).

Curtis, armed with the knowledge of Ritchey's discovery, immediately checked his own Crossley plates, and found several more novae on them, in NGC 4527 and M 100. He recognized at once that the faintness of the novae that Ritchey and he had discovered in the spirals in comparison with novae in the Milky Way proved conclusively that the spirals are at great distances from us, far outside our Galaxy. He stated in this first, quick paper that "[t]he occurrence of these new stars in spirals must be regarded as having a definite bearing on the 'island universe' theory of the constitution of the spiral nebulae" (Curtis 1917b). In an immediate follow-up paper he gave a very clear, quantitative statement of just how the magnitudes of the novae prove the very large distances of the spirals, and that they are constituted of stars (Curtis 1917c). He wrote these papers in the time he could spare from his World War I work; Curtis was gone from Lick Observatory from the summer of 1917, when he set up a navigation school for merchant marine officers in San Diego, until mid-1919, when he returned from his duties as an optical expert designing range finders and fire control telescopes at the National Bureau of Standards in Washington.

Many of Curtis' ideas on the spiral nebulae as galaxies were first published by Roscoe F. Sanford in his Ph.D. thesis, done at Lick under Curtis' supervision (fig. 14.7). Sanford was an older student, who had worked as an assistant at Mount Hamilton, San Luis, Argentina, and the Lick southern station in Chile for nine years before he returned to California in 1915 (Wilson 1958). In his thesis, Sanford (1917) clearly explained that the spectra, velocities, and apparent distribution in the sky of the spiral "nebulae" all showed that they are really "other universes of stars, comparable with our own." The apparent absence of spiral nebulae from the Milky Way regions he correctly attributed to the "intervening absorbing or occulting matter" known in its most extreme form as "dark nebulae" from Barnard's work. Sanford (1917) even recognized in primitive form the two stellar populations made famous much later by Walter Baade, the yellow and red stars more evenly spread in a roughly spherical distribution about the center of "the [our] Galaxy," and the earlier-type blue stars in the outer parts of the plane, as in the spirals. After finishing this fantastically modern-sounding thesis, Sanford took a job at Mount Wilson in which he spent the rest of his life studying the spectra of carbon stars, hardly ever returning to problems of the structure of galaxies and the universe.

Campbell, in a lecture he delivered as the retiring president of the American Association for the Advancement of Science in December 1916, surveyed the entire field of study of the nature of the nebulae. Of the spirals he concluded "*We are not certain how far they are; we are not certain what they are.* However, the hypothesis that they are enormously distant bodies, that they are independent systems in different degrees of development, is the one which seems to be in best harmony with the known facts." He cited the evidence from their spectra, obtained by Slipher, that the spirals "consist chiefly of multitudes of stars," and from the direct

Fig. 14.7 Lick Observatory group in 1907, including W. W. Campbell (seated on chair at left), John C. Duncan, Edward A. Fath, Roscoe F. Sanford, and Charles D. Perrine (standing fourth, fifth, seventh, and eighth from left, respectively), all early observers of spiral "nebulae" and the stars within them. (Mary Lea Shane Archives of Lick Observatory Photograph)

photographs of Sanford and Curtis that "our stellar system" as a whole is most probably a spiral "nebula," or has analogies to a spiral nebula. He stated that the Chamberlin–Moulton picture of a spiral nebula as the result of a close passage of one star past another was "possible," but the very large number of spirals, and their avoidance of the Milky Way, rather than being in it where the stars are, seemed to negate this idea (Campbell 1917). By this time Curtis (1918c) had raised the estimate of the total number of spirals within reach of the Crossley reflector to at least 700,000, and possibly over a million, but the ever-conservative Campbell gave it as "tens of thousands."

However, one scientist who held out against the idea of spiral "nebulae" as distant galaxies was Mount Wilson astronomer Harlow Shapley, soon thereafter to become director of Harvard College Observatory. Almost immediately after Curtis' (1917c) second paper on the novae in spirals, Shapley (1917) using essentially the same data, was much more skeptical of the idea that "a spiral nebula, NGC 224 [M 31] for instance, is a remote stellar system." In a more complete paper two years later, Shapley (1919) summarized the evidence that "apparently leads to the rejection of the hypothesis that the spiral nebulae should be interpreted as separate stellar systems." This evidence was chiefly his own work on the distances of globular clusters, clearly in our Galaxy, which gave it such large dimensions that Shapley could not believe that the spirals were outside it. His distances were somewhat in error because he did not realize the importance of extinction by interstellar dust, making the cluster stars fainter and hence apparently more distant, but basically he had the right size. Our Galaxy *is* large. Shapley also constantly emphasized the bright "nova" that had appeared in M 31 in 1885 (which we today understand was a supernova, and hence intrinsically much more luminous than "ordinary" novae), and the apparent relative proper motions within the spirals measured by van Maanen, which we shall discuss below.

Curtis (1919), on the other hand, in a very perceptive paper, summarized all the evidence stated above, that now seems so irrefutable to us, that the spiral "nebulae" are actually "island universes" composed "of hundreds of millions of stars, . . . similar to our [G]alaxy in extent and number of stars." He mentioned the 1885 [super]nova, but treated it as an aberrant case. In 1920, Shapley and Curtis gave invited lectures summarizing these rival pictures before the National Academy of Sciences, in what came to be known later as the "Great Debate" (Curtis 1921; Shapley 1921). Actually it was not presented in a debate format and nobody "won," but Curtis was basically right and Shapley was not. By that time the bulk of informed astronomical opinion strongly favored the spirals-as-galaxies idea.[8]

[8] This section is based on the biographies mentioned in the text, and on correspondence between Campbell and Curtis (1918–1922), SLO. The "Great Debate" and the research leading up to it are described in greater detail by Hoskin (1976a, 1976b) and Smith (1982), but they tend to overlook much of the early Lick work.

However, one remaining obstacle to the universal acceptance of this picture was provided by the measurements of apparent interval motions within spirals by Adriaan van Maanen. He was born and educated in Holland, where he was a protégé of J. C. Kapteyn, the Groningen pioneer of the star-count method of studying galactic structure. After completing his doctoral degree at Utrecht in 1911, van Maanen came to the United States and worked one year as a volunteer assistant at Yerkes Observatory. Kapteyn was a Research Associate at Mount Wilson Observatory, where he spent each summer doing his own research, and advising Director George Ellery Hale and his staff on worthwhile projects to pursue. Kapteyn recommended van Maanen to Hale, and Fath, who had been working as Kapteyn's assistant, was eased out of his job, and van Maanen appointed in his place in 1912. Van Maanen's training had been in measuring small proper motions on photographic plates, and he continued this type of work all his life at Mount Wilson (Seares 1946). He acted on the principle that if he made enough measurements of enough points he could determine any quantity, no matter how small it was. He thus underestimated the effects of systematic errors, which are not random and do not average out, and also the human tendency we all have, no matter how much we try to conquer it, to get subconsciously the "right" or expected answers in apparently blind measurements of small quantities. As an assistant at Mount Wilson van Maanen measured hundreds of solar spectrograms in attempts to find the very small Zeeman effect that would indicate the general magnetic field of the Sun for which Hale was searching; several times Hale thought his assistant had found it but their further measurements did not confirm it.[9]

On his own van Maanen began measuring plates of spiral nebulae, taken with the 60-inch reflector, to see if he could detect proper motions of rotation within these objects. Slipher and Curtis had both independently tried to make such measurements earlier, but abandoned their efforts because they found the apparent movements so small, and the possible systematic errors so large. However, from his first measurement of M 101, based on two plates taken five years apart by Ritchey, van Maanen (1916) appeared to find motions that combined rotational and radial components, in the form of a streaming motion outward along the spiral arms. The mean motion, he stated, was $0''.022^{-1}$ (that is, about $0.''1$ over the five-year interval the plates spanned) at 5′ radius from the nucleus.

Van Maanen continued this program, measuring apparent proper motions of the same general form and amount in many other large spiral "nebulae," using second-epoch plates taken by John C. Duncan or himself after Ritchey left Mount Wilson (van Maanen 1921a, 1921b). As van Maanen realized from the start, and as he made explicit in a paper

[9] As described in Hale's letters to H. M. Goodwin (1912–1915), Huntington Library, San Marino, California.

summarizing his work up to that time, his measured proper motions were quite inconsistent with the spirals having the large distances derived by Curtis (van Maanen 1921c). Combined with available spectroscopically determined internal radial-velocity differences, they instead indicated distances for the spirals of at most 10^5 light years, or diameters of at most 300 light years, far smaller than our Galaxy. Van Maanen (1922) wrote: "This material does not seem to warrant the acceptance of the island-universe theory. It is possible of course that later results may contradict this statement; but it seems well to point out that the evidence based on the best material at present available does not indicate that the spiral nebulae are comparable with our Milky Way system." In a later summarizing paper (van Maanen 1923) he stated that it was "very improbable" that the displacements he had measured were instrumental, not real, and he concluded that the spirals must show a rotary or streaming motion. Furthermore, he said, these observational data confirmed "[t]he beautiful [theoretical] work of [Sir James] Jeans on the evolution of a highly compressible mass of rotating gas, as a gaseous nebula undoubtedly is." Van Maanen's observational result thus proved the correctness of his own mistaken preconceptions, at least in his own mind.

The Swedish astronomer Knut Lundmark, who came to the United States on a scientific visit a few years after World War I, studied very carefully the best photographs of the spiral "nebula" M 33. He noted, as other observers had before him, that it is "resolved into numerous star-like objects" in its outer parts, especially in the spiral arms. Some of them looked exactly like stars, he said. Others were slightly fuzzy; for this reason Ritchey, who had taken the best 60-inch plates of the spirals, had called them "nebulous stars." Lundmark (1921) pointed out that even on the largest scale photographs, most of the star images in a distant galaxy would be crowded together and appear as these slightly fuzzy, nebulous objects. Furthermore, the background of unresolved, fainter stars in such a system would appear as a nebulosity, adding to the fuzziness of the images of groups of stars. Lundmark recognized NGC 604, a large, nebulous object in M 33 with an emission-line spectrum, as a gaseous nebula similar to the Orion Nebula in our Galaxy. At Lick he obtained spectra of the spiral arms and of a few of the brighter objects in them with a small spectrograph on the Crossley reflector. From all this spectral evidence he concluded that it was "probable that the spiral nebula consists of ordinary stars, clusters of stars and some nebular matter." Combining these results with results he had obtained at Uppsala, that ordinary stars in our Galaxy are not at all clustered to M 33, he stated that together, all the observations of M 33 "speak for a large distance."

However, completely convincing evidence could only come from a larger telescope. This was the 100-inch reflector, which went into operation on Mount Wilson in 1919. Nearly simultaneously Edwin Hubble arrived on the scene as a new staff member. Within a few short years he was to become the outstanding investigator of "extragalactic nebulae" in the world.

Hubble was born in Missouri in 1889 of parents who moved to Wheaton, Illinois, but he always considered himself a displaced Kentuckian. He had a romantic, larger-than-life quality that is rare in practicing scientists (Robertson 1954; Mayall 1970a). At the University of Chicago he majored in physics and astronomy, worked part-time one year as a laboratory assistant to Millikan, won his varsity letters on the basketball and track teams, and earned his B.S. degree in 1910. He took a Rhodes Scholarship and spent three years at Oxford studying law and modern languages. Back in America he taught school for one year in Indiana, but he decided to return to the University of Chicago as a graduate student at Yerkes Observatory (Allen 1954). In August 1914, just before starting there, he attended the American Astronomical Society meeting at nearby Northwestern University at which Slipher presented his early paper on the spectra and radial velocities of the spiral "nebulae" (Hall 1970) (fig. 14.8).

Fig. 14.8 V. M. Slipher (second row from rear, second from left), Edward A. Fath (standing fifth from right, wearing hat), Edwin Hubble (standing third from right, bareheaded), and others at the AAS meeting, Northwestern University 1914, at which Slipher reported the large radial velocities of the spirals. (Dearborn Observatory Photograph)

Yerkes Observatory was then in its doldrums, and Hubble did his thesis on "Photographic Investigations of Faint Nebulae" under the supervision of Director Edwin B. Frost, who had never worked in the subject, and whose research career was quite pedestrian.

Hubble was a self-starter though, and his thesis was very creditable, considering the little help he got and the instrument he was using, the 24-inch reflector which Ritchey had built at Yerkes Observatory in 1901. On the plates Hubble took with it, he found several *clusters* of small, faint "nebulae." He commented on the large number of small, faint nebulae that are *not* spirals, but that are "round or nearly so, brightening more or less gradually toward the center, and devoid of detail." The brightest member of this class was M 60, and he said their spectra (which he had observed only for a few bright ones such as it, with a very low-dispersion objective prism system) were "continuous, and probably of the same type as those of the spirals and the globular clusters." Thus though Keeler and Curtis had supposed that all the small "nebulae" were unresolved spirals, Hubble had correctly recognized that they are not, and had defined the class of objects which he was later to name elliptical galaxies. He tried, without much success, to derive information on the distances of the nebulae from his meager observational data combined with half-understood theoretical ideas. His paper was not very good, but he was groping with inadequate instruments toward the solution of great problems (Hubble 1920). It foreshadowed much of his later career.

As Hubble was finishing his thesis Hale offered him a position on the Mount Wilson Observatory staff. In April 1917 the United States entered World War I on the side of Great Britain, however, and the former Rhodes Scholar immediately volunteered for the Army (Mayall 1970a). He was commissioned a captain, promoted to major, came under fire in France in November 1918, and was mustered out in San Francisco in August 1919. He went straight to Mount Wilson, and immediately plunged into work with the 60-inch and 100-inch reflectors (fig. 14.9).

In a masterful observational study of galactic nebulae, Hubble (1922a, 1922b) recognized and explained the fundamental difference between emission nebulae and reflection nebulae as the presence or absence of hot stars, and that their sources of luminosity are the stars involved in them. He defined the "non-galactic nebulae" as those whose "members tend to avoid the galactic plane and to concentrate in high galactic latitudes." He emphasized that they are not all spirals as Curtis had thought, but rather could be classified as spiral, elongated, globular, or irregular (Hubble 1922b). The "globular [elliptical] nebulae" were the most numerous, he stated.

In 1922 John C. Duncan, an Indiana graduate who had worked at Lowell Observatory, and then earned his Ph.D. at Lick Observatory in 1909, spent a year at Mount Wilson. He was the professor of astronomy at

Fig. 14.9 Edwin Hubble in his early years at Mount Wilson Observatory, about 1920. (From the archives of the Observatories of the Carnegie Institution of Washington)

Wellesley College, but he liked to visit Mount Wilson, where he specialized in nebular photography. On this visit he found three faint variable stars in the spiral "nebula" M 33, by intercomparing plates taken over many years, one with the 100-inch, several with the 60-inch, and others with the smaller telescopes at Lick, Yerkes, and Lowell Observatories. Duncan (1922) could not determine their periods, but he was sure that they were variables.

Hubble began his detailed study of "non-galactic nebulae" with the irregular NGC 6822, a system similar to the Magellanic Clouds. By 1923 he had, with the 100-inch reflector, found several small nebulae within it and twelve variable stars. The following year he turned his attention to M 31 and found, in addition to the novae of which by then a total of over thirty were known, six variable stars. For the brightest of these, only about eighteenth magnitude at maximum light, Hubble had 83 separate measurements of brightness. From them he determined its period, 31.4 days, and the form of its light curve, both of which showed it to be a Cepheid variable. This was the first clear and definite identification of a type of star other than a nova in a spiral "nebula." It definitely confirmed M 31 as a remote star system, and by analogy other fainter, smaller (in angular dimension) spirals as even more remote galaxies. Hubble also found four additional variable stars, although he had not by then determined the periods or light curves of any of the variables in this irregular system. He was a demon of energy, observing and making new discoveries in quick succession. By the end of 1924 he had identified several more Cepheid variables in M 31 and M 33, and sent a paper giving his results to the joint American Astronomical Society–American Association for the Advancement of Science meeting in Washington, D.C. Russell read it for him there on January 1, 1925. In this paper Hubble (1925a) derived from the Cepheid variables distances of about 285,000 pc for M 31 and M 33, thus, in the words of AAS Secretary Joel Stebbins, "bringing confirmation to the so-called island universe theory." The AAAS awarded Hubble a prize for this work as a noteworthy contribution to science. Later that year Hubble (1925b) published his definitive paper on NGC 6822, which included the light curves of several Cepheids in it which, by their apparent magnitudes, made it "the first object definitely assigned to a region outside the galactic system." He thought of the Magellanic Clouds as within our Galaxy, and evidently regarded the prior evidence from novae in several spirals described above as unconvincing.

The next year he published his long paper on M 33, "a spiral nebula as a stellar system" giving the evidence from the Cepheid variables that it is "an isolated system of stars and nebulae far outside the limits of the galactic system" (Hubble 1926a). In another outstanding paper that same year he summarized his statistical investigation of what he now called "extragalactic nebulae," systems complete in themselves and different from galactic nebulae, clouds of dust and gas, mingled with stars within a particular stellar system (Hubble 1926b). A few years later he wrapped up the first phase of this study with his long paper on M 31 (Hubble 1929a) (fig. 14.10). Spiral "nebulae" were "non-galactic" or "extragalactic

Fig. 14.10 Edwin Hubble at his desk, in the early 1930s. (From the archives of the Observatories of the Carnegie Institution of Washington)

nebulae" or, as we say today, galaxies.[10] Campbell (1926) by now was convinced that the spirals were galaxies, and that our Milky Way was one of them. His presidential address to the American Astronomical Society in

[10] This section is based on the Annual Reports of the Director of Mount Wilson Observatory 1921–1925, published in the *Carnegie Institution of Washington Year Books*, and on *Publications of the Astronomical Society of the Pacific*, **37**, 100, 1924, and *Popular Astronomy*, **33**, 252, 1925.

1926, endorsing this view, no doubt persuaded many of the remaining skeptics.

As Hubble carried out his direct photographic work on the "extragalactic nebulae" with the 100-inch, Slipher continued his spectroscopic work. In 1921 he measured the highest velocity then known, +1800 km/sec for the spiral NGC 584. This spectrogram required an exposure of 28 hours, spread over the clear, moonless parts of all the nights from December 31 through January 14. For another spiral, NGC 936, he obtained an only slightly smaller velocity from a 34-hour exposure (Slipher 1921a). Clearly, he was reaching the limits of what his telescope, spectrograph, and method could contribute. For some reason now unknown, Slipher (1921b) reported the result for NGC 584 in the *New York Times*. In this newspaper article he stated that if NGC 584 had been in the region of the Sun "at the beginning of the earth," during the age of the Earth (then taken on geological evidence as 3×10^9 yr) it would have traveled to a distance of "many millions of light years." This was one of the first published statements by an observational astronomer on the expansion of the universe or the separation of the galaxies.

Wirtz, who had survived on the losing side of World War I and returned to Kiel, collected all the radial velocities of spiral "nebulae" available in the literature, a total of 29, and solved again for the solar motion. With this large number of objects, he found an even larger value of the K term, 840 km/sec ± 140 km/sec, so much larger than its probable error that it "could not be disregarded." Furthermore, he found a correlation of radial velocity with the magnitude of the object, in the sense that the fainter the spiral, the larger its radial velocity tended to be. He also found an analogous correlation with angular dimension: the smaller the spiral, the larger its velocity tended to be (Wirtz 1922). Again he concluded more data were needed. Two years later, Wirtz realized that Willem de Sitter's theoretical paper on expanding cosmology required that the radial velocity increase with the distance. Wirtz (1924) used better measurements of the angular diameters of the spirals, from papers by Curtis and Pease, and found them to be well correlated with distance; he concluded that de Sitter's cosmology was confirmed. Wirtz used velocity data mostly from Slipher's measurements, but the Lowell observer had reached the limit of his telescope.

Similarly Lundmark, by now in England, sought evidence for or against de Sitter's ideas in observational data. Like Paddock and Wirtz before him, Lundmark (1924) analyzed the radial velocities of the spirals and found a large positive K term, 793 ± 88 km/sec. A year later, then in Kiev, Lundmark (1925) reanalyzed the radial velocities in terms of not just a constant K term, but a quadratic expression in the distance D,

$$K = a + bD + cD^2.$$

He used relative distances, in units of the distance to M31, based on relative angular diameters, and solved for the individual coefficients a, b, and c. Though the first two terms dominated for the forty-three galaxies for which measured radial velocities were available, mostly from Slipher, the solution gave a negative value of c and predicted a maximum radial velocity of 2250 km/sec. Again, more data were needed, specifically radial velocities of more distant galaxies.

Further progress required a bigger telescope. The biggest telescope in the world was the Mount Wilson 100-inch. Hubble had his hands full with direct photographic work, and Pease, who had gone to Washington as a technical weapons designer during World War I, had been shifted to nearly full-time work with the stellar interferometer at Mount Wilson after 1920.

The person who took over the spectroscopy of "extragalactic nebulae" at Mount Wilson and made it his life work was Milton L. Humason, probably the last highly successful research astronomer to base his career on an eighth-grade education (Mayall 1973). Born in 1891, Humason had been taken by his parents to Strain's Camp on Mount Wilson for their summer vacation in 1905, just as George Ellery Hale was starting the observatory there. Young Milton hated school but he loved the mountain and the astronomy center being erected amidst the pine trees on its peak. His parents let him quit high school after only a few days that fall, stay with family friends on Mount Wilson, and work at the camp and at the hotel which was then in operation on the summit. Later he became a mule driver on the Sierra Madre trail, along which the materials for the observatory's buildings and instruments were brought to the top by wagon. In 1911 he married Helen Dowd, the daughter of Jerry Dowd, the chief engineer and electrician at Mount Wilson. Humason got a better paying, more socially acceptable job as the foreman of a ranch a few miles east of Pasadena, but he yearned to get back to Mount Wilson and astronomy.

In 1917, with his father-in-law's support, Humason applied for the job of janitor at the observatory and was hired. This was only a stepping stone to the position of relief night assistant, which opened up when the new 100-inch went into operation in 1919. Humason proved so careful, conscientious, and skillful with the mechanical tasks of observing that he was soon assigned to work with the smaller telescopes when he was not assisting at the 60-inch or 100-inch, and in 1921 he was promoted to the research staff. He did all kinds of direct and spectroscopic observing for the other, better educated astronomers. Gradually, as Hubble's published results demonstrated the importance of the study of the "extragalactic nebulae," Humason was assigned to work more on these objects. Almost the exact opposite of Hubble, the quiet, careful, self-effacing Humason was outstanding as a spectroscopic observer who studied and understood the mechanical details of his instruments and techniques completely, and could sit for endless hours in the dark, quiet telescope dome, concentrating his entire attention on getting all the light of faint "nebulae" through the slit and onto the plate in perfect focus (fig. 14.11).

Fig. 14.11 Milton L. Humason, observer of the redshifts of the galaxies. (From the archives of the Observatories of the Carnegie Institution of Washington)

In 1927 Humason published his first paper on the radial velocities of "nebulae." One, the spiral M 101, had only a faint nucleus; the other, the nearby irregular NGC 6822 which Hubble had studied, had no nucleus at all. With the Cassegrain prism spectrograph Humason was using, it was impossible to get spectrograms of either of these objects well enough exposed to show measurable absorption lines on a strong continuous spectrum. In each, however, Humason (1927) obtained (in four to six hours) spectra of emission nebulae which, with all their light concentrated in a few bright lines, were observable representatives of the entire system. Humason was to use this technique time after time to push out to the limits of his successively improved spectrographs and telescopes, as did many other galaxy observers since. Although he measured the spectroscopic plates himself, Humason knew hardly any mathematics beyond the simplest arithmetic, and the reductions of his measurements to numerical values of radial velocity were done for him by the computers on the Mount Wilson Observatory staff. Hubble wrote most of the text of Humason's early papers for him, and is probably responsible for the statement in this one that the measured radial velocities of NGC 6822 (+216 km/sec) and M 101 (−133 km/sec) were "unusually low for non-galactic objects. This is consistent with the marked tendency already observed for the smaller velocities to be associated with the larger (and hence probably closer) nebulae and those which are highly resolved" (Humason 1927). This is the first published statement in a scientific paper from Mount Wilson of what later came to be known as Hubble's law.[11]

What were needed were distances of many more "extragalactic nebulae." Hubble had determined the distances of three, M 31, M 33, and NGC 6822, from their Cepheid variables, and earlier, similar determinations were available for two others, the Large and Small Magellanic Clouds. All the others were so distant that their Cepheid variables were too faint to measure unambiguously. However, in his other outstanding paper published the same year as his study of M 33, Hubble (1926b) had outlined his system of the classification of the forms of the "extragalactic nebulae." It had many roots in earlier work by H. Knox-Shaw, W. J. Reynolds, Curtis, and especially Lundmark, as described by Sandage (1977). Hubble (1926c) presented this paper orally, with many illustrative slides, at the meeting of the Astronomical Society of the Pacific at Mills College, Oakland, that year. His classification was a one-dimensional sequence Elliptical–Sa–Sb–Sc–Irregular (the spirals could also be barred, SBa–SBb–SBc). According to Hubble he had consciously attempted to find a descriptive classification independent of any theoretical ideas, but the result turned out to be essentially almost identical with the path of development of a rotating mass derived by Jeans, with increasing condensation to a plane with time. Within each type, Hubble (1926b) demonstrated, the magnitudes and angular diameters obeyed a simple relationship (of constant surface brightness) indicating that the fainter objects were physically similar to the brighter ones, but at larger distances

[11] This section is based on the biography of Humason by Mayall (1973), on correspondence between Mayall and Humason (1933–1965) and Helen D. Humason (1972), SLO, on my own memories of many conversations with Humason (1953–1958), and my interviews of Allan Sandage (1988), who knew him longer and better than I did.

from the Sun. Furthermore, within each type the difference in apparent magnitude between the entire "nebula" and its brightest stars was a constant. Thus evidently the absolute magnitude of each type of "extra-galactic nebula" was apparently approximately a constant, and the absolute magnitude of its brightest stars as well.

From the few galaxies with distances known from Cepheids, Hubble (1926b) derived the mean absolute magnitude of the brightest stars $M_S = -6.3$ (our Galaxy was included in this average), and of the "nebula" as a whole, $M_T = -15.3$.

Now Hubble was ready to study cosmology. In his first paper on a velocity-distance relation (Hubble 1929b), he used the distances determined by this method and the available radial velocities, chiefly from the work of Slipher. Hubble argued, mostly from his idea that his observational classification sequence mimicked Jeans' evolutionary theory for a rotating mass that all the galaxies, that even the ellipticals and Sa spirals in which he had no resolved stars and hence no distance or absolute magnitude determination, had the same mass and hence the same absolute magnitude. Hence he used $M_T = -15.2$ (a value only very slightly different from his earlier value) as the mean absolute magnitude of *all* "extragalactic nebulae." He thought the galaxies' absolute magnitudes had a range of 4 or 5 magnitudes about this mean; the brightest stars had a smaller dispersion, he said, but could be resolved only in the nearer objects.

In addition to Slipher's and other early observers' velocities, Hubble had one value determined by Humason (1929). It was for NGC 7619, the brightest "nebula" in a cluster, and its value, +3779 km/sec, was over twice as large as the highest previously known velocity (for NGC 584) and well above the maximum predicted by Lundmark's formula. Humason had taken two spectrograms of NGC 7619, with exposure times of 33 and 45 hours. The reason for observing a cluster was that its brightest member would be the easiest (actually least difficult) to observe, but its velocity would be representative of the entire cluster; for the apparent magnitude he used the mean of all the cluster members.

In his paper, Hubble (1929b) mentioned earlier determinations of "a K-term of several hundred kilometers [per second] which appear to be variable," and said that "[e]xplanations of this paradox have been sought in a correlation between apparent radial velocities and distances, but so far the results have not been convincing," but he did not refer to Wirtz' paper at all, and only very briefly to Lundmark's. In contrast, he said, his own paper used "only those nebular distances which are believed to be reliable." These were the twenty-four "nebulae" in which the individual brightest stars had been resolved and measured. From these objects he found that a linear velocity-distance relation

$$V = HD$$

(in our present notation) was a sufficiently good representation of the data, and derived $H = +465 \pm$ km/sec/Mpc or $+513 \pm 60$ km/sec/Mpc, depending on how he grouped the data (Hubble 1929b). At the conclusion of his paper, Hubble stated that its outstanding result was that the observational velocity-distance relation might represent the "de Sitter effect" (expanding universe).[12]

Hubble (1929b) then used the velocity-distance relation he had derived to calculate the distances and hence the absolute magnitudes of the other galaxies, for which no individual distances were available. He demonstrated in this way that the same mean absolute magnitude M_T approximately fitted them all.

Within less than a year Humason and Pease (1929) again doubled the highest radial velocity known, measuring +7300 km/sec for NGC 4835 and +7800 km/sec for NGC 4860, two of the brightest "extragalactic nebulae" in the Coma Berenices cluster. Hubble (1929c) obtained the magnitudes of these two and many other "nebulae" in the cluster, and using the same mean absolute magnitude estimated its distance as 16×10^6 pc. These numbers lead to nearly the same value of the Hubble constant, $H = 470$ km/sec/Mpc.

Soon after these papers were completed, Humason put into operation a specially designed, fast spectrograph based on an F/0.6 "Rayton" (microscope-objective type) lens, procured by Caltech for ultimate use in the 200-inch telescope, then only being planned. It was somewhat more than twice as fast as the previous spectrograph Humason had used at the 100-inch. With the new instrument he measured a radial velocity of +11,500 km/sec for the brightest member of a cluster of "nebulae" in Ursa Major in 1930. The exposure time required for this spectrogram was 45 hours (Humason 1930). He speeded up the spectrograph still further by removing one of the two prisms, reducing the dispersion to 875 Å/mm, still quite sufficient for measuring the large velocities of distant galaxies. Within a year and a half Humason (1931) could report radial velocities of 46 galaxies measured at Mount Wilson (a few of them by Pease), about half in clusters and half individual galaxies. The largest measured velocity by now was +19,700 km/sec for the brightest member of a distant cluster in Leo.

Hubble and Humason (1931) used these velocities in their next study of the velocity-distance relation. Here they adopted $M_S = -6.1 \pm 0.4$ as the absolute magnitude of the brightest stars, and $M_T = -14.7 \pm 1.8$ as the absolute magnitude of a galaxy. With the much larger body of data then available, they confirmed the linear velocity-distance relation, and found

[12] Hubble never completely accepted that the redshifts necessarily resulted from the radial velocities of the objects alone. From this first redshift paper (Hubble 1929b) to his last, the George Darwin Lecture presented in 1953 and published posthumously (Hubble 1954), he consistently maintained that they could not be distinguished from Doppler shifts but in principle should be described as "apparent velocities."

an only slightly different value of the Hubble constant, $H = 558$ km/sec/Mpc (Hubble and Humason 1931). In retrospect we can see that all these values are much too large chiefly because they used too faint an absolute magnitude for the brightest stars (the stellar spectroscopists at the time all believed Hubble had adopted far too *bright* an absolute magnitude), in many cases what they interpreted as "brightest stars" were actually much more luminous clusters or groups of stars and the nebulae in which they were involved, and the apparent magnitudes of the nearby "nebulae" they used, such as M 31 in particular, were too faint. In addition, several of the nearby nebulae they used for their calibration, such as NGC 6822, the Large and Small Magellanic Clouds, and M 33, are much less luminous than the typical more distant galaxies which they observed for radial velocity. Nevertheless, it was a brilliant first effort.

Within a few more years Humason (1934) had pushed out the highest velocity measured to $+39{,}200 \pm 500$ km/sec, for a 17.5 magnitude galaxy which was the brightest member of a faint, distant cluster in Bootis that Hubble had discovered. The exposure time for this spectrogram was nearly 18 hours. Hubble (1934) was investigating the distribution of "extragalactic nebulae" in the sky, and in space by counting numbers as functions of magnitude in selected areas distributed over the sky. By now he gave many references to papers by others, including theoretical papers. From his counts he concluded that the distribution of the galaxies in space was approximately uniform. Using an average mass of a galaxy $6 - 10 \times 10^8 M_\odot$, he estimated the mean mass density in the universe as approximately 10^{-30} gm/cm^3.

By this time it was clear to everyone that the spirals really were distant galaxies. Van Maanen was forced to recant his earlier proper-motion measurements of their internal motions. Hubble remeasured four of the objects in which van Maanen had found streaming motions, M 33, M 51, M 81 and M 101, now using plates taken with the 60-inch with a time interval of 20 years. In none of them did Hubble find any appreciable motion. Seth Nicholson and Walter Baade of the Mount Wilson staff independently measured these plates of M 81 and M 51, respectively, and agreed with Hubble (Hubble 1935). Van Maanen measured pairs of plates of M 33 and M 74 taken with the larger scale of the 100-inch and nine-year time interval, and found only much smaller motions than he previously had. Still, he found the motions in the same direction! Also he measured some pairs of plates of M 33 and M 101 taken at the Cassegrain focus of the 60-inch over a long time interval, and did not confirm his earlier results. He was forced to publish a statement that "[i]n consideration of the difficulty of avoiding systematic errors in this special problem, these results together with the measures of Hubble, Baade, and Nicholson which are given in the preceding article, make it desirable to view the motions with reserve" (van Maanen 1935).

Humason (1936) extended the limit of largest measured velocity to +42,000 km/sec in the Ursa Major No. 2 cluster, discovered by Baade. In it the brightest "nebula" was fainter than in the Bootis cluster, and the velocity was "somewhat uncertain." In both these distant clusters the brightest galaxies were so faint that even with the 100-inch telescope they could not be seen visually on the slit of the spectrograph. To get their spectra Humason (1936) had to develop an offset guiding method using a brighter star, whose angular distance from the object he had accurately measured on a long-exposure plate. Again, Hubble (1936a, 1936b) discussed these measured velocities and rederived very nearly the same values of M_S, M_T, and H. In these papers Hubble used for data, besides the velocities measured by Slipher and Humason, some of the first measurements by Nicholas U. Mayall at Lick Observatory.

Mayall had done his undergraduate work and one year of graduate study at the University of California, and then had worked for two years (1929–31) as an assistant at Mount Wilson. He was especially inspired by Hubble and Humason, and when he returned to Lick to complete his Ph.D. in 1931, he did his thesis on the "extragalactic nebulae," working on a problem that Hubble had suggested to him. In it, Mayall (1934) discussed and analyzed the Crossley plates taken by Keeler, Perrine, and Curtis in terms of nebulae counts. There were 305 such plates which he could use, taken between 1899 and 1920, and Mayall himself added 194 in 1932–33. He reduced all the counts to a common system, and derived the final result that the total number of "extragalactic nebulae" over the whole sky down to apparent photographic magnitude 19.0 (the approximate limit of the Crossley) corrected for the extinction in the Milky Way, was approximately 4.8×10^6. When Mayall got his Ph.D. in the middle of the Great Depression, the only way Director Robert G. Aitken could keep him on the Lick staff was by appointing him as a janitor, with additional duties as an assistant, but within a year he was able to promote him to a research job (Mayall 1970b). Mayall designed and supervised the building of an especially fast nebular spectrograph which, at the Crossley, was ideally suited for obtaining redshifts of faint, extended low surface-brightness spiral galaxies, and he worked in very close cooperation with Hubble and Humason on his observing program.[13]

Hubble's further work concentrated more and more on observational tests of theoretical cosmological predictions. He collaborated with physicists Richard C. Tolman and H. P. Robertson at Caltech, who were at the forefront of research in this field. This work has been excellently reviewed very recently by Sandage (1987, 1988). Hubble was a vivid writer and from his first popular article (Hubble 1929d) he demonstrated his mastery of the

[13] This section is based on the correspondence between Mayall and Humason mentioned above, correspondence between Mayall and Hubble (1933–1953), SLO, and the autobiographical chapter by Mayall (1970b).

use of heightened language to describe his research on the problems of the universe. His Bruce Medal lecture is an especially good example of Hubble's (1938) eloquent style. Hubble's books, *The Realm of the Nebulae* (1936) and *The Observational Approach to Cosmology* (1937), inspired a generation of young astronomers and physicists. Yet observational cosmology required big telescopes, and until well after World War II, Hubble and Mount Wilson completely dominated the field (fig. 14.12).

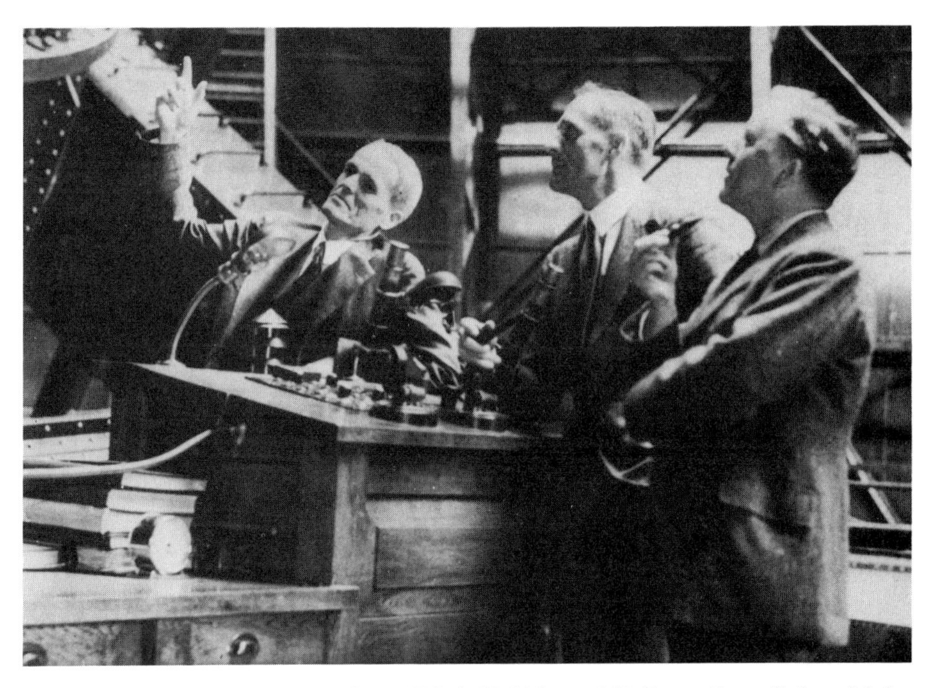

Fig. 14.12 Walter S. Adams, Edwin Hubble, and Sir James Jeans (left to right) at the Mount Wilson Observatory 100-inch telescope. (From the archives of the Observatories of the Carnegie Institution of Washington)

From Keeler to Hubble and Humason, observational astronomers, probing "what is out there," had measured and recognized galaxies as luminous markers in the universe. They had recognized their large, systematic velocities of recession, and had derived the first crude velocity-distance relation. At any particular time only a very few astronomers were working on these problems, a tiny minority of the much greater number working on stars. Practically all their discoveries had been made without the benefit of any theoretical ideas, and in several cases in spite of the incorrect theoretical or "common-sense" pictures the observers had had in their minds. All the big steps had required big telescopes or fast spectrographs. And most of them had been made by the astronomers who had almost unlimited observing time at their disposal.

Acknowledgements
I am most grateful to John S. Hall, Arthur A. Hoag, Allan Sandage, and Sidney van den Bergh for the information, suggestions, thoughts, comments, and criticisms they provided for this paper. I am also indebted to Dorothy Schaumberg and Irene H. Osterbrock for helping me to locate relevant source material in the Mary Lea Shane Archives of Lick Observatory, McHenry Library, University of California, Santa Cruz. Finally, I wish to thank John D. R. Bahng, Richard Dreiser, Anthony Misch, Steven Padilla, E. Robert Wilson, and the late William G. Hoyt for their assistance and cooperation in obtaining the photographs which illustrate this paper.

References
Aitken, R. G. (1943). Heber Doust Curtis 1872–1942. *Biographical Memoirs of the National Academy of Sciences*, **22**, 275.
Allen, C. D. (1954). Edwin Powell Hubble (1889–1953). *American Oxonian*, **41**, 101.
Campbell, W. W. (1917). The Nebulae, *Science*, **45**, 413.
(1926). Do We Live in a Spiral Nebula? *Publications of the Astronomical Society of the Pacific*, **38**, 75.
Curtis, H. D. (1917a). A Study of the Absorption Effects in the Spiral Nebulae. *Publications of the Astronomical Society of the Pacific*, **29**, 145.
(1917b). New Stars in Spiral Nebulae. *Publications of the Astronomical Society of the Pacific*, **29**, 180.
(1917c). Novae in Spiral Nebulae and the Island Universe Theory. *Publications of the Astronomical Society of the Pacific*, **29**, 206.
(1918a). Description of 762 Nebulae and the Clusters Photographed with the Crossley Reflector. *Publications of the Lick Observatory*, **13**, 9.
(1918b). A Study of Occulting Matter in the Spiral Nebulae. *Publications of the Lick Observatory*, **13**, 45.
(1918c). The Number of the Spiral Nebulae. *Publications of the Astronomical Society of the Pacific*, **30**, 159.
(1919). Modern Theories of the Spiral Nebulae. *Journal of the Washington Academy of Science*, **9**, 217.
(1921). The Scale of the Universe, Part II. *Bulletin of the National Research Council*, **2**, 194.
Duncan, J. C. (1922). Three Variable Stars and a Suspected Nova in the Spiral Nebula M 33 Trianguli. *Publications of the Astronomical Society of the Pacific*, **34**, 290.
Fath, E. A. (1909). The Spectra of Some Spiral Nebulae and Globular Star Clusters. *Lick Observatory Bulletin*, **5**, 71.
(1911). The Spectra of Spiral Nebulae and Globular Star Clusters. *Astrophysical Journal*, **33**, 58.
(1912). The Integrated Spectrum of the Milky Way. *Astrophysical Journal*, **36**, 362.
(1913a). The Spectra of Spiral Nebulae and Globular Star Clusters. Third Paper. *Astrophysical Journal*, **37**, 198.

(1913b). The Integrated Spectrum of the Milky Way. *Popular Astronomy*, **21**, 262.
(1914). A Study of Nebulae. *Astronomical Journal*, **28**, 75.
Hall, J. S. (1970). Vesto Melvin Slipher (1875–1969). *American Philosophical Society Yearbook*, p. 161.
Harrison, E. (1987). *Darkness at Night: A Riddle of the Universe*. Cambridge: Harvard University Press.
Hoskin, M. A. (1976a). Ritchey, Curtis and the Discovery of Novae in Spiral Nebulae. *Journal for the History of Astronomy*, **7**, 45.
(1976b). The "Great Debate": What Really Happened. *Journal for the History of Astronomy*, **7**, 169.
Hoyt, W. G. (1980). Vesto Melvin Slipher 1875–1969. *Biographical Memoirs of the National Academy of Science*, **52**, 411.
Hubble, E. (1920). Photographic Investigations of Faint Nebulae. *Publications of the Yerkes Observatory*, **4**, Part 2, 1.
(1922a). A General Study of Diffuse Galactic Nebulae. *Astrophysical Journal*, **56**, 162.
(1922b). The Source of Luminosity in Galactic Nebulae. *Astrophysical Journal*, **56**, 400.
(1925a). Cepheids in Spiral Nebulae. *Popular Astronomy*, **33**, 252.
(1925b). NGC 6822, A Remote Stellar System. *Astrophysical Journal*, **62**, 409.
(1926a). A Spiral Nebula as a Stellar System: Messier 33. *Astrophysical Journal*, **63**, 236.
(1926b). Extra-Galactic Nebulae, *Astrophysical Journal*. **64**, 321.
(1926c). Non-Galactic Nebulae I. Classification and Apparent Dimensions II. Absolute Dimensions and Distribution in Space. *Publications of the Astronomical Society of the Pacific*, **38**, 258.
(1929a). A Spiral Nebula as a Stellar System, Messier 31. *Astrophysical Journal*, **69**, 103.
(1929b). A Relation between Distance and Radial Velocity among Extra-Galactic Nebulae. *Proceedings of the National Academy of Sciences*, **15**, 168.
(1929c). Preliminary Estimate of the Distance of the Coma Cluster of Nebulae. *Publications of the Astronomical Society of the Pacific*, **41**, 247.
(1929d). The Exploration of Space. *Harper's Monthly*, **158**, 732.
(1934). The Distribution of Extra-Galactic Nebulae. *Astrophysical Journal*, **79**, 8.
(1935). Angular Rotations of Spiral Nebulae. *Astrophysical Journal*, **81**, 334.
(1936a). The Luminosity Function of Nebulae I. The Luminosity Function of Resolved Nebula as Indicated by Their Brightest Stars. *Astrophysical Journal*, **84**, 158.
(1936b). The Luminosity Function of Nebula II. The Luminosity Function as Indicated by Residuals in Velocity-Magnitude Relations. *Astrophysical Journal*, **84**, 270.
(1938). The Nature of the Nebulae. *Publications of the Astronomical Society of the Pacific*, **50**, 97.
(1954). The Law of Red-Shifts. *Monthly Notices of the Royal Astronomical Society*, **113**, 658.

Hubble, E. and Humason, M. L. (1931). The Velocity-Distance Relation among Extra-Galactic Nebulae. *Astrophysical Journal*, **74**, 43.
Humason, M. L. (1927). Radial Velocities in Two Nebulae. *Publications of the Astronomical Society of the Pacific*, **39**, 317.
(1929). The Large Radial Velocity of NGC 7619. *Proceedings of the National Academy of Sciences*, **15**, 157.
(1930). The Rayton Short-Focus Spectrographic Objective. *Astrophysical Journal*, **71**, 351.
(1931). Apparent Velocity-Shifts in the Spectra of Faint Nebulae. *Astrophysical Journal*, **74**, 35.
(1934). The Apparent Velocity of a Nebula in the Bootis Cluster No. 1. *Publications of the Astronomical Society of the Pacific*, **46**, 290.
(1936). The Apparent Radial Velocities of 100 Extra-Galactic Nebulae. *Astrophysical Journal*, **83**, 10.
Humason, M. L. and Pease, F. G. (1929). High Velocities of Faint Nebulae. *Publications of the Astronomical Society of the Pacific*, **41**, 246.
Keeler, J. E. (1899). On the Predominance of Spiral Forms among the Nebulae. *Astronomische Nachrichten*, **151**, 2.
(1900). The Crossley Reflector of the Lick Observatory. *Astrophysical Journal*, **11**, 325.
(1908). Photographs of Nebulae and Clusters Made with the Crossley Reflector. *Publications of the Lick Observatory*, **8**, 1.
Lundmark, K. (1921). The Spiral Nebula Messier 33. *Publications of the Astronomical Society of the Pacific*, **33**, 324.
(1924). The Determination of the Curvature of Space-Time in de Sitter's World. *Monthly Notices of the Royal Astronomical Society*, **84**, 747.
(1925). The Motions and the Distance of Spiral Nebulae. *Monthly Notices of the Royal Astronomical Society*, **85**, 865.
Mayall, N. U. (1934). A Study of the Distribution of Extra-Galactic Nebulae Based on Plates Taken with the Crossley Reflector. *Lick Observatory Bulletin*, **16**, 177.
(1970a). Edwin Powell Hubble. *Biographical Memoirs of the National Academy of Sciences*, **41**, 175.
(1970b). *Then There Was Light, Autobiography of a University. Berkeley 1868–1968*, ed. I. Stone. Garden City, N.Y.: Doubleday & Co., p. 107.
(1973). Milton L. Humason – Some Personal Recollections. *Mercury*, **2**, 3.
McMath, R. R. (1942). Heber Doust Curtis (1872–1942). *Publications of the Astronomical Society of the Pacific*, **54**, 69.
Newcomb, S. (1901a). *The Stars: A Study of the Universe*. New York: G. Putman & Sons.
(1901b). A Rude Attempt to Determine the Total Light of All the Stars. *Astrophysical Journal*, **14**, 297.
(1902). The Problem of the Universe. *International Monthly*, **5**, 395.
(1904). The Extent of the Universe. *Harper's Monthly*, *109*, 795.
Osterbrock, D. E. (1984a). The Rise and Fall of Edward S. Holden: Parts 1 and 2. *Journal for the History of Astronomy*, **15**, 81, 151.
(1984b). *James E. Keeler: Pioneer American Astrophysicist: And the Early*

Development of American Astrophysics. Cambridge: Cambridge University Press.

Paddock, G. F. (1916). The Relation of the System of Stars to the Spiral Nebulae. *Publications of the Astronomical Society of the Pacific*, **28**, 109.

Pease, F. G. (1915a). The Radial Velocity of the Nebula NGC 1068. *Publications of the Astronomical Society of the Pacific*, **27**, 133.

(1915b). Radial Velocity of the Andromeda Nebula. *Publications of the Astronomical Society of the Pacific*, **27**, 134.

(1915c). Radial Velocitics of Six Nebulae. *Publications of the Astronomical Society of the Pacific*, **27**, 239.

(1916). The Spiral Nebula Messier 33. *Publications of the Astronomical Society of the Pacific*, **28**, 33.

Perrine, C. D. (1904). The Number of the Nebulae. *Lick Observatory Bulletin*, **3**, 47.

(1916). The Number of the Nebulae: Remarks on a Paper by Dr. E. A. Fath in "Astronomical Journal," No. 658. *Astronomical Journal*, **29**, 70.

Ritchey, G. W. (1917a). Novae in Spiral Nebulae. *Publications of the Astronomical Society of the Pacific*, **29**, 210.

(1917b). Another Faint Nova in the Andromeda Nebula. *Publications of the Astronomical Society of the Pacific*, **29**, 257.

Robertson, H. P. (1954). Edwin Powell Hubble 1880–1953. *Publications of the Astronomical Society of the Pacific*, **66**, 120.

Sandage, A. (1977). Classification and Stellar Content of Galaxies Obtained from Direct Photography. In *Galaxies and the Universe*, ed. A. Sandage, M. Sandage, and J. Kristian, 1. Chicago: University of Chicago Press.

(1987). Observational Cosmology 1920–1985: An Introduction to the Conference. In *Observational Cosmology (IAU Symposium No. 124)*, ed. A. Hewitt, G. Burbidge, and L. Z. Fang. Dordrecht: D. Reidel Publ. Co.

(1988). Observational Tests of World Models. *Annual Review of Astronomy and Astrophysics*, **26**, 561.

Sanford, R. F. (1917). On Some Relations of the Spiral Nebulae to the Milky Way. *Lick Observatory Bulletin*, **9**, 80.

Scheiner, J. (1899). Über das Spectrum des Andromedanebels. *Astronomische Nachrichten*, **148**, 325. (English translation in *Astrophysical Journal*, **9**, 149.)

Seares, F. H. (1925). Note on the Distribution and Number of Nebulae. *Astrophysical Journal*, **62**, 168.

(1946). Adriaan van Maanen 1884–1946. *Publications of the Astronomical Society of the Pacific*, **58**, 89.

Shapley, H. (1917). Note on the Magnitudes of Novae in Spiral Nebulae. *Publications of the Astronomical Society of the Pacific*, *29*, 213.

(1919). On the Existence of External Galaxies. *Publications of the Astronomical Society of the Pacific*, **31**, 261.

(1921). The Scale of the Universe, Part I. *Bulletin of the National Research Council*, **2**, 171.

Slipher, V. M. (1913a). On the Spectrum of the Nebula in the Pleiades. *Lowell Observatory Bulletin*, **2**, 26.

(1913b). The Radial Velocity of the Andromeda Nebula. *Lowell Observatory Bulletin*, **2**, 56.
(1914). Spectrographic Observations of Nebulae. *Popular Astronomy*, **22**, 146.
(1915). Spectrographic Observations of Nebulae. *Popular Astronomy*, **23**, 21.
(1917). Nebulae. *Proceedings of the American Philosophical Society*, **56**, 403.
(1921a). Two Nebulae with Unparalleled Velocities. *Lowell Observatory Observation Circular*, Jan. 17.
(1921b). Dreyer Nebula No. 584 Inconceivably Distant. *New York Times*, Jan. 19.
Smith, R. W. (1982). *The Expanding Universe: Astronomy's "Great Debate," 1900–31*. Cambridge: Cambridge University Press.
Sullivan, R. (1916). Celestial Motions in the Line of Sight. *Popular Astronomy*, **24**, 109.
Townley, S. D. (1903). The Total Light of the Stars. *Publications of the Astronomical Society of the Pacific*, **15**, 13.
Truman, O. H. (1916). The Motions of the Spiral Nebulae. *Popular Astronomy*, **24**, 111.
van Maanen, A. (1916). Preliminary Evidence of Internal Motion in the Spiral Nebula Messier 101. *Astrophysical Journal*, **44**, 331.
(1921a). Investigations on Proper Motion. Fourth Paper: Internal Motion in the Spiral Nebula Messier 51. *Astrophysical Journal*, **54**, 237.
(1921b). Investigations on Proper Motion. Fifth Paper: The Internal Motion in the Spiral Nebula Messier 81. *Astrophysical Journal*, **54**, 347.
(1921c). Internal Motion in the Spiral Nebula Messier 33. Preliminary Results. *Proceedings of the National Academy of Sciences*, **7**, 1.
(1922). Investigations on Proper Motion. Eighth Paper: Internal Motion in the Spiral Nebula M 94 = NGC 4736. *Astrophysical Journal*, **56**, 208.
(1923). Investigations on Proper Motion: Tenth Paper: Internal Motion in the Spiral Nebula Messier 33, NGC 598. *Astrophysical Journal*, **57**, 264.
(1935). Internal Motions in Spiral Nebulae. *Astrophysical Journal*, **81**, 336.
Wilson, R. E. (1958). Roscoe Frank Sanford 1883–1958. *Publications of the Astronomical Society of the Pacific*, **70**, 360.
Wirtz, C. (1918). Über die Bewegungen der Nebelflecke. Vierte Mitteilung. *Astronomische Nachrichten*, **206**, 109.
(1922). Einiges zur Statistik der Radialbewegungen von Spiralnebeln und Kugelsternhaufen. *Astronomische Nachrichten*, **215**, 349.
(1924). De Sitters Kosmologie und die Radialbewegungen der Spiralnebel. *Astronomische Nachrichten*, **222**, 21.
Young, R. K. and Harper, W. E. (1916). The Solar Motion as Determined from the Radial Velocities of Spiral Nebulae. *Journal of the Royal Astronomical Society of Canada*, **10**, 134.

Discussion

Bertotti:

American observational astronomy had a much more prominent role in early cosmology than European astronomy. Is there an explanation for this?

Osterbrock:
Yes, I believe it is because galaxies are faint, and hence to get good observational data required large telescopes at good observing sites. These existed in the United States, partly because of geography and climate (the clear weather, good-seeing sites in California and Arizona), and partly because of a very few American millionaires, such as James Lick and Andrew Carnegie.

McCrea:
How did Hubble and Humason select the particular objects for redshift studies?

Osterbrock:
Their aim was to obtain spectra of the most distant "extragalactic nebulae" they could find. Chiefly this meant finding *small* (in angular diameter), *rich* clusters of *faint* galaxies, and selecting the brightest member or members to observe spectroscopically. They found these clusters on long exposure direct photographs taken with the 100-inch telescope.

15

Discovery of the cosmic microwave background

Robert W. Wilson

In the late 1950s plans were made to start working on communication satellites at Bell Laboratories, particularly at Holmdel. The first satellite tests were planned with NASA's Echo balloon. It was known that the return signal from Echo would be very weak because a sphere scatters the incoming radiation in all directions. In contemplating the weak signal, they decided that one should have a very low noise receiver system to receive that signal. It was very convenient that there were two Bell Laboratories devices which would go together and should make a very low noise receiving system. One of them was the traveling wave maser which Derrick Scovil and his group at Murray Hill were making (De Grasse et al. 1959a). This worked at liquid helium temperatures and had a noise temperature of a few Kelvin. Even after making a room temperature connection to it, you could have a receiver with a noise temperature of 10 K.

The other device which seemed to fit with the traveling wave maser was a horn-reflector antenna. The horn-reflector was invented by Al Beck and Harold Friis for use in a microwave relay system. In addition to turning the corner between the waveguide going up a tower and the horizontal communication path, it has the distinct advantage that when two of them are put back-to-back on a tower and have a very weak signal coming in on one side, a strong regenerated signal can be transmitted from the other side without interference. Its front-to-back ratio is very high. The corollary of this is that if a horn-reflector is put on its back, it will not pick up much radiation from the Earth and will be a very low noise antenna. Therefore, Art Crawford (1961) built a large (20-foot aperture) horn-reflector to be used with a traveling wave maser to receive the weak signals from Echo.

Figure 15.1 shows the 20-foot horn-reflector with its parabolic reflector on the left and cab on the right. Since the cab does not tilt, almost any kind of receiver can be conveniently put at the focus of this antenna (apex of the horn). It is fairly obvious that the horn shields the receiver from the ground, especially when it is looking up.

Fig. 15.1

Just to convince the engineers how good the shielding is, fig. 15.2 shows a polar diagram of a rather smaller horn-reflector antenna compared with the gain of a theoretical isotropic (uniform response) antenna. If we take an isotropic antenna and put it on a field with the 300° ground down below and zero degree sky up above, we expect it to pick up 150°; half of its response comes from the ground. Now look at this horn-reflector. It is certainly at least 35 db less responsive to the ground than the isotropic antenna would be. So we take 150° and divide it by a few thousand and we expect under a tenth of a Kelvin for the ground pickup from the horn-reflector.

In 1963, knowing of the existence of this antenna, I accepted a job at Bell Laboratories, leaving a postdoctoral position at Caltech. Arno had been there a year or so at that time, and obviously the only two radio astronomers in the place were going to work together because making a radio telescope do any kind of an observation is a job for at least two people. One might ask why two young astronomers wanted to work with such a tiny little antenna, with a collecting area of maybe 25 square meters, when there were much bigger antennas around. We knew it had very special properties. First, it is a small enough antenna that one could measure its gain very accurately. It was necessary only to be about a kilometer away to be in the far field, for making an accurate gain

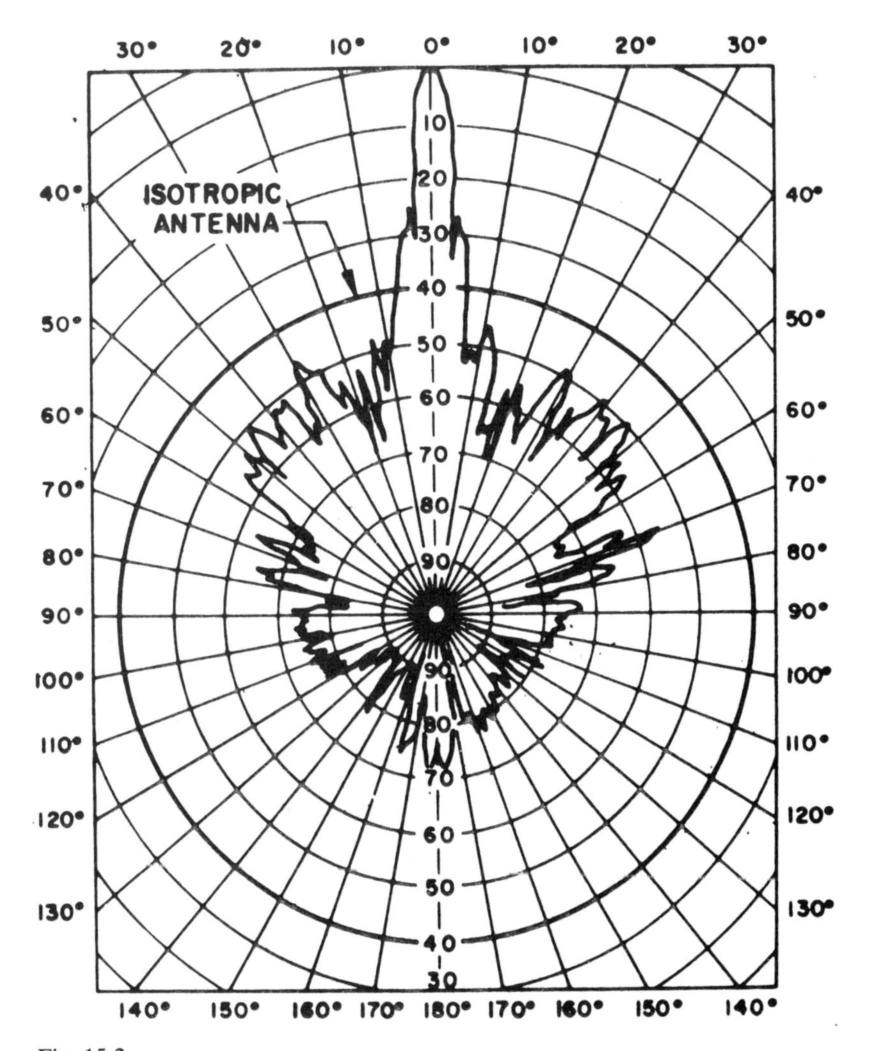

Fig. 15.2

measurement. And that, in fact, had already been started by Dave Hogg.

The availability of the traveling wave masers at several frequencies made this small antenna sensitive enough for work even with small diameter sources, and for sources which were of a diameter large enough to fill its beam, it would have been the most sensitive radio telescope in existence at the time. The other important thing about horn-reflectors I have pointed out before; the good shielding of the antenna leads to the conclusion that we ought to be able to understand all the sources of noise. Radio astronomers don't often understand the background temperature when they do the usual on–off experiment (subtracting a measurement pointing away from the source from the measurement on the source), but the 20-foot horn-reflector offered the possibility of absolute temperature measurement.

My interest in that possibility came directly from my thesis work at Caltech. There, with John Bolton, I had made a survey of the Milky Way. We had done it in the standard way of doing such things at the time; we pointed the antenna to the west of the Milky Way and let the Earth's rotation sweep the antenna beam through it and to the other side. Of course, we are inside the Galaxy, and there is no possibility of pointing completely away from the Galaxy to do an on–off measurement. What we observed was a simple curve in which the power goes up as we approach the Milky Way and comes back down on the other side, and has a fairly steady level more than 10° from the plain. I did the ususal thing; I drew a linear baseline from one side to the other and measured levels above it. That was enough to see the radiation from the plane of the galaxy but it was clearly somewhat unsatisfying because I knew that there was a possible large-scale component, the halo that people were talking about, which my measurement was not at all sensitive to.

After I went to Bell Laboratories, the 20-foot horn-reflector was released from the various satellite jobs it was doing. It had been designed for the Echo experiment which required operation at 13-cm wavelength, but it had later been used to receive a beacon from the Telstar satellite, so when Arno and I inherited it there was a 7.3-cm. maser receiver on it (Tabor and Sibilia 1963). It had a communications receiver which a radio astronomer would find hard to believe. The maser was followed by a low-noise nitrogen-cooled parametric amplifier which was followed by a low-noise traveling wave tube amplifier, and the gain stability was unbelievably bad. Our jobs were to turn all of this into a radio telescope by making a radiometer, finish up the gain measurement, and then proceed to do some astronomy projects.

We thought about what astronomy we ought to do and laid out a plan, which would take a few years. The first project was the absolute flux measurement. If we could measure the gain of the antenna to a few percent, which we thought we could, and then understand the temperature scale of the receiver very well, we could measure the standard calibration sources more accurately than had been done before. I planned to follow up on my thesis by taking a few selected cuts across the Milky Way and then confirm the spectrum of some of the sources that I had looked at. Next we wanted to check our ability to measure the galactic halo radiation. Extrapolating from a lower frequency, we did not expect to see any galactic halo at 7 cm. We wanted to prove that when we did try to make such a measurement we got a null result. After doing these projects, our plan was to build a 21-cm receiver scaled from our 7-cm receiver. We already had the maser in hand. We would then make the halo measurement and do a number of 21-cm line projects including reworking Arno's thesis of looking for hydrogen in clusters of galaxies.

At one point during that time John Bolton came for a visit and we laid out this plan of attack in front of him and asked his opinion, and he said, "Well, obviously the most important thing to do in that list is the 21-cm background measurement." He thought that it was an unexplored area and something that we really ought to do.

Actually, by the time I joined Bell Laboratories, Arno had started making a liquid helium-cooled noise source (Penzias 1965). In fig. 15.3 we see it; a piece of ordinary C-band waveguide, 90% copper, which runs down inside the dewar which has a 6-inch inside diameter. About halfway down, the waveguide is thinned to reduce its heat conductivity and finally there is a carefully designed absorber in the bottom. There's a sheet of mylar in the flange near the bottom which keeps the liquid helium out of the upper part of the waveguide. There are some holes in the bottom section so that the liquid helium can surround the absorber and there will be no question of the physical temperature of the absorber. The cryogenics has been taken care of with the baffles which make a heat exchange between the cold helium gas and the waveguide. We realized that we had to know the radiation from the walls of the waveguide, so there are a series of diode thermometers on the waveguide for measuring its physical temperature distribution. We calculated the radiation of the walls using these temperatures and the measured loss in the waveguide.

When we first transferred the contents of a 25-liter dewar of liquid helium into the cold load, it would fill up to a fairly high level and we calculated the radiation temperature at the top to be approximately 5 K – just eight-tenths of a Kelvin above the temperature of the liquid helium. Fifteen hours later or so (we usually ran down before the helium did), the

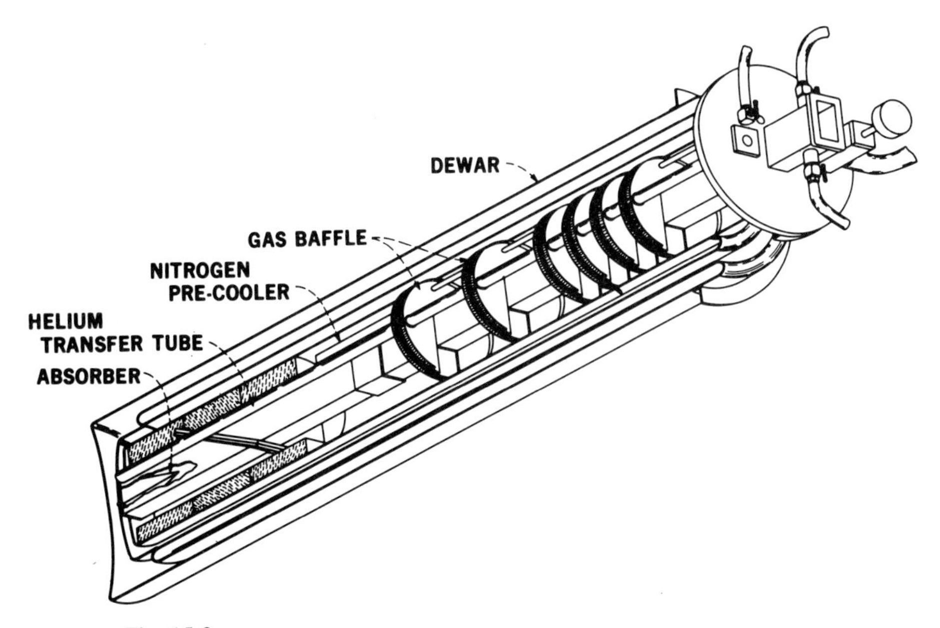

Fig. 15.3

liquid helium level would be down near the absorber and we would calculate the flange temperature to be about 6°. Comparing it to the horn-reflector, the change agreed within something like a tenth of a degree over that period, so we felt we had a reasonably good calibration of what was going on in our cold load.

While Arno was doing that, I set up the radiometer shown in fig. 15.4 (Penzias and Wilson 1965a). As with most of our astronomical equipment at Bell Laboratories, this is somewhat unusual. The horn-reflector has a circular waveguide at its output, and so we decided to use that property in a scheme which Doug Ring and others at Crawford Hill had used in the past. It takes advantage of the fact that two orthogonal polarizations will pass through round waveguide. The polarization coupler near the antenna couples the signal from the reference noise source into the horizontal polarization mode traveling toward the maser and allows vertical polariza-

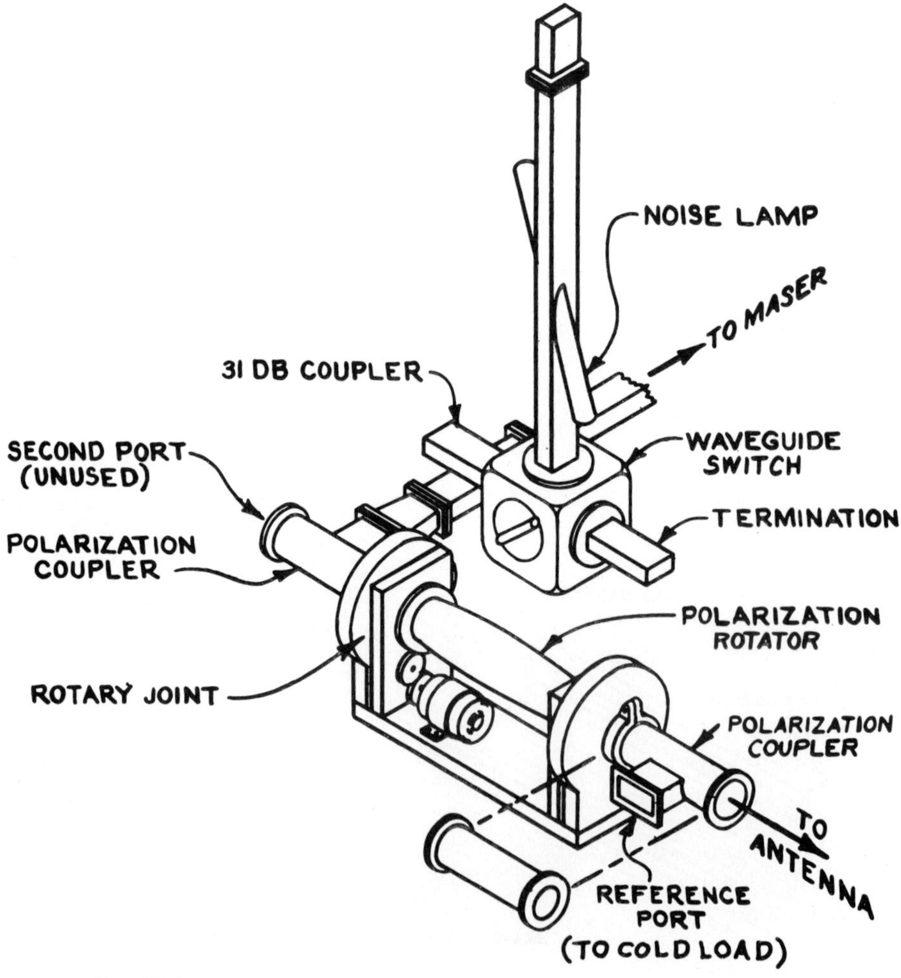

Fig. 15.4

tion from the antenna to go straight through. The polarization rotator is the equivalent of a half-wave plate. It is a squeezed piece of waveguide with two rotary joints; another polarization coupler at the back picks one polarization off and sends it over to the maser. By rotating the squeezed waveguide, we could switch between the reference noise and the antenna. The noise tube was a secondary temperature standard in our measurement of Cas A and other sources. An important aspect of this radiometer design is that except for the unused port, all ports of the waveguide were terminated at approximately the same low radiation temperature. Thus small reflections would not have a large effect. All parts of the system, however, were well matched and the unused port could be opened to room temperature with no effect on measurements.

Figure 15.5 shows a picture of the actual installation including the rotary joint which allowed the horn-reflector to turn while the receiver stayed stationary in the cab. The cold load was connected to the reference port through an adjustable 0.11 db attenuator which could add a well-calibrated additional noise. You can see that having the output of the horn-reflector in a cab which moves only in azimuth was very important for this experiment.

Before we started making measurements with this system, there had been careful measurements of horn-reflectors with traveling wave masers at Bell Laboratories. First, before going to the trouble of building a 20-foot

Fig. 15.5

horn-reflector, the antenna and maser groups had put together a test system (De Grasse et al. 1959b). They had a 6 GHz maser and a small horn-reflector antenna. They hooked the two up with a calibrating noise lamp and saw that indeed they got a system temperature of something like $18\frac{1}{2}$ K which was very nice, but they had expected to do a little better. You see in fig. 15.6 that contrary to the prediction I made before, they have assigned 2 K to the antenna for the back lobe and other pickup from the antenna, $2\frac{1}{2}$ K for atmosphere, and 10.5 K for the temperature of the maser. The makers of the maser were not very happy with that number. They thought they had made a better maser than that, but within the accuracy of what they knew about all the components, they solved the problem in making things add up by assigning additional noise to those components. Arno had used this horn-reflector for another astronomy project and was aware of the extra 2 K that had been assigned to it. One of the reasons that he built the cold load was to improve on their experiment.

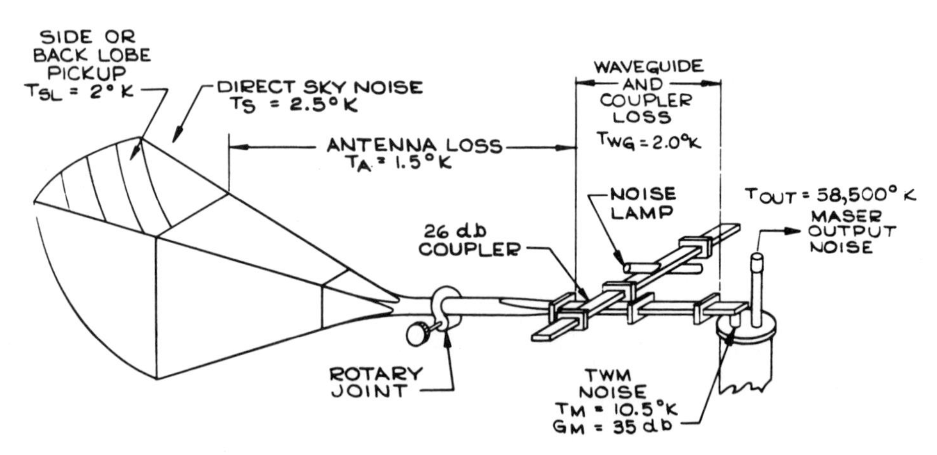

Fig. 15.6

This group had measured the atmospheric radiation (sky noise) by the same technique that Dicke had first reported on in 1946. Figure 15.7 shows a chart of a measurement such as Arno and I made with the 20-foot horn-reflector. It shows the radiometer output as the antenna is scanned from a fairly high elevation angle down to 10° elevation angle. This is a chart with power increasing to the right, and you can see what the power out of the receiver did. The circles correspond to the expected change if the zenith sky brightness is 2.2 K and the crosses to 2.4 K. You can see that the curve is a very good fit to the expected curve down to at least 10° elevation, so a well-shielded antenna makes the measurement of the atmospheric radiation very easy.

After the 20-foot horn-reflector was built and was being used with the Echo satellite, Ed Ohm, who was a very careful experimenter (Ohm

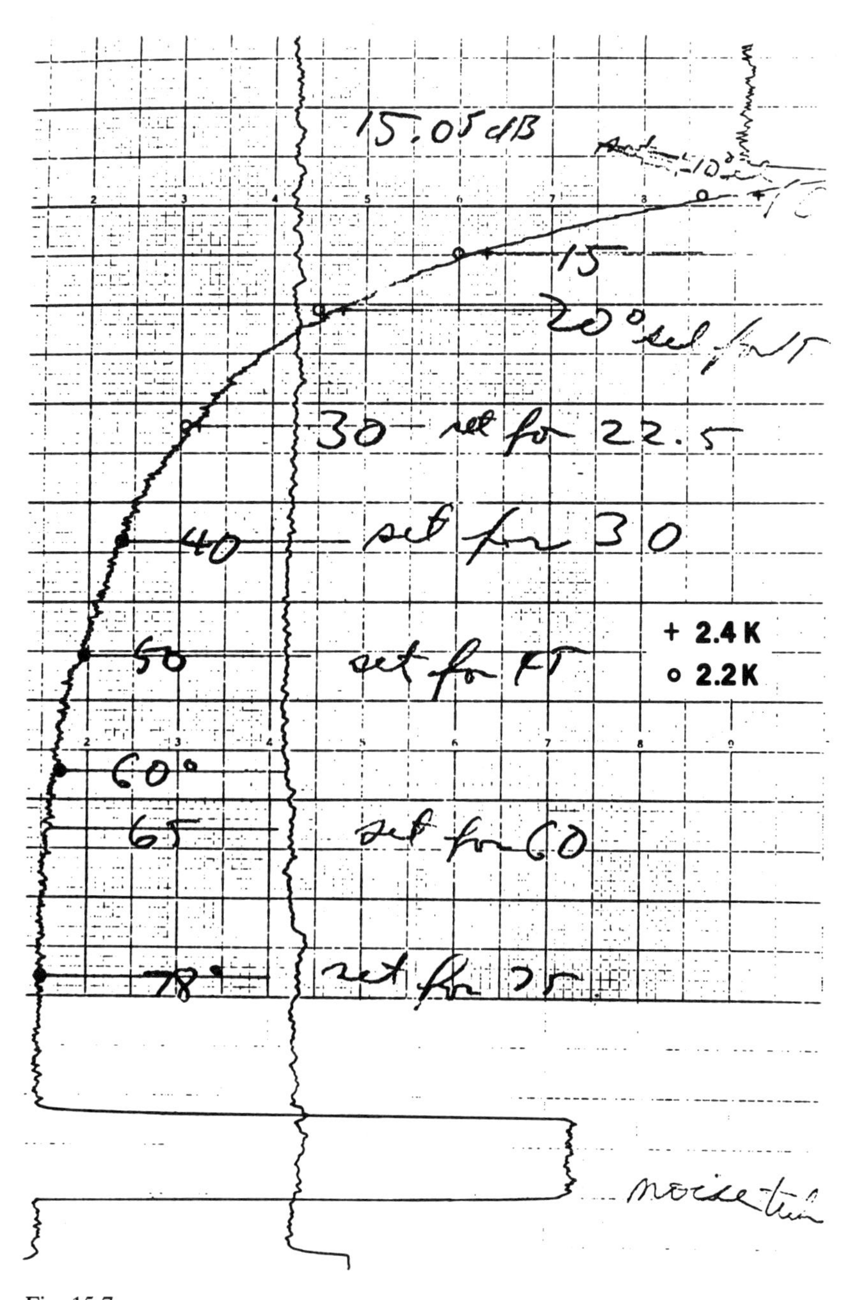

Fig. 15.7

1961), added up the noise contribution of all the components of the system and compared it to his measured total. In fig. 15.8 we see that from the sum of the components he predicted a total system temperature of 18.9 K, but he found that he consistently measured 22.2, or 3.3 K more than what he had expected. However, that was within the measurement errors of his summation, so he did not take it to be significant.

TABLE II – SOURCES OF SYSTEM TEMPERATURE

Source	Temperature
Sky (at zenith)	2.30 ± 0.20°K
Horn antenna	2.00 ± 1.00°K
Waveguide (counter-clockwise channel)	7.00 ± 0.65°K
Maser assembly	7.00 ± 1.00°K
Converter	0.60 ± 0.15°K
Predicted total system temperature	18.90 ± 3.00°K

the temperature was found to vary a few degrees from day to day, but the lowest temperature was consistently 22.2 ± 2.2°K. By realistically assuming that all sources were then contributing their fair share (as is also tacitly assumed in Table II) it is possible to improve the over-all accuracy. The actual system temperature must be in the overlap region of the measured results and the total results of Table II, namely between 20 and 21.9°K. The most likely minimum system temperature was therefore

$$T_{system} = 21 \pm 1°K.*$$

The inference from this result is that the "+" temperature possibilities of Table II must predominate.

Fig. 15.8

Well, our first observations were somewhat of a disappointment because we had naturally hoped that these things I have mentioned were just errors in the experiments. Figure 15.9 is the first measurement with our receiver. At the bottom and top, the receiver is switched to the antenna and in between to the cold load. The level from the antenna at 90° elevation matched that from the cold load with 0.04 dB of attenuation (~7.5 K total radiation temperature). At the bottom I recorded measurements of the temperature-sensing diodes on the cold load.

That was a direct confrontation. We expected 2.3 K from the sky, 1 K from absorption in the walls of the antenna, and we saw something that was obviously considerably more than that. It was really a qualitative thing rather than just quantitative because the antenna was hotter than the helium reference and it should have been colder. But we knew that the problem was either in the antenna or beyond. Arno's initial reaction was "Well, I made a pretty good cold load!" The most likely problem in such an experiment is that you do not understand all the sources of extra radiation in your reference noise source, but it is not possible to make it have a lower temperature than the liquid helium.

Initially, it seemed likely that we could not do the Galactic halo experiment, but at that time our measurements of the gain of the antenna

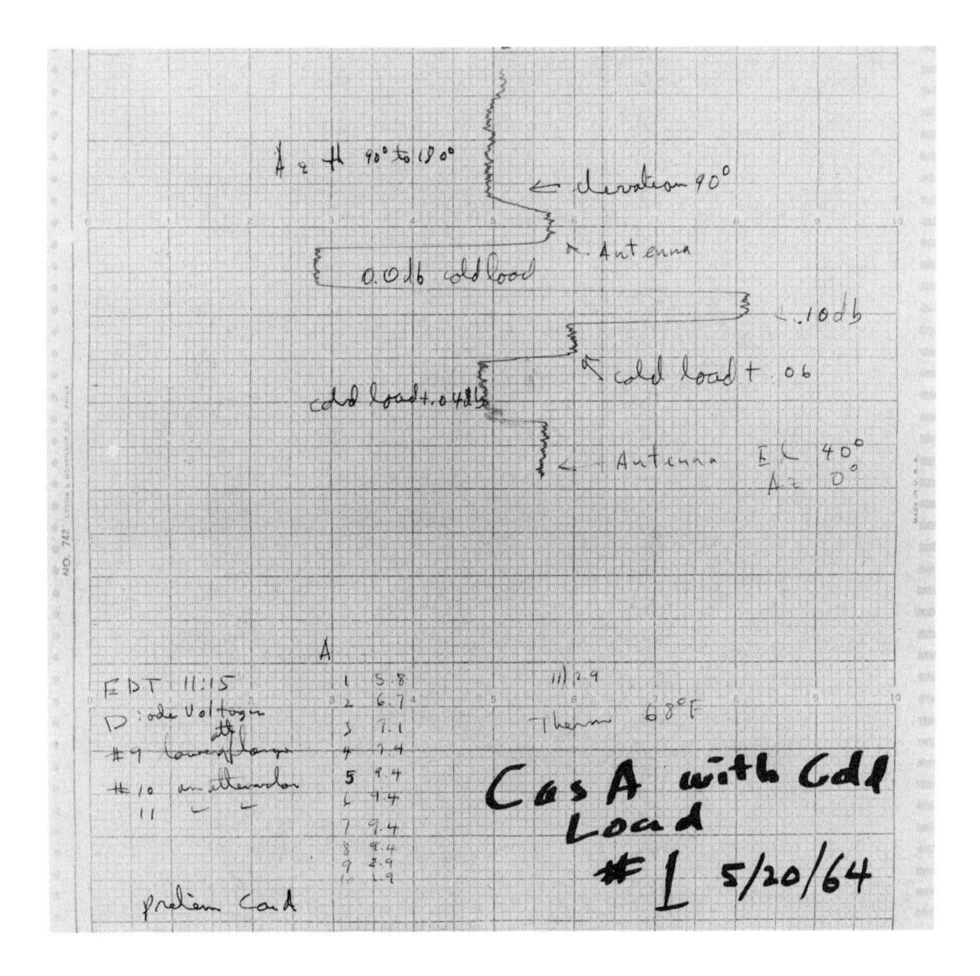

Fig. 15.9

had started (Hogg and Wilson 1965) and we wanted to go on with the absolute flux measurements before taking anything apart or trying to change anything. So we ended up waiting for almost a year before doing anything about our antenna temperature problem; however, we were thinking about it all that time.

We thought of several possible explanations of the excess antenna temperature. Many radio astronomers at the time thought the centimeter wave atmospheric radiation was about twice what we were saying. That would have gone a long way toward explaining our problem. However, the curve for the zenith angle dependence in fig. 15.7 indicates that we were measuring the atmospheric absorption or emission correctly. It turned out later that the centimeter astronomers had applied refraction corrections to their measurements of radio sources in the wrong sense. John Shakeshaft finally straightened this out.

Crawford Hill overlooks New York City; perhaps man-made interference was causing trouble. Therefore we turned our antenna down and scanned around the horizon. We found a little bit of superthermal radiation, but, given the horn-reflector's rejection of back radiation, nothing that would explain the sort of thing that we were seeing.

Could it be the Milky Way? Not according to extrapolations from low frequencies. The galactic poles should have a very small brightness at 7 cm and our actual measurements of the plane of the Milky Way did fit very well with the extrapolations. Perhaps it was discrete sources. The strongest discrete source we could see was Cas A and it had an antenna temperature of 7 K. Point sources extrapolate in frequency in about the same way as the radiation from the Galaxy, so they seemed a very unlikely explanation. That left radiation from the walls of the antenna itself. We calculated nine-tenths of a degree Kelvin for that, taking into account the actual construction of the throat section of the antenna which is the most important. It was a piece of electroformed copper and we could measure similar sorts of waveguides in the laboratory. We had to wait sometime to finish the flux measurement, but in the spring of 1965, almost a year later, we had completed it (Penzias and Wilson 1965a). The Earth had made a complete cycle around the Sun and nothing had changed in what we were measuring. We pointed to many different parts of the sky, and unless we had a known source or the plane of the Galaxy in our beam, we had never seen anything other than the usual antenna temperature. In 1962 there had been a high altitude nuclear explosion over the Pacific which had filled up the van Allen belts. We were initially worried that something strange was going on there, but after a year, the population of van Allen belts had gone down considerably and we'd not seen any change.

There was a pair of pigeons living in the antenna at the time, and they had deposited the usual white diaelectric in the part of the horn where they rousted. So we cleaned up the antenna, disposed of the pigeons, and put some aluminium tape over the joints between the separate pieces of aluminium that made it up. All of this made only a minor improvement.

We were really scratching our heads about what to do until one day Arno happened to be talking to Bernie Burke about other matters. After they had finished talking about what Arno had called for, he mentioned our problems, our dilemma that the Galactic halo experiment was not ever going to work, and that we could not understand what was going on. Bernie had heard about Jim Peebles' calculations of microwave radiation from a hot big bang and suggested that we get in touch with Dicke's group at Princeton. So of course Arno called Dicke. Dicke was thinking about oscillating big bangs which he concluded should be hot. After a discussion on the phone, they sent us a preprint and agreed to come for a visit. When they came and saw what we had done, I'm sure they were thoroughly disappointed, but agreed that what we had done was probably right.

Afterward the two groups wrote separate letters to the *Astrophysical Journal* (Dicke et al. 1965; Penzias and Wilson 1965b).

We made one last check before actually sending off our letter for publication. We took a signal generator, attached it to a small horn and took it around the top of Crawford Hill to artificially increase the temperature of the ground and measure the back lobe level of the 20-foot horn, maybe there was something wrong with it. But the result was as low as we expected. So we sent the letter in!

Arno and I of course were very happy to have any sort of an answer to our dilemma. Any reasonable explanation would have probably made us happy. In fact, I do not think that either of us took the cosmology very seriously at first. We had been used to the idea of steady-state cosmology; I had come from Caltech and had been there during many of Fred Hoyle's visits. Philosophically, I liked the steady-state cosmology. So I thought that we should report our result as a simple measurement; the measurement might be true after the cosmology was no longer true!

After our meeting, the Princeton experimental group returned to complete their apparatus and make their measurement, with the expectation that the background temperature would be about 3 K.

The first confirmation of the microwave cosmic background that we knew of, however, came from a totally different, indirect measurement. This measurement had, in fact, been made thirty years earlier by Adams (1941, 1943), Dunham (1937, 1939, 1941), and Dunham and Adams (1937) had discovered several faint optical interstellar absorption lines which were later identified with the molecules CH, CH^+, and CN. In the case of CN, in addition to the ground state, absorption was seen from the first rotationally excited state. McKellar (1941) using Adams' data on the populations of these two states calculated that the excitation temperature of CN was 2.3 K. The rotational transition occurs at 2.64 mm wavelength, near the peak of a 3 K blackbody spectrum. Shortly after the discovery of the background radiation, G. B. Field et al. (1965, 1966), Field and Hitchcock (1966), I. S. Shklovsky (1966), and P. Thaddeus (1966) (following a suggestion by N. J. Woolf) independently realized that the CN is in equilibrium with the background radiation. (We now know enough about the interstellar medium to say that there is no other signficant source of excitation where these molecules are located.) In addition to confirming that the background was not zero, this idea immediately confirmed that the spectrum of the background radiation was close to that of a blackbody source for wavelengths larger than the peak. It also gave a hint that, at short wavelengths, the intensity was departing from the $1/\lambda^2$ dependence expected in the long wavelength (Raleigh–Jeans) region of the spectrum and following the true blackbody (Planck) distribution.

In December 1965 Roll and Wilkinson (1965) completed their measurement of (3.0 ± 0.5) K at 3.2 cm, the first confirming microwave measurement. This was followed shortly by Howell and Shakeshaft's (1966) value of (2.8 ± 0.6) K at 20.7 cm and then by our measurement of (3.2 ± 1) K at 21.1 cm (Penzias and Wilson 1967). (Half of the difference between these two results comes from a difference in the corrections used for the galactic halo and integrated discrete sources.) By mid-1966, the intensity of the microwave background radiation had been shown to be close to 3 K between 21 cm and 2.6 mm, almost two orders of magnitude in wavelength.

Earlier theory

I have mentioned that the first experimental evidence for cosmic microwave background radiation was obtained (but unrecognized) long before 1965. We soon learned that the theoretical prediction of it had been made at least sixteen years before our detection. George Gamow (1948) had made calculations of the conditions in the early universe in an attempt to understand galaxy formation. Although these calculations were not strictly correct, Gamow and his collaborators calculated that the density of radiation in the hot early universe was much higher than the density of matter. In this early work, the present remnants of this radiation were not considered. However, in 1949, Alpher and Herman (1949) followed the evolution of the temperature of the hot radiation in the early universe up to the present epoch and predicted a value of 5°K. They noted that the present density of radiation was not well known experimentally. In 1953, Alpher, Follin, and Herman (1953) reported what has been called the first thoroughly modern analysis of the early history of the universe, but failed to recalculate or mention the present radiation temperature of the universe.

In 1964, Doroshkevich and Novikov (1964) had also calculated the relic radiation and realized that it would have a blackbody spectrum. They quoted E. A. Ohm's article on the Echo receiver, but misunderstood it and concluded that the present radiation temperature of the universe is near zero.

Cosmology is a science which has only a few observable facts to work with. The discovery of the cosmic microwave background radiation added one – the present radiation temperature of the universe. This, however, was a significant increase in our knowledge since it requires a cosmology with a source for the radiation at an early epoch and is a new probe of that epoch. More sensitive measurements of the background radiation in the future will allow us to discover additional facts about the universe.

References

Adams, W. S. (1941). Results with the Coudé Spectrograph of the Mt. Wilson Observatory. *Astrophysical Journal*, **93**, 11–23.

(1943). The Structure of Interstellar H & K lines in 50 Stars. *Astrophysical Journal*, **97**, 105.

Alpher, R. A. and Herman, R. C. (1949). Remarks on the Evolution of the Expanding Universe. *Physical Review*, **75**, 1089–1095.

Alpher, R. A., Follin, J. W., and Herman, R. C. (1953). Physical Conditions in the Initial Stages of the Expanding Universe. *Physical Review*, **92**, 1347–1361.

Crawford, A. B., Hogg, D. C., and Hunt, L. E. (1961). A Horn-Reflector Antenna for Space Communication. *Bell System Technical Journal*, **40**, 1095.

DeGrasse, R. W., Schultz-Dubois, E. O., and Scovil, H. E. D. (1959a). Three-Level Solid-State Travelling-wave Maser. *Bell Systems Technical Journal*, **38**, 305.

DeGrasse, R. W., Hogg, D. C., Ohm, E. A., and Scovil, H. E. D. (1959b). Ultra Low Noise Antenna and Receiver Combination for Satellite or Space Communication. *National Electronics Conference*, **15**, 370.

Dicke, R. H., Peebles, P. J. E., Roll, P. G., and Wilkinson, T. D. (1965). Cosmic Black-Body Radiation. *Astrophysical Journal*, **142**, 414–419.

Doroshkevitch, A. G. and Novikov, I. D. (1964). Mean Density of Radiation in the Metagalaxy and Certain Problems in Relativistic Cosmology. *Doklady Akademii Nauk SSSR*, **154**, 809; also in *Soviet Physics Doklady*, **9**, 111–113.

Dunham, T. (1937). Interstellar Neutral Potassium and Neutral Calcium. *Publications of the Astronomical Society of the Pacific*, **49**, 26.

(1939). The Material of Interstellar Space. *Proceedings of the American Philosophical Society*, **81**, 277.

(1941) Concentration of Interstellar Molecules. *Publications of the American Astronomical Society*, **10**, 23.

Dunham, T. Jr. and Adams, W. S. (1937). New Interstellar Lines in the Ultraviolet Spectrum. *Publications of the American Astronomical Society*, **9**, 5.

Field, G. B. and Hitchcock, J. L. (1966). Cosmic Black-Body Radiation at $\lambda = 2.6$ mm (from Observations of Interstellar CN). *Physical Review Letters*, **16**, 817–818.

Field, G. B., Herbig, G. H., and Hitchcock, J. L. (1965). Radiation Temperature of Space at λ 2.6 mm. *Astronomical Journal,* **71**, 161.

Gamow, G. (1948). The Evolution of the Universe. *Nature*, **162**, 680–682.

Hogg, D. C. and Wilson, R. W. (1965). A Precise Measurement of the Gain of a Large Horn-Reflector Antenna. *Bell System Technical Journal*, **44**, 1019.

Howell, T. F. and Shakeshaft, Jr. R. (1966). Measurement of the Minimum Cosmic Background Radiation at 207-cm Wavelength. *Nature*, **210**, 1318–1319.

McKellar, A. (1941). Molecular Lines from the Lowest States of the Atomic Molecules Composed of Atoms Probably Present in Interstellar Spaces. *Publications Dominion Astrophysical Observatory Victoria B.C.*, **7**, 251–272.

Ohm, E. A. (1961). Project Echo Receiving System. *Bell System Technology Journal*, **40**, 1065.

Penzias, A. A. (1965). Helium-Cooled Reference Noise Source in a 4-kMc Waveguide. *Review of Scientific Instruments,* **36**, 68.

Penzias, A. A. and Wilson, R. .W. (1965a). Measurement of the Flux Density of CasA at 4080 MHz. *Astrophysical Journal*, **142**, 1149.

(1965b). A Measurement of Excess Antenna Temperature at 4080 Mc/s. *Astrophysical Journal*, **142**, 419.

(1967). Measurement of Background Temperature at 1415 MHz. *Astronomical Journal*, **72**, 315.

Roll, P. G. and Wilkinson, D. T. (1966). Measurement of Cosmic Background Radiation at 3.2 cm Wavelength – Support for Cosmic Black-Body Radiation. *Physical Review Letters*, **16**, 405–407.

Shklovsky, I. S. (1966). Astronomical Circular No. 364. Academy of Sciences of the U.S.S.R.: Moscow.

Tabor, W. J. and Sibilia, J. T. (1963). Masers for the Telstar® Satellite Communications Experiment. *Bell System Technical Journal*, **42**, 1863.

Thaddeus, P. and Clauser, J. F. (1966). Cosmic Microwave Radiation at 263 mm from Observation of Interstellar CN. *Physical Review Letters*, **16**, 819–822.

Discussion

Sullivan:

Bob, you were too modest to mention the amazing parallels between your discovery and that of Karl Jansky thirty years earlier, but they are worth mentioning. First of all, besides the fact that they both occurred at Bell Laboratories, there is the connection provided by two men: (a) Harald Friis, who was Jansky's boss and who was important in the invention of the parabolic horn such as you used, and (b) Art Crawford, who was a close colleague of Jansky's as well as intimately linked with your antenna.

More importantly, what were the characteristics of the men and machines that allowed these two fundamental discoveries of an all-sky, weak background to be made? In each case, the antenna was the best designed and the most sensitive in the world for its intended use in communications research. In each case, the receivers were carefully constructed so as to achieve unparalleled sensitivity and stability. In each case, the men involved hammered away at trying to understand the origin of a small, but persistent effect that simply would not go away – was it patience or just plain stubbornness? It appears that Bell Laboratories was indeed hiring the same kind of researchers in the early 1960s as it did in the late 1920s!

Wilson:

Thank you, Woodie. Yes, there are amazing parallels. It is also interesting that those discoveries were made in a part of Bell Laboratories research which was (and is) rather goal oriented. I believe that the combination of communications goals which forced the exploration of new science and an attitude of wanting to understand new things in as fundamental a way as possible is responsible for many of the successes at Bell Laboratories.

Sullivan:
In the design of your experiment and in its initial observations, how much were you motivated by the previously reported (although hardly certain) excess system temperature found by Ohm and others?

Wilson:
Arno was aware of the earlier work when he started to design the reference noise source. The expectation was that there would be no problem when we made a proper measurement. I do not remember being influenced by it very much during the year that we were puzzling over the problem. We had a much more direct measurement.

16

The entry of radio astronomy into cosmology: radio stars and Martin Ryle's 2C survey

Woodruff T. Sullivan, III

The influence of radio astronomy on the cosmology of the mid twentieth century has been enormous and in this brief paper I give an account of how the fledgling technique first impinged on the study of the universe as a whole. In doing so, one is inevitably led to the person of Martin Ryle (1918–84) (fig. 16.1). It was he who, in 1955, first carried radio astronomy into cosmology in an important way, and it was also he who triggered over the following years a great debate about the evidence for steady state versus evolving worlds. This is particularly remarkable because for many years previous it was also he who had most forcefully argued that radio stars were a strange breed of star in our own Milky Way, having nothing to do with cosmology. But over the period 1952–54 he changed his mind and one of the main aims of this paper is to investigate how this switch came about.

In this paper I will sketch the radio star scene as it developed over 1946–52 (largely from the point of view of Ryle's group), present the available evidence as to how and why Ryle dramatically changed his view on the radio stars, and finally gave some idea of the reception accorded the first announcement of these results.[1] The following paper by Peter Scheuer will then carry on from that point.

16.1 1946–51: First work on radio stars

Those few radio engineers and physicists who worked immediately after World War II on "extraterrestrial noise," as the phenomenon was then known, primarily studied the Sun. A few projects, however, were undertaken on the strong galactic background radiation that Karl Jansky had originally discovered in 1932. It was during such a survey that a group led by James Stanley Hey in England noticed that one particular region of the Milky Way – in the constellation of Cygnus – seemed to vary in

[1] A complete account of the pre-1953 period will appear in a forthcoming monograph on the history of radio astronomy. For a detailed analysis of radio source work throughout the 1950s, examining in particular the radio source count controversies after 1955, see Edge and Mulkay (1976).

Fig. 16.1 Martin Ryle (*c.* 1960).

intensity. They could only determine that the varying region was less than 2° in size, but nevertheless deduced that this discrete source, later to be known as Cygnus A, must have extremely small physical dimensions in order to be fluctuating on a time scale of minutes (Hey, Parsons, and Phillips 1946). At the Radiophysics Laboratory in Sydney John Bolton and Gordon Stanley (1948) then used their sea-cliff interferometer to show that the source was in fact less than 8′ in size. Moreover, Bolton (1948) discovered six other "radio stars," as they soon came to be known, and it became clear that one was dealing with an entirely new class of object. At the Cavendish Laboratory in Cambridge Martin Ryle and his student Graham Smith (1948) first became involved with the radio stars when they set up a two-element interferometer to study Cyg A, only to stumble upon a second, yet stronger source, Cassiopeia A.

To understand the radio stars in general, it was deemed vital to understand the nature of their intensity scintillations. During the period 1947–50, there was a great deal of argument (mostly unpublished) as to whether the scintillations originated in the radio star itself, as Hey had originally assumed, or during passage through the Earth's ionosphere. Or were both mechanisms operating, perhaps to differing degrees in different sources? Studies by Bolton's group in 1949 (from both Sydney and New Zealand) convinced them that the scintillations were ionospheric (Stanley and Slee 1950), and Bolton so wrote Ryle. But Ryle felt that Cyg A's maximum altitude of only 15° as seen from the Antipodes made these observations unreliable. He therefore persuaded Bernard Lovell at Jodrell Bank, University of Manchester (then devoted to meteor radar studies), to undertake joint, simultaneous observations at a variety of spacings. These were carried out during the second half of 1949 and resulted in convincing evidence that the scintillations were indeed ionospheric (Little and Lovell 1950; Smith 1950). The Cambridge group, however, had observed a second type of variation, namely strong 15-second-long bursts simultaneously observed at sites 160 km apart. These they took to be intrinsic to the radio stars, thereby indicating that these enigmatic objects could be no more than 15 light-seconds in size, that is, about ten times the size of the Sun. Although Ryle never regarded these bursts as "proven" (they had been rare during the original experiments and in fact were never again observed), they were nevertheless frequently cited over the next two years as contributing evidence that most sources were indeed radio *stars*.

The notion of radio stars was first brought to the fore independently by the German astrophysicist Albrecht Unsöld (1949a, 1949b) and Ryle (1949). Unsöld suggested that the radio stars could be low-mass dwarf stars of low optical luminosity and yet of considerable activity in terms of giant flares or similar phenomena. He was also attracted to the idea that these same radio stars might be the powerhouses for the galactic cosmic rays, whose origin was unknown. Ryle too liked the idea of solving the cosmic ray puzzle at the same time as the radio star one, but he in fact had a *third* phenomenon that could be understood by this class of dark stars. This was the galactic background radiation, whose origin had originally been taken to be thermal emission from a hot interstellar medium. But by 1949 it was becoming clear that one could not explain the spectrum and sky distribution by a thermal model. Ryle argued that the galactic background was simply the integrated radiation of all the radio stars that could not be individually resolved.

At Leiden Observatory the eminent theorist Jan Oort had long been interested in radio astronomy and trying to launch an observational effort in Holland. From discussions and correspondence with Ryle and Bolton he grew enthusiastic about explaining the galactic background as the integrated effect of radio stars and developed a detailed model with his student Gart Westerhout. In this model the radio stars represented a class of star intermediate between Walter Baade's Populations I and II. The one discrepancy, however, was at the north galactic pole, where observations at 100 MHz by Bolton and Westfold (1950) revealed an excess intensity of 600 K in brightness temperature. Bolton and Westfold (1951: 484) had stated that this excess possibly pointed to an isotropic component, which might "be extragalactic in origin or even originate in the solar system." Westerhout and Oort (1951: 327), however, favored the idea that this isotropic component originated as the integrated radiation from distant *galaxies*. Noting this possibility, they wrote that it "would be of the utmost interest, inasmuch as it would give an entirely new datum regarding the structure of the universe." This was not, however, the first suggested link between radio astronomy and cosmology, for Oort in fact was building on earlier work he had assigned to his colleague and former student Hendrik Van de Hulst. It is little known that in the famous 1944 colloquium in Nazi-occupied Holland in which Van de Hulst (1945) predicted the 21 cm hyperfine line of hydrogen he also examined the problems of (a) the galactic background, (b) whether high-n hydrogen recombination lines might be detectable in the radio spectrum (concluding no), and (c) radio cosmology! The last was based on the galactic background measures of Jansky and of Grote Reber, as well as on Reber's (1940) claimed detection of the Andromeda nebula.[2] After supposing that all galaxies were similar to our own and to Andromeda in the respect that their radio spectra were much flatter than their optical spectra, Van de Hulst then used an Olbers'-paradox type argument to show (using Otto Heckmann's recent [1942] monograph) that the universe must indeed be expanding and of finite age, not static and of infinite age, or else the brightness of the radio sky would be far larger than observed. As we shall see, this type of argument (although not applied to the nature of the redshifts) was to be central in later debates about the nature of radio stars. It is not known what influence Van de Hulst's work had on Ryle, but it may have been considerable, for in May 1951 Oort sent him an English

[2] This detection would seem to be incorrect, based on later measurements of the 160 MHz intensity of the Andromeda nebula (for example, by Hanbury Brown and Hazard [1951]), as well as the too-narrow width of the drift scan illustrated by Reber (1940).

translation of the relevant section.[3] Note, however, that in 1951 Van de Hulst (1951: 175) considered it "very unlikely" that most known discrete sources were extragalactic.

In mid-1951 the situation was that most people accepted the idea of radio *stars*, but that no one knew much about these putative objects, and even less about how they emitted. This was despite the fact that the only optical counterparts that had been suggested, based on positions of 10–15′ accuracy determined by Bolton, Stanley, and Slee (1949), were not at all stellar. The first association (Taurus A) was with the Crab nebula, the second (Virgo A) with the elliptical galaxy M87, and the third (Centaurus A) with the peculiar nebula NGC 5128. Although for the first time one had a plausible link between galactic noise and (optical) astronomy, over the next couple of years these identifications were hardly viewed as certain (even by Bolton). The Crab looked like the best bet, but if it was so strong, then why weren't the supernova remnants of Tycho and Kepler easily detected? As Ryle (1976: 27T)[4] recalled:

> Not being optical astronomers, we didn't know how remarkable the Crab was . . . At that stage we didn't know anything about it except that it was a little fuzzy thing on a photograph.

At this time (April 1951) a meeting was held in London on the "Dynamics of Ionized Media" and its proceedings (Boyd 1951), although only distributed in a mimeographed format, provide a valuable benchmark for the attitudes of the day toward the radio sources. The major astrophysical conundrum, no matter what the sources' distances, was to explain their extremely high brightness temperatures ($>10^8$ K). Ryle felt that a super-hot outer atmosphere of a dark star was a better bet than putting the sources at extragalactic distances where their luminosities would be horrendously large. And if the sources were to be entire galaxies, one would have to maintain such a temperature over a very large object. Furthermore, was it not suggestive that the ratio of the intensity of the strongest discrete radio source to that of the radio background was similar to that of the brightest star to the optical Milky Way? And there was the economy of hypotheses in explaining the discrete sources and the galactic background and the cosmic rays all with one type of object, albeit a mysterious one. Ryle (1950: 241) and Smith (1951a: 95) had also been able to show that the numbers of known radio stars could indeed be extrapolated to fainter levels to produce the observed amount of galactic background. This required the radio stars to have a number density of ~3 per cubic parsec (Ryle) or 0.004/pc^3 (Smith), roughly like that of the

[3] Oort:Ryle, 22 May 1951, file 12 (uncatalogued) and box 1/2, Ryle papers (= RYL), Churchill College, Cambridge University. Oort had also sent this translation a few months earlier to Hanbury Brown at Jodrell Bank.

[4] This citation refers to p. 27 of the transcipt of Ryle's 1976 interview with the author.

(a)

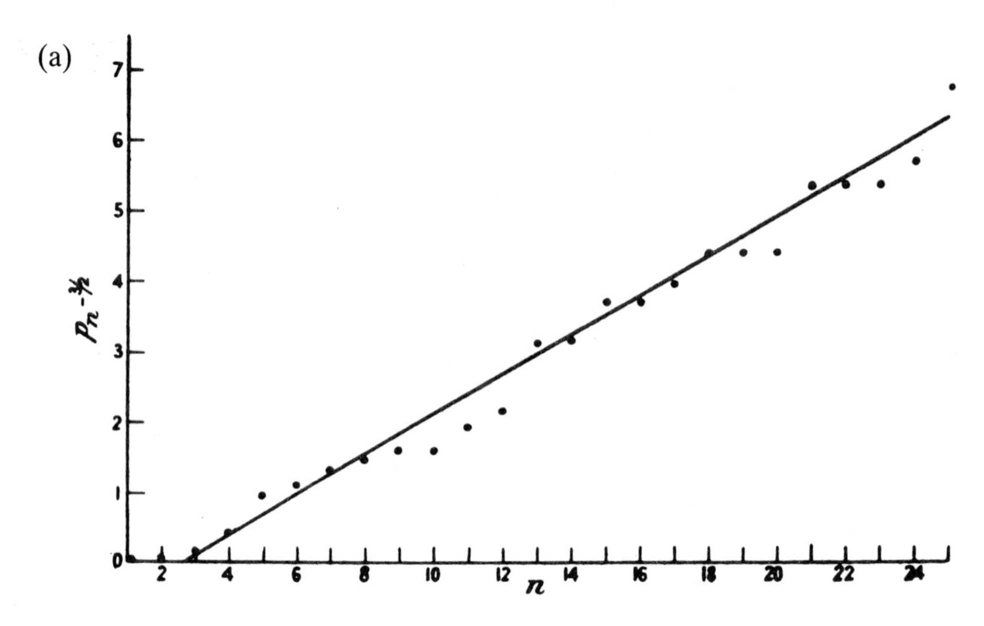

(b)

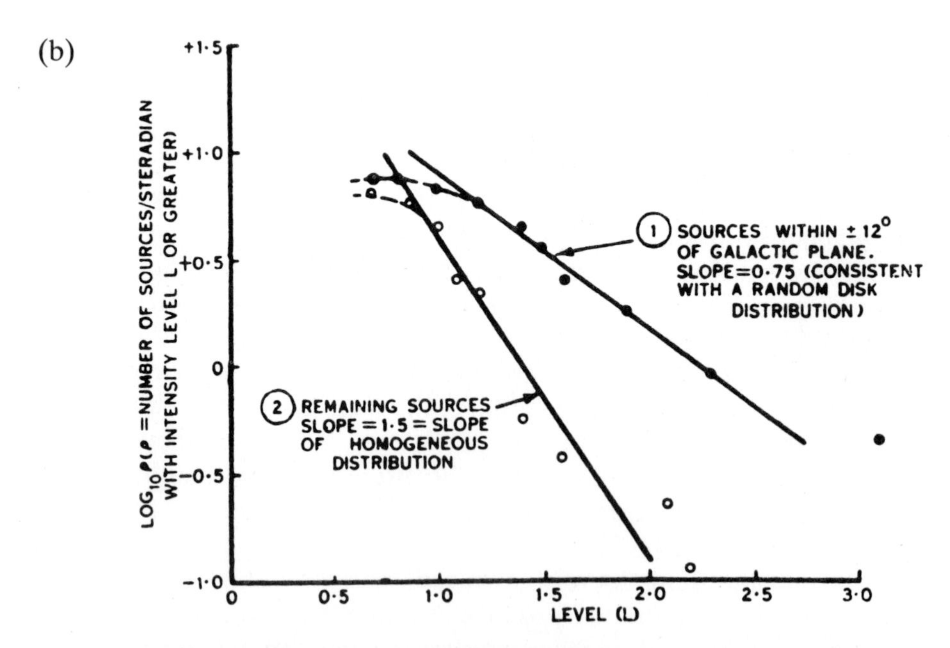

Fig. 16.2 (a) The first plot of radio source intensities analogous to a log N − log S plot, made by Ryle and Smith in 1949 and published by Ryle (1950: 241). The plot is based on the 25 most intense sources of a preliminary version of what eventually became the 1C survey. P_n is the power received from the nth most intense source.
(b) Mills's (1952: 275) first log N − log S plot, in the form that later became standard. The level L is the logarithm of the source flux density in units of 10 Jy. The different slopes for the (17) Class I and (60) Class II sources are emphasized.

known stars (other estimates of the time from Sydney, Jodrell Bank, and Leiden had number densities ranging as low as $5 \times 10^{-8}/\mathrm{pc}^3$). In the process Ryle and Smith produced the first diagram for radio sources analogous to what later became known as a "log N–log S" plot (fig. 16.2a), that is, a plot of (the logs of) the flux density S versus the number N of sources having a flux density $\geqslant$S. Their plot's constancy of slope implied a uniform distribution of sources.

Ryle's views did not go unchallenged at this meeting. The theorist Tommy Gold was unconvinced and reasoned, with his colleague Fred Hoyle assenting, that the isotropic distribution of the known ~75 radio stars could equally well place them very removed from us as very close (relative to the width of the Galactic disk). He furthermore pointed out that either hypothesis required an entirely new class of object, but that peculiar galaxies only had to have a radio luminosity of 10^3 times that of the Milky Way, whereas peculiar stars had to have one of 10^8 times that of the Sun. Sparks flew (fig. 16.3). This was not the first time that Hoyle and Gold had been at loggerheads with Ryle – previous occasions had centered

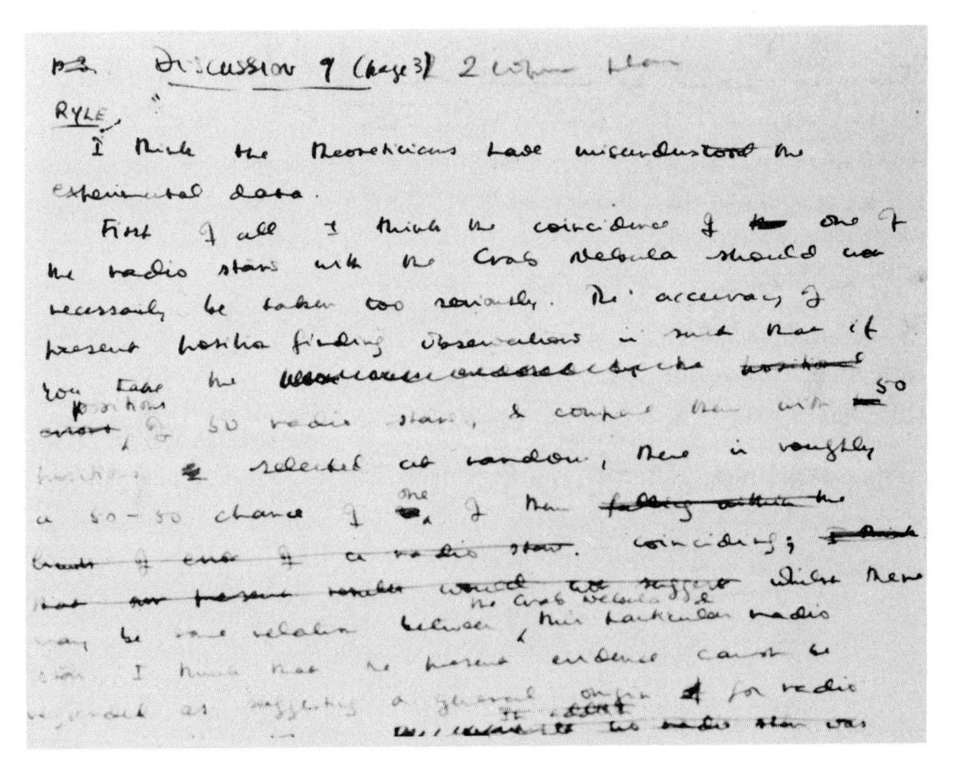

P3. Discussion 9 (page 3) 2 [illegible]

RYLE,

I think the theoreticians have misunderstood the experimental data.

First of all I think the coincidence of one of the radio stars with the Crab Nebula should not necessarily be taken too seriously. The accuracy of present position finding observations is such that if you take the positions of 50 radio stars, & compare them with 50 positions selected at random, there is roughly a 50-50 chance of one of them coinciding; [illegible] there may be some relation between the Crab Nebula & this particular radio [illegible] I think that the present evidence cannot be regarded as suggesting a general origin for radio [illegible]

Fig. 16.3 Ryle's manuscript version of discussion remarks submitted to the editor of the proceedings of the April 1951 conference on "Dynamics of Ionized Media" (Boyd 1951: 58). His opening remark that "I think the theoreticians have misunderstood the experimental data" hints at the liveliness of the discussion with Gold and Hoyle. [From box 1/2, RYL]

on the theory of solar radio bursts as discussed in Cambridge at evening meetings of the Kapitza Club.[5]

16.2 1951–52: Optical identifications

Shortly after the London conference Smith (1951b) finished a long series of observations that led to ~1′ accuracies for the positions of the four strongest sources in the northern sky. These confirmed Bolton's earlier suggestions for Tau A and Vir A, but disappointingly led to nothing obvious for the two strongest sources, Cas A and Cyg A. Smith wrote to Walter Baade and Rudolph Minkowski at Mt. Wilson Observatory, however, and within a year they had turned the 100-inch and 200-inch telescopes to the identification and study of what was to be found at Smith's positions (see Smith and Lovell 1983). For Cas A it was an unprecedented network of fast-moving filaments showing no particular pattern of motions, and for Cyg A it was a peculiar, faint, extragalactic nebula with a speed of recession of 17,000 km/s, implying the astonishing distance of 30 Mpc (300 times the distance to the center of our Galaxy) (Baade and Minkowski 1954a). Baade immediately interpreted its peanut-like shape and its highly excited spectrum as evidence that one was actually seeing the collison of *two* galaxies, a process that he had recently studied in the context of clusters of galaxies. The unexpected distance and optical faintness of Cyg A, combined with its great radio intensity, opened up the possibility of measuring multitudes of radio sources whose optical counterparts would be so distant as to be beyond the faintness limit of even the mighty 200-inch telescope. The central question remained, however, as to what fraction of the radio stars were indeed Cyg A's.

After many discussions with Baade at the September 1952 meeting of the International Astronomical Union (IAU) in Rome, Ryle was particularly impressed with the optical evidence. Back home, he described these developments to his Cavendish colleagues as "dramatic; a turning point in radio astronomy; the completion of the first stage of radio star observations." And to his fellow Cantabrigians in the Kapitza Club he said:

> Radio stars have provided the Kapitza Club on several previous occasions with real opportunities for speculation – [there was] so very little information available concerning their constitution. There were, I think, even those who were not really convinced of their reality. This stage, I am afraid, is now over, as a number of these have recently been identified beyond any reasonable doubt with visible objects.[6]

Much later, Ryle recalled:

> Perhaps the greatest discontinuity [in my career] was with the identification of Cygnus A. That showed that we were in the cosmology game . . .

[5] Ryle: Hoyle and Ryle: K. C. Westfold, 26 November 1949, box 1/1, RYL. Also Gold (1976: 6–7T).

[6] Notes for talks given to the Cavendish radio group and to the Kapitza Club ("Recent Results in Radio Astronomy"), 16 October 1952 and undated (about October 1952), file 8 (uncat.), RYL.

> To me that was the point where one said, "Well now, if other things like Cygnus exist, here is something which we can likely see, even with our little instrument, as far away as the 200-inch can see. This is something much more interesting than it might have been – much more interesting than [the radio sources] being galactic objects, much more interesting than M87's [fairly normal galaxies]"...
>
> Now of course Cygnus-type sources needed only represent one percent of the total source population. We didn't know that ... We didn't know anything, because here we'd got a new sky ... but there was now the possibility of being in the cosmology game.
>
> (Ryle 1976: 37–8, 49–50T)

The new optical identifications were undoubtedly exciting and promising, but they still did not settle the question of the distances to the bulk of the radio stars – the scorecard of galactic-to-extragalactic for identified objects was about even at 5 to 4 (not counting nearby "normal" galaxies). There is no doubt, however, that Ryle began at this time more seriously to consider the possibility that a good fraction of the radio stars might be extragalactic, even though he continued to favor their being in the Galaxy. Baade also was convinced that the unidentified sources represented a portion of his Population II (and thus were necessarily stars) and that the identified few were an unrepresentative "tip of the iceberg." First of all, he had preferentially searched for optical counterparts to those sources that were *extended* in the radio. Secondly, if one looked at the brightest optical objects, Baade argued, one found not only the familiar stars, but also the very distant Andromeda nebula and Large and Small Magellanic Clouds.[7]

16.3 Views outside Cambridge

This paper focuses on Ryle and his group, but it is important to note that the radio sky was perceived differently in Sydney and in Manchester (see Mills 1984 and Hanbury Brown 1984 for accounts by two key participants). Bernard Mills (1952) surveyed the southern sky and found 77 radio stars that he argued split up naturally into two groups. Class I sources, more intense and sometimes extended in size, concentrated to the plane of the Milky Way, while Class II sources were isotropically distributed. Mills took the former to be in our Galaxy and was uncertain about the latter, although leaning toward them being extragalactic. He also presented his first log N–log S plots (fig. 16.2b), which showed that Class I and Class II sources yielded distinctly different slopes, the value of −1.5 for Class II implying a uniform distribution of sources. Stating that "the definite identification of a class of source with rare extragalactic objects would be of great importance to cosmological theory," he then

[7] Baade:Ryle, 1 July 1952, file 4 (uncat.), RYL. Also see Baade and Minkowski (1954b).

explored the consequences of assuming that all Class II sources were of similar luminosity to Virgo A (taken to be in the Virgo cluster of galaxies). Mills found that the background radiation at the pole could indeed be accounted for by such radio sources if they occurred as far away as 1000 Mpc, close to the estimated size of the Hubble sphere (600 Mpc). He concluded: "Thus these assumptions lead to a reasonable result, but only further work can determine to what extent they are correct" (Mills 1952: 281).

Also in Sydney, Bolton, Stanley, and Slee (1954) published their own survey of 104 sources and broadly agreed with Mills. They obtained, however, a slope much steeper than −1.5 in their log N–log S plot for sources away from the galactic plane. This excess of faint sources was argued not to be an instrumental artifact, but they did not discuss it as a possible effect of cosmology. They concluded:

> A plausible explanation is that the Sun (if these sources are galactic) or the Galaxy (if the sources are extragalactic) is in a local region of low source density and that somewhere towards the limit of the survey we reach a region of much higher density ... However, there is not much point in speculating too far on this result as it could also be produced by a large dispersion in absolute magnitudes [intrinsic luminosities] amongst the sources of the survey.
>
> (Bolton, Stanley, and Slee 1954: 129)

Meanwhile, at Jodrell Bank, Robert Hanbury Brown and Cyril Hazard used a giant (218-ft diameter) fixed reflector first to detect the Andromeda nebula in 1950 and then to undertake a long series of studies of "normal" galaxies. Their first calculation of any extragalactic component within the radio background assumed all galaxies to emit like the Andromeda nebula. With a "very simple model of the universe" (for example, no effects of recession on observed intensities), they then found the expected contribution to be negligible (Hanbury Brown and Hazard 1951). This had changed, however, by the middle of 1953 when they had developed a model of the galactic background radiation that was roughly evenly split between an isotropic (and probably extragalactic) component and a galactic component arising from the integrated effect of Mills's Class I sources (Hanbury Brown and Hazard 1953a). They basically agreed with Mills's division and felt that Ryle was missing extended sources because of his interferometric technique. Furthermore, Hazard (1953: 165) concluded his Ph.D. thesis by stating that "the majority of the Class II sources so far observed are probably extragalactic objects." Working out their number density on his model, he found it not too different from that of normal spiral galaxies. In all of this, however, it was sobering that the four major surveys done before 1953 not only were yielding few optical identifications, but also profoundly disagreed with one other (fig. 16.4).

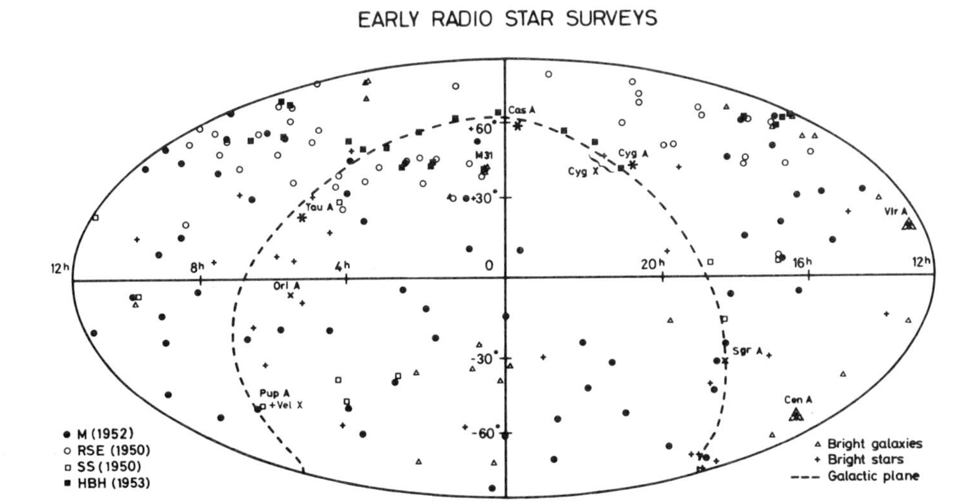

Fig. 16.4 An all-sky plot in equatorial coordinates of the positions of the discrete radio sources from the surveys of Stanley and Slee (1950), Ryle, Smith and Elsmore (1950), Mills (1952), and Hanbury Brown and Hazard (1953b). The fifty brightest optical stars and optical galaxies are also indicated, as well as the names of the strongest sources. Note the complete lack of correlation between the brightest optical objects and the radio sources, as well as the minimal agreement between different radio surveys in portions of the sky where they overlap.

16.4 The 2C survey and Ryle's change of view

Immediately upon completion of the first Cambridge ("1C") survey of 50 sources (Ryle, Smith, and Elsmore 1950), Ryle began planning for the next step, a giant antenna system that would produce far more sources and far better positions. The first objective would allow better statistics to check on whether the distribution of radio stars was compatible with their also accounting for the galactic background, and the second would lead to many more opitcal identifications. The antenna consisted of four parabolic cylinders with a total collecting area of one acre split among the corners of a 1900 × 200-foot field. During the design stage in 1950–51 and construction during 1952 it was hoped eventually to tabulate 500 to 1000 sources in this 2C survey, but nobody knew what exactly would obtain when the survey's strip chart recordings began to appear in March 1953. By July, Ryle's student John Shakeshaft, who was in charge of data reduction, was able to show to a Jodrell Bank conference a preliminary map of about 150 sources in a 10°-wide strip of declination. By October, the antenna had covered over two-thirds of the northern sky, but data reduction was falling well behind data collection. Observations were completed in May 1954 and the energies of the group were immediately redirected to producing a catalogue.

Because there is only one article from Ryle's group between 1951 and 1955 concerning the nature of radio stars, it is difficult to ascertain when and how Ryle became interested in cosmology and changed to the view that the radio stars were extragalactic and therefore constituted a cosmological probe. A thorough search of Ryle's personal papers has also not yielded as much evidence as one would like, but I now list what is known:

> *c*1942: A wartime radar colleague (D. J. Allen-Williams) recalls that Ryle often enjoyed discussing cosmological questions during his free time.[8]
> April 1951: Conference on "Dynamics of Ionized Media"; as related above, Ryle is adamantly in favor of radio stars being within the Milky Way.
> May 1951: Oort sends Ryle a translation of the cosmological portion of Van de Hulst's 1945 paper (as related above).
> December 1951: In his Ph.D. thesis Smith (1951a: 101) states that there is "no evidence for the existence of a type of extragalactic body whose radio emission greatly exceeds that from our own Galaxy" and that "the contribution from extragalactic sources, even from the polar regions, would on this basis be less than 2%."
> June 1952: Ryle receives a copy of Hermann Bondi's book *Cosmology* for review (but never reviews it).
> February 1953: Ryle mentions in a talk that the effect of the redshift on the radio intensity of any extragalactic source is much less than in the optical and that this may allow radio techniques to probe out to far greater redshifts.[9]
> Spring 1953: In a popular article Ryle is neutral about whether the majority of radio stars are galactic or extragalactic.[10]
> April 1953: Ryle states that his radio observations may now provide "important evidence for the cosmologist."[11]
> October 1953: First evidence of Ryle discussing the possibility that radio measurements might distinguish between cosmological theories. This is found at the conclusion of his notes for a talk given to the Cavendish Physical Society and is reminiscent of Van de Hulst's earlier argument:
>
> The relatively slow [flat] law of the radio spectrum [as opposed to the optical spectra of stars, which means that objects with redshifts beyond $\frac{1}{2} \cdot c$ will fast become too faint to observe optically] suggests another very important possibility. The integrated radiation of optical light converges rapidly due to the redshift and [one gets] a negligible contribution beyond

[8] Smith (1986: 502). This obituary of Ryle should also be consulted for further details of his life and career.

[9] Notes for a talk to the Cambridge Philosophical Society, 16 February 1953, pp. 15–16, file 8 (uncat.), RYL.

[10] "Radio Stars: Study of Their Nature and Identification," *The Times Science Review* (London), Spring 1953, p. 7.

[11] Notes for a talk to a symposium on radio astronomy held at the US National Academy of Sciences, Washington, DC, April 1953, p. 9, file 8 (uncat.), RYL.

> $\frac{1}{2}\cdot$c. But at radio [wavelengths] very slow [convergence]. Therefore integrated radiation depends mainly on very distant region of the universe.
>
> Therefore expect an isotropic component of radiation whose intensity depends on what happens when recession velocity becomes large – the actual value depends on what particular brand of cosmology you favour. The effects out to $\frac{1}{2}\cdot$c show little difference.
>
> Now it so happens that there *is* a large isotropic component of radiation which cannot be explained in terms of emission from galactic sources. If indeed it is extragalactic, it offers the possibility of being able to distinguish between some of the cosmological theories.
>
> Whether the observations will ever be sufficiently accurate one cannot say, but it is nice to think that the cosmologists may one day not lose complete freedom of choice of the conditions beyond the optical limit.[12]

This theme of reining in the cosmologists was also struck in notes for a course given by Ryle in about 1953:

> Cosmologists have always lived in a happy state of being able to postulate theories which had no chance of being disproved – all that was necessary was that they should work in the observable Universe out to regions where the velocity is about $\frac{1}{2}\cdot$c. Since everything goes like $1/(1 - v^2/c^2)$, nothing interesting happens 'til much further. Now we do seem to have some possibility of exploring these most distant regions. Even if we never actually succeed in measurements with sufficient accuracy to disprove any cosmological theory, the threat may discourage too great a sense of irresponsibility.[13]

It can be seen that Ryle's first method of using radio data to test cosmology was not to count sources, but to use the integrated background radiation, the same isotropic excess component that Westerhout and Oort had discussed over two years earlier. This led directly to a short paper by Shakeshaft (1954), submitted in June 1954, in which he considered whether the polar component could arise from the integrated radiation of a population of extragalactic sources (taken to be colliding galaxies).[14] He concluded that if the entire, measured intensity at the north galactic pole were extragalactic, then "Relativistic Cosmology" could account for it and steady-state theory (see below) could not. But a final decision was withheld because it was still possible that the Milky Way was surrounded by a very large radio halo, as recent measurements by his fellow student John Baldwin had shown for the Andromeda nebula.

[12] "Recent Advances in Radio Astronomy," 14 October 1953, notes for a talk to the Cavendish Physical Society, p. 17, file 8 (uncat.), RYL.

[13] Notes for a (probably undergraduate) course, undated (about 1953), p. 55, file 35 (uncat.), RYL.

[14] Almost simultaneously with Shakeshaft's study, Wolfgang Priester, working at Kiel under Unsöld, also investigated whether the polar component of the background radiation could be explained by extragalactic radio sources. Priester (1954), however, used only cosmologies as found in Heckmann (1942); nor did he attempt to test cosmological models using the radio data. He concluded that sources that arose from collisons of galaxies in clusters could supply the polar component.

Most of a first complete reduction of the 2C survey appears to have been accomplished within two months of its completion – quite a feat when one remembers that hundreds of feet of strip chart recordings were being measured by hand. A "computer" in those days referred to a *person*, not a machine. (For example, in 1952 Van de Hulst complained about a four-week slowdown in the reduction of Leiden's 21-cm hydrogen-line survey because their *computer* had been injured playing football!) In any case it was shortly after this first reduction, in the early summer of 1954, that Ryle became convinced that almost all the radio stars were extragalactic.[15] The first archival evidence is a notebook entry dated 30 July 1954:

> If new statistics result is right and majority of radio stars are extragalactic, there are three main problems:
> 1. *Study of extragalactic sources* – cosmological application – little hope of much identifications other than what should come out of present 1000 [sources]. Main object will be to see further – even if area of sky has to be restricted . . .
> 2. *Study of large-diameter, galactic [plane] concentrated sources* . . .
> 3. *Origin of galactic radiation* [with no entry following]

And for 3 August:

> Plot of log N – log Intensity (N is no. brighter than I) gives curve which is approximately of slope 3, indicating that density [of radio stars] must increase approximately as r^3 [distance cubed] . . . On relativistic cosmology, best fit (assuming $\varrho \propto \varrho^2_{gal}$ [density of sources proportional to the square of density of galaxies] and using α [slope] = -3 . . .[16]

16.5 The announcement

For the next six months Ryle and his group worked on the details of the survey and its interpretation. In February 1955 the survey itself was submitted for publication (Shakeshaft, Ryle, Baldwin, Elsmore, and Thomson 1955) and in the end it contained 1936 sources, about ten times more than previously known (fig. 16.5). The number of possible new optical identifications that turned up, however, was exceedingly disappointing – at most only ten or twenty had been found, even with Minkowksi searching on plates of the then-in-progress Palomar Observatory Sky Survey. But it was the companion paper that caused great commotion.[17] Entitled "The Spatial Distribution and the Nature of Radio Stars" and submitted in March to the *Proceedings of the Royal Society* by Ryle and his student Peter Scheuer (1955), this paper gave various log N–log S plots (fig. 16.6) and results from the so-called P(D)

[15] Peter Scheuer recalls that Ryle came to the view that the sources had to be extragalactic (a) rather suddenly, and (b) during the period February–April 1954 (letter to author, 14 November 1988).

[16] Dated pages from laboratory notebook of Ryle's, location now not known (photocopies made in 1981).

[17] Also see Edge and Mulkay (1976: 152–9) for a discussion of the 2C results and their reception.

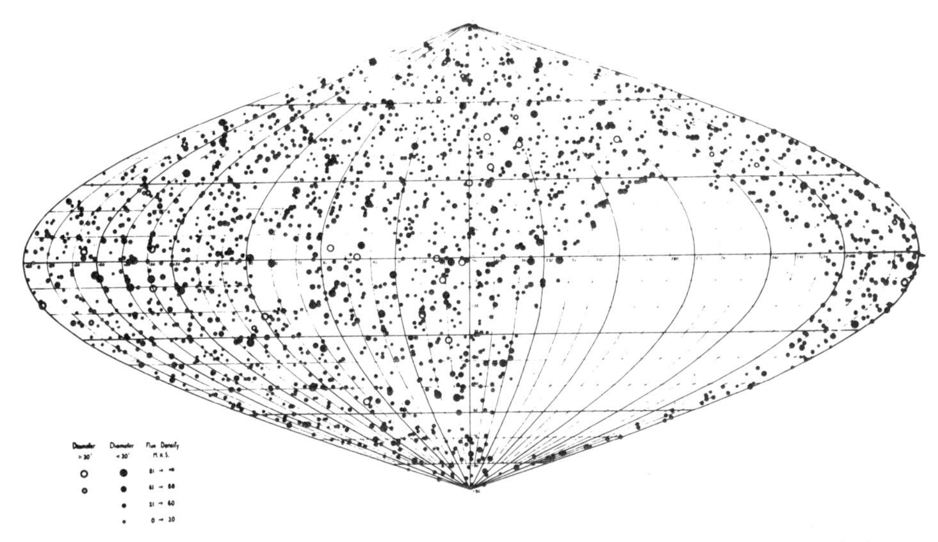

Fig. 16.5 The 1936 radio sources of the 2C survey (Shakeshaft et al. 1955) shown in galactic coordinates. Source flux densities are indicated by the sizes of dots; open circles indicate sources of angular size $>20'$.

analysis (see the following paper by Scheuer). Ryle and Scheuer argued that the steep slope ($\ll -1.5$) of the source counts at the faint end, together with limits placed by the brightness of the background radiation (whose cause now was even less well understood), could only be reasonably explained if the sources were (a) extragalactic, (b) of similar luminosity to Cyg A, and (c) of much greater number density at larger distances than nearby. This immediately meant that everyone lived in an evolutionary universe, not a steady-state one where no large-scale properties can vary with distance or time. As Ryle (1955: 146) put it in his Halley Lecture delivered in Oxford on 6 May:

> This is a most remarkable and important result, but if we accept the conclusion that most of the radio stars are external to the galaxy, and this conclusion seems hard to avoid, then there seems no way in which the observations can be explained in terms of a Steady-State theory.

The steady-state theory had been invented in the late 1940s at Cambridge by Bondi, Gold, and Hoyle and provided an attractive and testable alternative to the various older "big bang" models developed in the decade following the introduction of general relativity. Ryle's sweeping disproof of steady-state theory naturally caught the imagination and attention of both the general public and the scientific community. Here was no less than a major theory of the universe being overthrown. One Cavendish Laboratory press release of the time said that "by penetrating further into space than any previous observations made by mankind ... deductions can be made about that most fundamental of astronomical

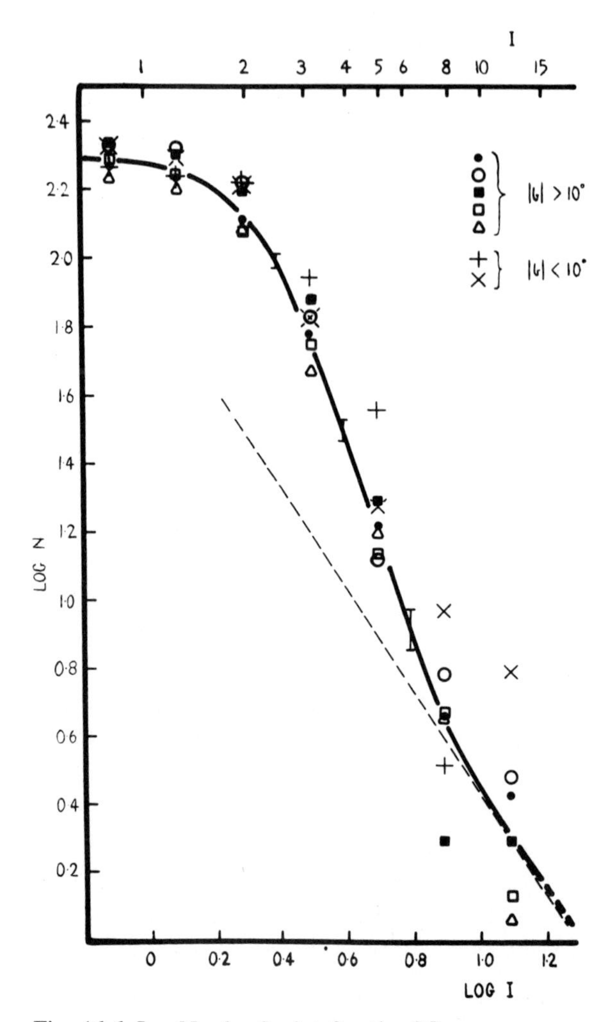

Fig. 16.6 Log N − log S plot for the 2C survey sources (Ryle 1955). N is the number of sources per steradian and I is the intensity in units of 10 Jy. The different symbols correspond to different regions of the sky, as shown in the upper right. The full curve is a fit for all regions of galactic latitude >10°, while the dashed line is a slope of −1.5, clearly incompatible with the observations.

problems, the origin of the universe."[18] All of this caught the steady-staters by surprise – until the spring of 1955 they had had no idea that Ryle's latest survey would be put to any such purpose (Gold 1976: 16T). As Ryle explained his policy at the time:

> Before the Halley lecture I am afraid we had been somewhat cagey because of the large number of hungry cosmologists we have prowling around – we wanted to be able to collect our thoughts a little on the next stage before they pounced![19]

[18] "The Cambridge Survey of Radio Stars," March 1955, box 1/3, RYL.

[19] Ryle: J. L. Pawsey, 27 June 1955, file A1/3/1, archives of the CSIRO Radiophysics Division, Epping, New South Wales.

Given the long, checkered history of (optical) galaxy counts applied to cosmology by Edwin Hubble and others, as well as the fact that one still did not even understand what the individual radio sources were, Hoyle and Gold were astounded at Ryle's conclusions and the confidence with which they were given. The strained relations of the past now moved into a new arena. After Ryle presented these results at a meeting of the Royal Astronomical Society, Gold commented:

> I have been greatly impressed by this magnificent survey, which has exceeded all expectations of a few years ago. I am also glad to see that there is now agreement that many of these sources are likely to be extragalactic, as I suggested here and elsewhere, with much opposition, four years ago. Mr. Ryle then considered that such a suggestion must be based on a misunderstanding of the evidence . . . But on present evidence it is very rash to regard the great majority of weak sources as extremely distant.[20]

Gold then asked for a pre-publication list of all the sources, and Ryle sent it to him, but only reluctantly:

> Having spent 4½ rather hectic years dealing with mud, steel and concrete and with much tedious analysis, I think that [my students] are entitled to six months' consideration of the results after completing the survey paper and before its publication!. . . Would you therefore please regard this list as pre-publication data for your own use in disproving our interpretation of the observations?[21]

Hoyle for his part wrote to Mills and asked whether his southern survey were sufficiently advanced that it could provide a check on Ryle's statistics (Mills 1976: 60–1T). Ryle was also unprepared for the intensity of the opposition from the steady-state theorists. As he recalled much later:

> [The intense controversy] was a considerable shock, because of course the trouble with cosmology up till then was that it had been a playground for mathematicians – "Is space curved this way or that way?" and all these things. It was nothing very much to do with the real world and observations had never, and apparently would never, make any effect on it. It was a game mathematicians could play, safe from all possible attack.
>
> But the development of the steady-state model was an important break-through. Here was something that made specific predictions in a wide range of not-necessarily-thought-of possible measurements. It said that the universe was in a state that could remain the same through time as well as space . . . And as soon as you know you can detect sources at redshifts large enough for things to happen on other cosmolgies, then you can detect a difference . . . It was remarkable what an absolute storm it provoked. Well of course, it wasn't helped by the fact that the press got hold of the story.
>
> (Ryle 1976: 53T)

[20] Proceedings of the RAS Meeting of 13 May 1955, *Observatory*, **75**, 106.
[21] Ryle : Gold, 28 May 1955, box 1/3, RYL.

The reactions of those who received a pre-print of the paper allows us today to gauge its impact over thirty years ago. The distinguished British cosmologist William McCrea wrote:

> I am filled with admiration for your work and obviously it is of first-rate importance . . . I cannot see anything clearly wrong with your general interpretation. Moreover, it would be extraordinarily satisfactory if radio observations could so seemingly definitely decide between models when a decision by optical methods appears to be so far out of reach.
>
> In spite of not seeing anything obviously wrong, one has the irrational impression that it is all too easy! This may be simply because we have grown used to the idea that any discrimination between different models or different theories can be made only on the basis of "second order" effects. It seems then astonishing if it can be made just by noticing the difference between 1.5 and 3.0.

Oort, a most conservative man in his judgements of scientific worth, was beside himself with excitement:

> I have just read your and Scheuer's article on the radio stars, and do not want to wait another hour to congratulate you with the splendid discovery you have made. To me it is the most fascinating discovery I have ever seen.

Finally, Joseph Pawsey, the scientific leader of the Sydney radio astronomers, wrote on 6 July:

> The content of the paper is revolutionary. If the data are correct, I think your explanation the most plausible (i.e. the best) and the implications in cosmology immense.

But Pawsey then went on and gave some preliminary results from a survey that Mills had been undertaking with his own new large antenna, a 1500-foot long, cross-shaped array of dipoles. This was the first hint of what would soon grow into a second, full-scale attack on Ryle's results:

> I immediately checked on the current state of the 85 Mc/s [Mills] Cross results . . . There are 500 or 600 [sources] on the list and when these are plotted, $\log I - v - \log N$, they fall on a slightly irregular line of slope -1.5 . . . Thus the observations appear not to agree with yours, but we must investigate the situation much more carefully before there is any certainty . . . This is all very mysterious. We shall do our best to clarify the issue as soon as possible.[22]

In August 1955 Pawsey came to England for an IAU symposium on radio astronomy held at Jodrell Bank, but further checks and a larger sample had only confirmed the statistical disagreement between the two

[22] McCrea : Ryle, 4 March 1955, file 4 (uncat.); Oort : Ryle, 1 June 1955, box 1/3; Pawsey : Ryle, 6 July 1955, box 1/3 – all RYL.

surveys (compare Pawsey 1957 with Ryle 1957). A detailed comparison of a region visible from both Hemispheres was needed and this cross-check was at first attempted through the mails, but not to Mills' satisfaction:

> We had some correspondence about this and Ryle was quite unshakable . . . So we felt the only thing to do was to do a thorough analysis, as far as we could, of comparing the two catalogues . . . But our results were, as far as we could determine, just completely dismissed by the Cambridge group. So we had no alternative but to come out with [a very critical] sort of paper.
>
> (Mills 1976: 63–5T)

The paper was by Mills and Slee (1957) and it concluded:

> We have shown that in the sample area, which is included in the recent Cambridge catalogue of radio sources, there is a striking disagreement between the two catalogues. Reasons are advanced for supposing that the Cambridge survey is very seriously affected by instrumental effects which have a trivial influence on the Sydney results. We therefore conclude that discrepancies, in the main, reflect errors in the Cambridge catalogue, and accordingly deductions of cosmological interest derived from its analysis are without foundation.
>
> An analysis of our results shows that there is no clear evidence for any effect of cosmological importance in the source counts.

The battle was joined.

Acknowledgements

This study would not have been possible without the generous access to Martin Ryle's papers allowed me by Churchill College, Lady Rowena Ryle, and Professor Antony Hewish of the Cavendish Laboratory. I thank them all for their help. Peter Scheuer has been indispensable in helping me decipher and understand parts of the Ryle papers. I am grateful to Professor Martin Rees, Director of the Institute of Astronomy, Cambridge University, for his support during the period of research for this paper, and to the Observatoire de Meudon, in particular Dr. Chantal Balkowski, for further sabbatical support. I acknowledge permissions to quote from published passages and from interviews, as well as to use the illustrations, from the individuals and journals cited. Finally, I thank the History and Philosophy of Science Program of the U.S. National Science Foundation for its consistent funding over the years of my research on the history of radio astronomy.

References

Baade, W. and Minkowski, R. (1954a). Identification of the Radio Sources in Cassiopeia, Cygnus A, and Puppis A. *Astrophysical Journal*, **119**, 206–14.

(1954b). On the Identification of Radio Sources. *Astrophysical Journal*, **119**, 215–37.

Bolton, J. G. (1948). Discrete Sources of Galactic Radio Frequency Noise. *Nature*, **162**, 141–2.
Bolton, J. G. and Stanley, G. J. (1948). Variable Source of Radio Frequency Radiation in the Constellation of Cygnus. *Nature*, **161**, 312–13.
Bolton, J. G., Stanley, G. J., and Slee, O. B. (1949). Positions of Three Discrete Sources of Galactic Radio-Frequency Radiation. *Nature*, **164**, 101–2.
(1954). Galactic Radiation at Radio Frequencies. VIII. Discrete Sources at 100 Mc/s between Declinations +50 Degrees and −50 Degrees. *Australian Journal of Physics*, **7**, 110–29.
Bolton, J. G. and Westfold, K. C. (1950). Galactic Radiation at Radio Frequencies. I. 100 Mc/s. Survey. *Australian Journal of Scientific Research*, **A3**, 19–33.
(1951). Galactic Radiation at Radio Frequencies. IV. The Distribution of Radio Stars in the Galaxy. *Australian Journal of Scientific Research*, **A4**, 476–88.
Boyd, R. L. F. (ed.) (1951). *Proceedings of the Conference on Dynamics of Ionized Media*. Dept. of Physics, University College, London (mimeographed, unbound).
Edge, D. O. and Mulkay, M. J. (1976). *Astronomy Transformed: The Emergence of Radio Astronomy in Britain*. New York: John Wiley and Sons.
Hanbury Brown, R. (1984). Paraboloids, Galaxies and Stars: Memories of Jodrell Bank. In *The Early Years of Radio Astronomy*, ed. W. T. Sullivan, III, pp. 213–35. Cambridge: Cambridge University Press.
Hanbury Brown, R. and Hazard, C. (1951). Radio Emission from the Andromeda Nebula. *Monthly Notices of the Royal Astronomical Society*, **111**, 357–67.
(1953a). A Model of the Radio-Frequency Radiation from the Galaxy. *Philosophical Magazine*, **44**, 939–63.
(1953b). A Survey of 23 Localized Radio Sources in the Northern Hemisphere. *Monthly Notices of the Royal Astronomical Society*, **113**, 123–33.
Hazard, C. (1953). An Investigation of the Extra-Galactic Radio Frequency Emissions. Ph.D. thesis, Manchester University.
Heckmann, O. (1942). *Theorien der Kosmologie*. Berlin: Springer.
Hey, J. S., Parsons, S. J., and Phillips, J. W. (1946). Fluctuations in Cosmic Radiation at Radio-Frequencies. *Nature*, **158**, 234.
Little, C. G. and Lovell, A. C. B. (1950). Origin of the Fluctuations in the Intensity of Radio Waves from Galactic Sources – Jodrell Bank Observations. *Nature*, **165**, 423–4.
Mills, B. Y. (1952). The Distribution of the Discrete Sources of Cosmic Radio Radiation. *Australian Journal of Scientific Research*, **A5**, 266–87. Erratum (1953). *Australian Journal of Physics*, **6**, 125.
(1984). Radio Sources and the log N–log S Controversy. In *The Early Years of Radio Astronomy*, ed. W. T. Sullivan, III, pp. 147–65. Cambridge: Cambridge University Press.
Mills, B. Y. and Slee, O. B. (1957). A Preliminary Survey of Radio Sources in a Limited Region of the Sky at a Wavelength of 3.5 Meters. *Australian Journal of Physics*, **10**, 162–94.
Pawsey, J. L. (1957). Preliminary Statistics of Discrete Sources Obtained with

"Mills Cross." In *Radio Astronomy*, ed. H. C. Van de Hulst, pp. 228–32. Cambridge: Cambridge University Press.
Priester, W. (1954). Zur Deutung der extragalaktischen Radiofrequenz-Strahlung. *Zeitschrift für Astrophysik*, **34**, 283–94.
Reber, G. (1940). Cosmic Static. *Astrophysical Journal*, **91**, 621–4.
Ryle, M. (1949). Evidence for the Stellar Origin of Cosmic Rays. *Proceedings of the Physical Society*, **A62**, 491–9.
(1950). Radio Astronomy. *Reports on Progress in Physics*, **13**, 184–246.
(1955). Radio Stars and Their Cosmological Significance. *Observatory*, **75**, 137–47.
(1957). The Spatial Distribution of Radio Stars. In *Radio Astronomy*, ed. H. C. Van de Hulst, pp. 221–7. Cambridge: Cambridge University Press.
Ryle, M. and Scheuer, P. A. G. (1955). The Spatial Distribution and the Nature of Radio Stars. *Proceedings of the Royal Society*, **230**, 448–62.
Ryle, M. and Smith, F. G. (1948). A New Intense Source of Radio-Frequency Radiation in the Constellation of Cassiopeia. *Nature*, **162**, 462–3.
Ryle, M., Smith, F. G., and Elsmore, B. (1950). A Preliminary Survey of the Radio Stars in the Northern Hemisphere. *Monthly Notices of the Royal Astronomical Society*, **110**, 508–23. Erratum (1951). *Monthly Notices of the Royal Astronomical Society*, **111**, 641.
Shakeshaft, J. R. (1954). The Isotropic Component of Cosmic Radio-Frequency Radiation. *Philosophical Magazine*, **45**, 1136–44.
Shakeshaft, J. R., Ryle, M., Baldwin, J. E., Elsmore, B., and Thomson, J. H. (1955). A Survey of Radio Sources between Declinations −38 Degrees and +83 Degrees. *Memoirs of the Royal Astronomical Society*, **67**, 97–154.
Smith, F. G. (1950). Origin of the Fluctuations in the Intensity of Radio Waves from Galactic Sources – Cambridge Observations. *Nature*, **165**, 422–3.
(1951a). Some Studies of Radio Stars. Ph.D. thesis, Cambridge University.
(1951b). An Accurate Determination of the Positions of Four Radio Stars. *Nature*, **168**, 555.
(1986). Martin Ryle (1918–1984). *Biographical Memoirs of Fellows of the Royal Society*, **32**, 497–524.
Smith, F. G. and Lovell, A. C. B. (1983). On the Discovery of Extragalactic Radio Sources. *Journal of the History of Astronomy*, **14**, 155–65.
Stanley, G. J. and Slee, O. B. (1950). Galactic Radiation at Radio Frequencies. II. The Discrete Sources. *Australian Journal of Scientific Research*, **A3**, 234–50.
Unsöld, A. (1949a). Über den Ursprung der Radiofrequenzstrahlung und der Ultrastrahlung in der Milchstrasse. *Zeitschrift für Astrophysik*, **26**, 176–99.
(1949b). Origin of the Radio Frequency Emission and Cosmic Radiation in the Milky Way. *Nature*, **163**, 489–91.
Van de Hulst, H. C. (1945). Radiogolven uit het wereldruim. I. Herkomst der radiogolven. *Nederlandsch Tijdschrift v. Natuurkunde*, **11**, 210–21. Available in English in *Classics in Radio Astronomy* (1982), ed. W. T. Sullivan, III. Dordrecht: Reidel.
(1951). *A Course in Radio Astronomy*. Leiden: Sterrewacht (bound mimeograph format).
Westerhout, G. and Oort, J. H. (1951). A Comparison of the Intensity Distribu-

tion of Radio-Frequency Radiation with a Model of the Galactic System. *Bulletin of the Astronomical Institutes of the Netherlands*, **11**, 323–33.

Discussion

Trimble:

Do we correctly understand you to say that Van de Hulst in effect worried about a radio version of Olbers' Paradox and proposed the "Bondi solution" for it before Bondi?

Sullivan:

Yes, indeed. His 1944 paper, published in Dutch in 1945 in the *Nederlandsch Tijdschrift voor Natuurkunde*, contains far more than the well-known prediction of the 21-cm hydrogen line: (1) calculations concerning the galactic background as free–free emission, (2) a consideration of the possibility of detecting hydrogen (high-n) recombination lines in the radio spectrum, and (3) a comparison of the predictions of expanding- and static-universe models for the radio sky brightness, based on Reber's claimed detection of M31. For (3), basing his models on Heckmann's treatise, he concluded that a static universe would be too bright.

Note added in proof: One further early study relating radio astronomy and cosmology has recently come to light. Although unknown in the West until years later, a pioneering 1953 paper by the Soviet astrophysicist Iosif Shklovsky considered many cosmological problems to which radio data could be applied: "The photometric paradox for the radio emission of the metagalaxy [in Russian]", *Astron. Zh.*, **30**, 495–507 (1953). From existing surveys, Shklovsky determined that there existed a 150 K extragalactic component to the radio background (at 100 MHz). On the basis of this and cosmological models applied to Mills's (1952) Class II radio sources (assumed to be extragalactic), he concluded that (1) the radio Olbers's paradox could not be solved by a Charlier-type hierarchical cosmology, (2) the best solution was that the redshift must operate at radio wavelengths as well as at optical, and (3) the redshift was due to a Doppler effect, not "tired light". Shklovsky considered these the "first concrete cosmological results obtained with the help of radio astronomy".

17

Radio source counts

Peter Scheuer

17.1 Summary

The story of the radio source counts from 1953 to 1965 consists of many intertwined themes. The first was the nature of radio sources: were they galactic or extragalactic? If galactic, did the source counts disprove the steady-state theory or were they consistent with it? There were conflicting catalogues of sources and conflicting claims about the distribution of angular diameters of sources. No less important were much more subjective attitudes, including optical astronomers' profound distrust of any data from an interferometer, as opposed to something with a 'pencil beam' that they could recognise as a real telescope, as well as the changing attitudes of radio astronomers to each other, as individuals and as members of groups (particularly, of course, those at Cambridge and at Sydney), and the response of steady-state theorists to the threat of a conflict with observation.

By now we all (or nearly all) agree on the answers. The Second Cambridge Catalogue of radio sources was mostly wrong, but the conclusion drawn from it, that powerful radio sources were more abundant in the past, was right. The Mills Cross catalogue was largely right, but the conclusion drawn from it, that the source counts were after all compatible with steady-state theory, was wrong. The worries and debates about the bias of interferometers against sources of large diameter turned out to be largely irrelevant, for very few of the sources were in fact resolved by the early interferometers. The real weakness of the earliest results was their poor statistical significance, but that did not become a central issue in the debate until the closing stages of the controversy, when that argument had already lost much of its force.

At the time, these ingredients were mixed up to an extent which a future historian of science will not necessarily see from the fairly carefully worded published papers. Some confusion between issues was inevitable when we did not know which points of detail would turn out to be the crucial ones. Some confusion was no doubt allowed to persist and exploited when it suited polemical purposes. I draw no moral from this story, except that it is a particularly fine counter-example to the notion that scientists proceed in a logical step-by-step sequence called scientific method; I expect we shall continue to proceed by a series of more or less large leaps into the dark followed by partial retreats.

17.2 Prologue

This talk will not be an elegant presentation of great themes, such as we heard yesterday. Historians are better and more persistent than I am in probing the literature and other fossil evidence, so I think the most useful thing I can do is to try to give an impression of how one particular episode – the controversy over source counts and the steady-state theory – felt at the time. In doing so, I give notice that I was a student and then a postdoc in Ryle's group, much involved in the interpretation of the source counts, but one step away from the observations themselves; thus, while I shall try to be fair, the way I saw it will be different not only from the way that Hoyle or Mills saw it but also from the way that Ryle, Shakeshaft, Baldwin or Elsmore saw it.

In 1953 over 100 radio sources had been detected, but virtually nothing was known about their nature. Three were identified with nebulae known or suspected to be supernova remnants, three with peculiar galaxies – two close, but one (Cygnus A) surprisingly distant, with a redshift of 0.056. Of the rest, a significant number were very close to the Galactic Plane, but most were much too isotropically scattered to be supernova remnants. The growth in numbers of the high galactic-latitude population of radio sources, $N(>S)$, with the limiting flux density S, was also known to be very roughly consistent with the law $N(>S) \propto S^{-3/2}$ expected for a uniform distribution in space. They could either be other galaxies – but mysteriously failed to be identified as such, and therefore had to be some special rare or optically invisible kind – or they might be a very common type of galactic star, very dim in the optical band. The latter theory had the advantage (Westerhout and Oort 1951) that the radio stars could then also account for the galactic background radiation, which was clearly not free–free emission, but had no clear alternative explanation.

17.3 Two steps forward

It was against this background that Martin Ryle and much of his group began the second survey of 'radio stars'. For this purpose, Ryle had designed a double interferometer (Ryle and Hewish 1955) consisting of four parabolic cylinders, fed by rows of co-phased dipoles along their focal lines. The interference fringes in the N–S and E–W directions would not only discriminate against the galactic background radiation, but also permit accurate positions to be determined for the stronger sources, and thus greatly increase the chances of identifying more sources with optical objects. Though very cheap and simple by the standards of Big Science the new aerial represented a huge advance on the 'long Michelson' array of the first Cambridge survey (Ryle et al. 1950). It was our big instrument.

Ryle's choice of operating frequency had been determined (among the few channels relatively free from man-made interference) by considering the limitations in sensitivity and angular resolution. At lower frequencies the sensitivity would be better, because one could build better low-noise receivers, and the sources were stronger. At higher frequencies the resolution would be better since the beamwidth for a given telescope aperture is proportional to wavelength, in each direction; however, the sensitivity would be worse. Not knowing what to expect made the choice something of a leap in the dark. A rough extrapolation of the $N(>S)$–S relation indicated that the two limits would be about equal, i.e. there would be about one source above noise per beamwidth, at around 120 MHz (not a clear channel); the number of detectable sources would be limited by sensitivity at 160 MHz and by angular resolution at 81.5 MHz. Ryle decided to work with 81.5 MHz feeds first, perhaps because the technology would be simpler, safer and more reliable.

As soon as the new aerial came into operation, in the second half of 1953, it showed that indeed there were vast numbers of radio sources all over the sky. Almost wherever the beam pointed, there were very clear interferometer fringes, demonstrating the presence of a radio source of small angular diameter. The numbers were not thinning out with distance, as they should, sooner or later, if they were local galactic stars. On the contrary, when the first strips of sky were analysed, a plot of numbers indicated something quite astonishing. So far from turning over at weaker flux densities, the plot of numbers versus limiting flux density was much steeper than $N \propto S^{-3/2}$, clearly indicating a source density increasing with distance. Furthermore, the source counts seemed to be the same for different parts of the sky – though the statistics were poor, and needed to be confirmed on a much larger sample of the sky. At this point – it must have been in the winter of 1953–4 or the early spring of 1954 – Martin Ryle's attitude on 'radio stars' changed almost overnight. They simply had to be extragalactic; there was no way the solar system could be at the centre of a more or less spherical hole in a galactic population. That conclusion agreed with what Gold had been saying for some time (though of course for other reasons). However, simultaneously, there had to be another conclusion: the radio source population was increasing outwards, and that flatly contradicted the steady-state theory of Bondi and Gold, and of Hoyle, according to which the number density must be the same at all places and all epochs.

Martin Ryle pointed out that, since the survey records showed fringes almost everywhere, quite a lot of sources above the normal detection limit would lie buried in the aerial beams already occupied by stronger sources, and the loss of these weak sources would bias the statistics. To sidestep the problems due to such effects I proposed a radically different approach to the source statistics, in which (as one would now say) one models the probability distribution P(D) of deflection amplitudes D on the interferometer records to be expected for sources with a given N(>S)–S relation sprinkled in random positions over the sky, and compares this directly with the histogram of observed interferometer amplitudes. This eliminates the subjective element in extracting individual sources from the records. It turned out that, for a power-law N(>S)–S relation the calculation could be reduced to one nasty integral, which, even in 1954, could be computed numerically without too much effort. Long before the calculations were completed it became clear that 'confusion' between sources lying close together would affect the source statistics in quite the opposite sense. Underlying weak sources would more often boost than reduce the measured flux of the brightest source in the beam, and this effect far outweighed the loss of similar sources under really strong sources. On the one hand, the P(D) analysis confirmed that the observations required a log N(>S)–log S relation with a slope steeper than -1.5, and that was very important in the long controversy that followed. On the other hand, it indicated a slope much less steep than the slope of -3 that came from the source counts. The second conclusion was almost as unwelcome to the rest of the Cambridge group as the first was reassuring.

So far as I was concerned, the matter finished there for the next two years, as I had a thesis to write (on a different topic) by a deadline in October 1954, when I was due to report at Catterick Camp for National Service.

17.4 One step back

Meanwhile, Mills' group at the CSIRO Radiophysics laboratory in Sydney had built a prototype Mills Cross with an $8° \times 8°$ beam (Mills and Little 1953) and then built the full-scale version with a $50' \times 50'$ beam. Although even their preliminary survey was not published until 1957, some early results were reported by J. L. Pawsey at a symposium on radio astronomy at Jodrell Bank in 1955 (see Mills and Slee 1957). The Australians found a log N–log S relation consistent with a slope of -1.5, and quite inconsistent with a slope of -3. (It appears that a copy of the typescript of our paper [Ryle and Scheuer 1955] had been sent to Sydney, and Mills had done some quick preliminary analysis on his sources so far.)

The stories that reached Cambridge from Sydney over the next two years must have been very disturbing: Mills' group had found virtually no agreement between his sources and 2C sources in the region of overlap between the Northern and Southern observations. In 1956 the feeds and receivers of the Cambridge interferometer were changed to observe at

159 MHz (with a consequent 4-fold decrease in the solid angle of the primary beam), to make the 3C (third Cambridge) survey of sources. Two things soon became plain: (i) rather few of the 2C sources agreed with 3C sources, and (ii) even the 3C survey was limited more by confusion than by sensitivity. To see these differences between catalogues in proper perspective one must remember that none of the early catalogues were very reliable. For example, Bolton, Stanley and Slee (1954) made a survey of 101 sources using a cliff interferometer; they reported that 42 of the 69 sources from Mills (1952) that were in the common area, and 11 of the 17 1C (Ryle, Smith and Elsmore 1950) sources in the common area and above the flux limits of both surveys, agreed with sources in their list. It is also true that the overlap region between Northern and Southern surveys was at the greatest zenith distances for both, and thus showed neither at its best. Even so, the new instruments were an order of magnitude bigger than their predecessors, and one expected the agreement to become better rather than worse! Although it was rarely said in so many words (for none of us wanted to raise Martin's blood pressure above safe limits, for his sake even more than for our own), we all knew that the majority of the sources in the 2C catalogue did not exist as individual objects.

The essential point is that a peak in the amplitude on the chart records is unlikely to be dominated by a single source (rather than two or more weaker ones) except for peaks so large that there is only one in every 20 to 30 beamwidths. Confusion limits the reliability of a survey at that level, far above the level of one source per primary beam. Furthermore, a weak confusing source can easily shift the peak amplitude of quite a strong source by one interferometer fringe; thus a source detected by an interferometer lies with high probability in one of several small separated patches, instead of a single larger error ellipse. This difference makes it easier to understand the discrepant results reported by the Sydney and Cambridge groups when enough information had been published to make a comparison possible (e.g. Mills and Slee 1957; Archer at al. 1959).

In view of the disputes about the source catalogues, the P(D) analysis became important to us for establishing valid conclusions about the source counts, but this approach found little acceptance elsewhere for a long time. To begin with, the initial results of the P(D) approach had been published in the first paper on the source counts (Ryle and Scheuer 1955) but as an afterthought and with no details of the calculations. I did not find barrack rooms conducive to writing a paper, but even after it was written it was almost as hard to publish as a paper on perpetual motion engines. It finally came out (Scheuer 1957) in a largely mathematical journal not widely read by astronomers, and stripped of almost any mention of cosmology, but at least we could refer to a publication, and could no longer be accused of using a secret method.

17.5 Confrontation

The arguments came to a head at the Paris Symposium on Radio Astronomy in August 1958 (Bracewell 1959). Dr Pawsey (Sydney) chaired the session on 'Discrete Sources and Cosmology'. The mostly rather carefully worded written version of the discussion gives only a pale reminder of the heat of the argument.

Thoughts prevalent at the time were voiced by Hanbury Brown (Manchester) in his introductory lecture: 'It is generally believed that Class II sources' (i.e. those not concentrated to the galactic plane), 'which are by far the more numerous, are extragalactic', but 'Two obvious difficulties in drawing cosmological conclusions from the source counts are: (1) we have no distance calibration, and (2) there is no evidence that we are dealing with a homogeneous population', and later 'In conclusion I must emphasise how far we are at present from understanding the extragalactic radio-source population; we have in fact no reliable idea of the shape of the radio-source luminosity function.' Now, when pressed, well-informed radio astronomers would agree that (i) steady-state theory predicts unambiguously that sources of any given power P will have a source count N(>S) which may be as steep as $S^{-3/2}$ at high flux densities S and becomes less steep at lower S, and (ii) any weighted sum of such relations for different P must have these same properties; therefore the luminosity function is irrelevant to the particular conclusion that a source count steeper than $S^{-3/2}$ rules out the steady-state theory. However, the very general way in which such statements were habitually made (and Hanbury Brown's version was of course strictly correct) left the broader astronomical community with the impression that no cosmological conclusions whatsoever could be drawn from source-count data without extensive information on the distances of the sources.

That did not prevent source counts from being discussed. At this stage, a 'preliminary survey' of the region $-20° <$ declination $< +10°$ had been published by Mills and Slee (1957) and the 3C survey was essentially complete (see Archer et al. 1959); observations with the 4C aperture synthesis instrument had just begun (Ryle 1959, top of p. 527).

Here another prolific source of misunderstanding appears in the discussion of the interferometer data. The age of Fourier optics had hardly begun, so that, while most astronomers had a formal knowledge of Fourier transforms, very few had the kind of intuitive understanding that comes with regular use. Consequently, the defects of the 2C catalogue were associated with the use of interferometers which gave only certain Fourier components, as opposed to pencil beams which gave real knowledge about radio sources in the sky.

Indeed, some of the radio astronomers seemed to encourage the optical astronomers' intuitive distrust of interferometers. But Minkowski (1959) got it exactly right in his concluding lecture:

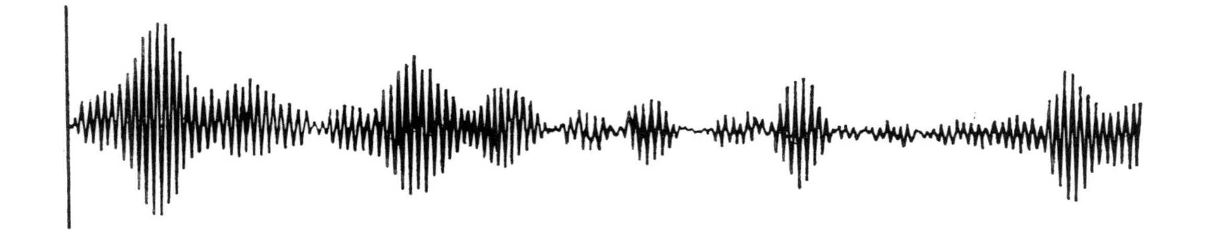

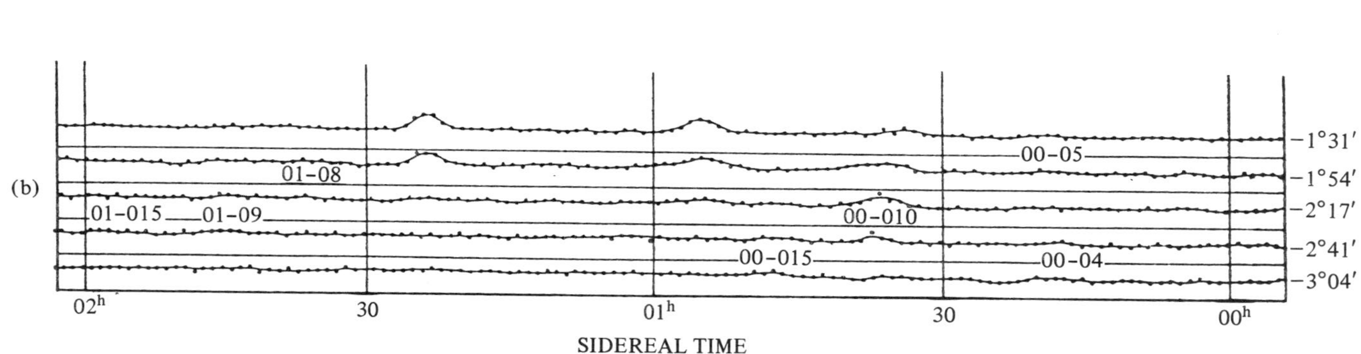

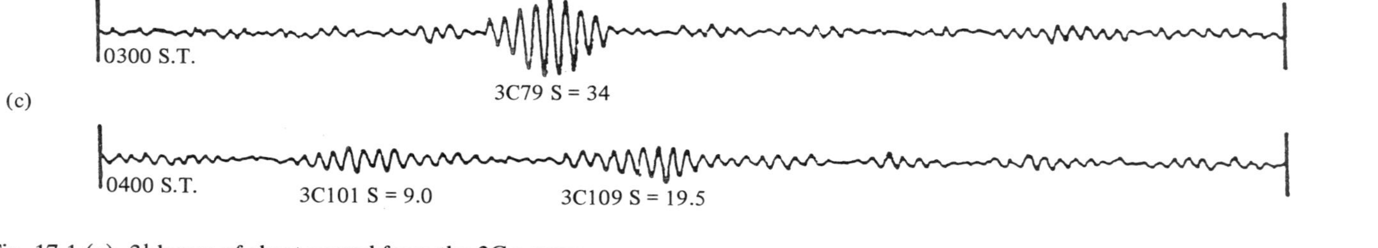

Fig. 17.1 (a) $3\frac{1}{2}$ hours of chart record from the 2C survey.
(b) $2\frac{1}{2}$ hours of chart record from the Mills Cross survey (five adjacent declination strips).
(c) 2×1 hour of chart record from the 3C survey.
In (b) and (c) the numbers of sources listed in the respective catalogues are marked. Figs. (a) and (c) are taken from Ryle & Hewish (1955) and Edge et al. (1959) respectively, and are reproduced by kind permission of the Royal Astronomical Society and the surviving authors: fig. (b), from Mills & Slee (1957), is reproduced by kind permission of the Editor of the *Australian Journal of Physics*.

> The discussion of the relative merits of the pencil beams and the interferometers for surveys of radio sources has tended to obscure the merits of the interferometer ... there is no basic difference between pencil beam and interferometer as regards the effects of confusion. The main difference between the Sydney and the Cambridge surveys is the fact that the Mills Cross is sensitivity limited, while the Cambridge interferometer is confusion limited. The inability of the interferometer to record sources beyond a certain size is an inherent difference, however.
> It is now generally recognised that the Cambridge survey at 81.5 Mc/s was severely affected by confusion.

Though disputes about catalogues were irrelevant to the P(D) analysis, it still bore guilt by association. In one discussion I was allowed to come to the front to answer questions about it, but the real attacks came after I had sat down again, and I was not allowed to reply to any of them. Thus, Gold: 'Scheuer's type of analysis cannot overcome the genuine lack of information, and if we cannot trust a particular assumption about the pattern in the sky we cannot derive any benefit from it. We are therefore left with a clear case for larger aperture antennas and we could not get the same information "on the cheap" by any device of analysis'; and Jennison: 'No applications of statistical analysis can fill in the gaps in the Fourier components.' I believe that some of the distrust arose from the mistaken notion that we were claiming to get more information than the observations contained. In fact, the statistics of record amplitudes are very insensitive to the source counts at levels below one source per primary beam, and about a page of my paper was devoted to explaining exactly how that arose in the mathematics. What we did claim (and has been confirmed by subsequent use in X-ray as well as radio astronomy) is that one can make effective use of observations down to about that level, as opposed to the level of one source per 20 to 30 beamwidths at which individual sources can be recognised reliably, and the even higher level at which flux densities cease to be biased towards high values.

However, there was also a much more substantial criticism of the interferometer data. Interferometers resolve out sources whose angular diameters exceed the fringe spacing. More distant sources of the same kind would have smaller angular diameters, and appear undiminished, leading to a spurious excess of faint sources. As Ryle suspected even then (see e.g. Ryle 1959: 526), the overwhelming majority of extragalactic radio sources have angular diameters too small to affect these early source counts noticeably, but given the state of knowledge in 1958 it was a legitimate scientific doubt. I certainly worried about it a lot, and we even contemplated surveys made with scaled interferometer systems, the deeper surveys having a correspondingly narrow fringe spacing. Some not very rigorous attempts of this sort were indeed made, using parts of existing aerials, and a remark by Ryle in the discussion of Paper 91 (Mills) probably refers to one of these. My worries about angular diameters are also reflected in a paper (Edge et al. 1958) which was a very uneasy compromise between the views of Ryle and my co-authors on the one hand and my (as it turned out, excessively) cautious views on the other. In 1958 the most striking evidence for an important effect due to angular resolution was Mills and Slee's (1957) detection of a large number of extended sources (i.e. sources comparable in size with their 50 arc–minute beam) even at high galactic latitudes. In the end, ironically, most of these turned out to be chance groupings of weaker sources – the small remnant of 'confusion' that remained even in the Mills Cross survey.

17.6 Three steps sideways

Many more surveys were made in the years that followed, and they became much better. With the development of aperture synthesis and the use of higher frequencies the problems of confusion went away, and different surveys found the same sources wherever they overlapped in sky coverage and flux range. Yet the statistical controversies did not go away, they merely shifted ground. The source counts of the early 1960s converged to a slope of about -1.8 at high flux densities S, but flattened out at lower flux densities [as had already been indicated by P(D) plots (Hewish 1961)]. The number of bright sources missing below a line $N(>S) \propto S^{-3/2}$ was only of the order of 100 over the whole sky, so it could never be increased by more or better measurements. While such a number is too large to be credible as a random fluctuation, it is also too small for assertions about isotropy over the sky to carry much statistical weight. So a number of 'local hole' theories were proposed, which invoked various kinds of large-scale clustering (e.g. Hanbury Brown 1962; Hoyle and Narlikar 1962). Much later, Kellermann (1972) gave a good and detailed discussion of such theories, in a broader context that includes evolving models.

The arguments therefore became more complex. The correct standard of comparison for the observed source counts is, of course, not the $N \propto S^{-3/2}$ law for a uniform distribution in Euclidean space, but the count predicted by the steady-state theory, which rapidly flattens for sources at significant redshifts; thus the deficit of bright sources is really much greater than 100. Unfortunately, this argument did require some information about the distances of sources, though even very crude lower limits were sufficient. Thus the fact that most radio sources could not be identified with galaxies on the Palomar Sky Survey, and had small angular sizes, indicated that most radio galaxies were powerful and distant. But the only piece of quantitative evidence independent of assumptions about the nature of radio galaxies was the total brightness of the radio sky: if, say, half of the observed sources were within a redshift of 0.01, then the sky brightness due to all such sources had to be at least 15 times the contribution of all the known sources. Since the observed sources already accounted for a substantial fraction of the radio background, the 'local hole' in the source distribution could not be so very local after all, and therefore the difference between steady-state and Euclidean cosmology was significant.

These arguments were already clearly stated by Ryle at the Paris Symposium (Ryle 1959), and were spelt out in detail, with quantitative estimates, in Ryle and Clarke (1961). Hanbury Brown (1962) produced a model distribution of radio galaxies clustered on a 50 Mpc scale, which used a luminosity function consistent with that of Ryle and Clarke based solely on the brightness of the radio background, and which, according to Hanbury Brown, could produce a sufficiently deep 'local hole'. Clarke et al. (1963) disagreed. It is a useful reminder of our general lack of understanding of radio sources at that time, and the scarcity of optical identifications in particular, that as late as 1963 Sciama could produce, quite seriously though perhaps not with conviction, a 'local hole' theory based on a hypothetical population of galactic radio sources (Sciama 1963).

Despite the enormous improvements in the observations, the first really deep surveys, and the discovery of quasars, that was still the state of source counts when, in 1965, Penzias and Wilson discovered the microwave background. After that, nearly all astronomers agreed that the universe must be evolving, even those who had been staunch supporters of steady-state theory, and those who wished could do so without conceding that radio source counts were relevant evidence. The phase of source counts versus steady-state theory was over. In a sense, the debate continued with many of the same participants, on ground that had shifted again; this time it was about the redshifts of quasars, and hence whether quasars could tell us anything about cosmology. Of course we shall never know, but I do wonder whether that debate would have been as long and as passionate if it had not been the direct successor to an earlier one.

Source counts were still made, but now to trace the evolution of the source population, not just to establish the existence of evolution, and for that purpose one needs comprehensive measurements of distances. It is quite a different game, and one I found less interesting. However, I did watch, fascinated, when in the early 1970s the X-ray astronomers went through many of the same arguments and even misunderstandings as the radio astronomers had fifteen years earlier. They did so much faster and more peaceably, and I hope in that they may have been helped by our experiences.

17.7 Epilogue

In 1959, to my immense surprise, Pawsey wrote to invite me to come to CSIRO Radiophysics Division, where I then spent three happy years. Bernie Mills met me at the docks and took me to a pub for an introduction to Australian beer. I got to know him a little (he left Radiophysics soon afterwards) and we got along pretty well, though we never did agree about radio sources.

References

Archer, S., Baldwin, J. E., Edge, D. O., Elsmore, B., Scheuer, P. A. G. and Shakeshaft, J. R. (1959). In *Paris Symposium on Radio Astronomy* (IAU Symposium No. 9, URSI Symposium No. 1), ed. R. N. Bracewell, pp. 487–491. Stanford University Press.

Bolton, J. G., Stanley, G. J. and Slee, O. B. (1954). *Australian Journal of Scientific Research*, **7**, 109.

Bracewell, R. N. (ed.) (1959). *Paris Symposium on Radio Astronomy* (IAU Symposium No. 9, URSI Symposium No. 1). Stanford University Press.

Clarke, R. W., Scott, P. F. and Smith, F. G. (1963). *Monthly Notices of the Royal Astronomical Society*, **125**, 195.

Edge, D. O., Scheuer, P. A. G. and Shakeshaft, J. R. (1958). *Monthly Notices of the Royal Astronomical Society*, **118**, 183.

Edge, D. O., Shakeshaft, J. R., McAdam, W. B., Baldwin, J. E. and Archer, S. (1959). *Memoirs of the Royal Astronomical Society*, **68**, 37.

Hanbury Brown, R. (1959). In *Paris Symposium on Radio Astronomy* (IAU Symposium No. 9, URSI Symposium No. 1), ed. R. N. Bracewell, pp. 471–474. Stanford University Press.

(1962). *Monthly Notices of the Royal Astronomical Society*, **124**, 35.

Hewish, A. (1961). *Monthly Notices of the Royal Astronomical Society*, **123**, 167.

Hoyle, F. and Narlikar, J. V. (1962). *Monthly Notices of the Royal Astronomical Society*, **125**, 13.

Kellermann, K. I. (1972). *Astronomical Journal*, **77**, 531.

Mills, B. Y. (1952). *Australian Journal of Scientific Research*, **A5**, 266; ibid., **5**, 456.

Mills, B. Y. and Little, A. G. (1953). *Australian Journal of Scientific Research*, **6**, 272.

Mills, B. Y. and Slee, O. B. (1957). *Australian Journal of Physics*, **10**, 162.

Minkowski, R. (1959). In *Paris Symposium on Radio Astronomy* (IAU Symposium No. 9, URSI Symposium No. 1), ed. R. N. Bracewell, pp. 536–538. Stanford University Press.

Ryle, M. (1959). In *Paris Symposium on Radio Astronomy* (IAU Symposium No. 9, URSI Symposium No. 1), ed. R. N. Bracewell, pp. 523–527. Stanford University Press.

Ryle, M. and Clarke, R. W. (1961). *Monthly Notices of the Royal Astronomical Society*, **122**, 349.

Ryle, M. and Hewish, A. (1955). *Memoirs of the Royal Astronomical Society*, **67**, 13.

Ryle, M. and Scheuer, P. A. G. (1955). *Proceedings of the Royal Society*, **A230**, 448.

Ryle, M., Smith, F. G. and Elsmore, B. (1950). *Monthly Notices of the Royal Astronomical Society*, **110**, 508.

Scheuer, P. A. G. (1957). *Proceedings of the Cambridge Philosophical Society*, **53**, 764.
Sciama, D. W. (1963). *Monthly Notices of the Royal Astronomical Society*, **126**, 195.
Westerhout, G. and Oort, J. H. (1951). *Bulletin of the Astronomical Institutes of the Netherlands*, **11**, 323.

Discussion

Hoyle:

I would like to mention a point that clears up an apparent difference between Peter Scheuer's paper and what I said concerning the good fit of the 3CR radio galaxies to the steady-state theory. The radio galaxies, although comprising the bulk of the sources, omit the quasar component. It is the latter which has the steep log N–log S slope. It was the combination of about −2.5 for the quasars with about −1.3 for the radio galaxies which led to −1.7 for the entire survey.

Bondi:

1) The story of the non-publication of your paper is surely ripe for historical investigation. Was it submitted and rejected, or was it never submitted? 2) When the radio astronomers disagreed, I felt I should wait until the dust settled before seriously studying whether the results were or were not compatible with the steady-state theory.

Scheuer:

I don't know the whole story. As you know, papers had to be submitted by a senior person (e.g. for *Monthly Notices*, a Fellow of the Royal Astronomical Society); furthermore, the story may have begun before I returned to Cambridge. But the details are irrelevant to my point. If the paper was submitted and rejected, I have no reason at all to suppose it was not done in good faith (though I should of course say, mistakenly!). If for any reason Martin did not submit it to one of the standard astronomy journals we shall probably not find out why, and it doesn't really matter why.* I mentioned the matter to show what a tense situation we had around 1957; the mere fact that my little paper was published in a surprising place shows that, in the prevailing climate of opinion, there was a problem of some sort. On balance I would rather not know the details;

* Following an enquiry by Bondi, a recent thorough search of the archives of the Royal Astronomical Society by its present executive secretary shows that the paper was never received by the society. Scheuer wishes to add that he has come across a letter (dated 12 June 1956) which Martin Ryle sent him at Catterick Camp, before the text had been put in its final form. The letter shows that even then it was clear that the 'P(D) paper' was to be submitted to some journal other than *Monthly Notices*, but does not explain why. (The editors.)

prodding around in such things might well create unpleasantness where now there is none.

Bertotti:
I would like to compare the quick development of the number counts problem for radio sources with the practically absent counterpart for nebulae in the nineteenth century. From W. Herschel's fundamental work in 1785 the subject remained dormant, in spite of the fact that several thousands of nebulae were available. The lack of a statistical approach to 'extragalactic' astronomy before Hubble is striking. Do you have any comment.

Scheuer:
Surely this was simply because there was no astronomical photometry even approaching the accuracy one needs. In radio astronomy, there was a quantitative measurement of flux density from the beginning, on a linear scale, even if it was sometimes not on an absolute scale.

Braccesi:
I do not agree with Professor Scheuer's statement that the debate about source counts came to an almost abrupt end in 1965 with the discovery of the 3° K cosmological background radiation.

In fact, even before those times, most of the radio astronomers had already agreed that source counts, plus the few sparse, but compelling, identifications with distant objects, in conjunction with the small number of identifications with optically relevant ones, did allow for the rejection of the steady-state cosmology, even if the slope of the source counts was as low as 1.5. If a date has to be put to this, one may refer to the 1961 Ryle paper delivered at the IAU Symposium on the 'Problems of Extragalactic Research' and the ensuing discussion, particularly to the comments by Mills, plus the summary talk of McVittie.

After that the accent shifted towards the problem of a quantitative appraisal of the characteristics of the cosmological evolution of the radio source population, with all the problems of identification, redshifts etc. which everyone knows.

One may also recall that in 1963, at the First Texas Symposium, the 'Quasar' symposium, neither Gold nor Hoyle made any claim in favour of the steady-state theory, nor any question concerning the nature of the quasar redshifts was raised, although Geoffrey Burbridge spoke at length there.

Thus, I think it safe to say that steady state had been generally abandoned in favour of Friedmann evolving models years before the discovery of the cosmological 3° K background radiation. That discovery, when the blackbody nature of the radiation was established, and thanks to the magnificent 1967 work of Wagoner, Fowler and Hoyle on the primeval nucleosynthesis, only led the community to agree on the particular model one had to consider. To put a date also to the moment of this generalised consent, I would refer to the Texas Symposium held in New York in December 1966, where the ageing Gamov was so warmly cheered.

Scheuer:
I did not want to suggest that the discovery of the microwave background caused an instantaneous change of attitude among the great majority of astronomers. Indeed, many had never really accepted steady-state theory. But in 1965 even most of those who had until then vigorously defended the steady-state theory recanted; the succession of models designed to save the theory, rebuttals, and rebuttals of rebuttals, stopped. I should also add that (as I warned at the start of my talk) I am describing a personal view, and the steady-state theory was strongest in Britain. The scene probably looked quite different when seen from Italy or the USA.

Sullivan:
(1) It has always seemed remarkable to me that, for the great majority of the astronomical community in the 1960s, the steady-state theory was killed off not by the direct approach of showing that the early universe was different in its population of objects, but by a serendipitous discovery that was taken to give great credence to the opposing big-bang cosmology. Many people were involved in trying to take advantage of the well-defined and much-publicised falsifiability of steady-state cosmology, but it did not die in such a Popperian manner. Rather, it died because of the fulfilment of a little-known, never-emphasised prediction by its rival theory. Perhaps the community was tired of a decade of haggling over source counts and, long before they were ever resolved, was only too pleased to move on to the entirely different sort of evidence provided by the background radiation.

Although really too young to be reminiscing in the present company, nevertheless I think it would be of interest to recall an amazing event, at least to one first-year graduate student at the University of Maryland. In 1966–67 our chairman, Dennis Sciama, gave a talk there in which he stated that he was *recanting* from the steady-state cosmology. I was astounded – I thought recantations had gone out of style over three centuries earlier in the country where we now speak! This was very soon after the Penzias–Wilson discovery and I would like to ask him now what reasons he had for his change of camp.

(2) It is important to note that while Ryle clearly embraced the idea of doing cosmology and that by 1954 he saw this as the most exciting advance of the 2C survey, the situation in Australia was quite different. It is clear

from the archival and the published evidence that Mills' work, at least in the 1955–58 period, was much more motivated by his disagreements with Ryle's source counts and the resulting interpretations than by any desire to do cosmology, for or against steady-state theory.

18

The discovery of quasars

Maarten Schmidt

The discovery of quasars in the early 1960s was the first of many discoveries that revolutionized astronomy over the next two decades. In presenting the history of the discovery of quasars as I remember it, I will first mention early work on radio galaxies, because it was in the course of studies of these objects that quasars were discovered. Then I will mention the radio source 3C 48 which provided early clues that were not understood until years later. Next, I will describe the search for identifications of radio sources of small angular diameter which culminated in the identification of 3C 273 and the discovery of its redshift. I will conclude with some remarks concerning the effect of this and other discoveries on astronomical research.

18.1 Radio galaxies

Following the classic studies by Baade and Minkowski in 1952 on the optical identification of radio sources such as Cas A, Cyg A, etc., Rudolph Minkowski continued his work on radio galaxies. This culminated in his spectacular determination of the redshift (0.46) of 3C 295, just months before his retirement in 1960.

Gradually, in 1961, I started to take up where Minkowski had left off. Tom Matthews (Caltech) would supply optical identifications of radio sources based on positions determined with the Owens Valley Radio Observatory twin 90-foot interferometer. I would take spectra with the nebular spectrograph at the prime focus of the 200-inch telescope at Palomar. With the photographic means available at the time, progress was slow, exposure times being in the range of 2 to 10 hours.

Through the work of Burbidge and others, the large energy output of the large double-lobed radio galaxies was explained through synchrotron models with an enormous energy content, typically around 10^{61} ergs. These large energies posed a major problem in radio astronomy in the early 1960s.

18.2 The radio source 3C 48

Soon after Minkowski's retirement, Matthews obtained an accurate position for the radio source 3C 48. It had a small angular diameter and was expected to be a distant cluster of galaxies. I was not involved in work on this source. As far as I know, Allan Sandage took a direct plate of the field in September 1960 which showed a stellar object with faint nebulosity at the radio position. Sandage obtained the first spectra of the stellar object a month later, which showed it to be extremely peculiar, the only prominent features being strong broad emission lines. Photometry of the stellar object by Sandage showed a strong ultraviolet excess, such as exhibited by white dwarfs. Guido Münch and Jesse Greenstein obtained further spectra in subsequent months. The results of the joint effort were presented in an unscheduled paper at the 107th meeting of the American Astronomical Society in New York in December 1960 (Matthews, Bolton, Greenstein, Münch, and Sandage 1961).

Further photometry of the 3C 48 stellar object showed that it was variable. There were no publications about the object in 1961 and 1962. It was generally thought at that time that 3C 48 was probably the first radio star (see Matthews and Sandage 1963).

18.3 Radio sources of small angular diameter

When I became involved in taking spectra of radio galaxies in 1961, Tom Matthews had become interested in radio sources of small angular diameter, in the expectation that they might be radio galaxies at large redshift. Angular diameters were derived by Manchester radio astronomers at Jodrell Bank based on interferometry at extended baselines. As many as seven radio sources that eventually turned out to be quasars were investigated before the redshift of quasars was discovered. Table 18.1 presents the chronology of the spectroscopic observations.

As far as I remember, the first three sources studied in 1961 were all identified with galaxies. The spectra did not support the identifications in the sense that they showed no strong emission lines, as usually exhibited by radio galaxies. The optical identifications of 3C 196 and 3C 286 were with star-like objects that eventually turned out to be quasars. My spectra of 3C 196 showed a continuum without convincing emission or absorption lines. 3C 286 exhibited a broad emission line at 5170 A, which I reported in August 1962 (Schmidt 1962). I also obtained in May 1962 a spectra of two galaxies mistakenly identified with 3C 273 and 3C 254, respectively.

3C 147 was the next source identified with a stellar object by Tom Matthews. As with 3C 48, 3C 196, and 3C 286, it eventually turned out to be a quasar. My optical spectra showed several emission lines in the red part of the spectrum. I discussed them at a conference on extragalactic radio sources held at the Goddard Space Science Institute in New York in

Table 18.1. *First spectroscopic observations of optical objects identified with quasi-stellar radio sources*

Date	Radio source	Optical object
April 1961	3C 286	misidentification
June 1961	3C 280	misidentification
June 1961	3C 298	misidentification
May 1962	3C 196	quasar
May 1962	3C 286	quasar
May 1962	3C 273	misidentification
May 1962	3C 254	misidentification
Oct. 1962	3C 147	quasar
Dec. 1962	3C 273	quasar

December 1962. I attempted to explain the spectrum in terms of helium emission from an expanding shell, but did not publish this interpretation.

18.4 The radio source 3C 273

The fifth radio source identified with an object of stellar appearance was 3C 273. Hazard, Mackey, and Shimmins (1963) had been observing lunar occultations of the source in April, August, and October 1962. John Bolton sent Tom Matthews the first accurate positions that were obtained in August 1962. The source was double, one of the components lying within a few seconds of arc from a star of magnitude 13. The second component, at an angular distance of about 20 seconds of arc, coincided with an optical wisp or jet pointing away from the star.

My initial suspicion was that the jet was a peculiar galaxy associated with the radio source and that the star of magnitude 13 was a foreground object. Since the jet was very faint and would require a long exposure, I decided to take a spectrum of the bright star, expecting that it could be eliminated from consideration with little expenditure of telescope time. Even then, when I took the first spectrum of the star at the end of the night of December 27/28, 1962, it was badly overexposed. I simply was not used to observing such bright objects. The spectrum did look peculiar. At the ultraviolet end, where the exposure was about right, a broad emission line appeared, at 3240 A. In addition, I noticed emission lines at 5630 and 5790 A and suspected the presence of other lines.

Two nights later I obtained a spectrum with the correct exposure, which showed six emission lines, including those already seen. In January 1963, Oke (1963) observed 3C 273 spectrophotometrically at the 100-inch telescope on Mount Wilson and detected a strong emission line at 7600 A.

With seven lines in hand, it seems odd in retrospect that no larger effort was undertaken to identify the lines. I showed the list of wavelengths to I. S. Bowen, who had identified the green nebular lines as forbidden lines in 1926 and to B. Baschek. Suggested identifications with helium lines were not convincing.

The puzzle was suddenly resolved in the afternoon of February 5, 1963, while I was writing a brief article about the optical spectrum of 3C 273. Cyril Hazard had written up the occultation results for publication in *Nature* and suggested that the optical observations be published in an adjacent article. While writing the manuscript, I took another look at the spectra. I noticed that four of the six lines in the photographic spectra showed a pattern of decreasing strength and decreasing spacing from red to blue. For some reason, I decided to construct an energy-level diagram based on these lines. I must have made an error in the process which seemed to contradict the regular spacing pattern. Slightly irritated by that, I decided to check the regular spacing of the lines by taking the ratio of their wavelengths to that of the nearest line of the Balmer series. The first ratio, that of the 5630 line to H-beta, was 1.16. The second ratio was also 1.16. When the third ratio was 1.16 again, it was clear that I was looking at a Balmer spectrum redshifted by 0.16.

Assuming that the remaining lines also had a redshift of 0.16, their rest wavelengths showed that the ultraviolet line was to be identified as the unresolved Mg II doublet at 2798 A; another line was identified as forbidden [O III] at 5007 A. Oke's line observed at 7600 A came close to the wavelength of H-alpha.

I was stunned by this development: stars of magnitude 13 are not supposed to show large redshifts! When I saw Jesse Greenstein minutes later in the hallway and told him what had happened, he produced a list of wavelengths of emission lines from a just completed manuscript about the spectrum of 3C 48. Being prepared to look for large redshifts, it took us only minutes to derive a redshift of 0.37. One of the emission lines turned out to be Mg II at 2798 A. The fact that both objects exhibited Mg II emission, which until that time had never been observed in extragalactic objects, provided strong support for the existence of the redshifts.

The interpretation of the large redshifts was an extraordinary challenge. Greenstein and I soon found that an explanation in terms of a gravitational redshift was essentially impossible on the basis of spectroscopic arguments. We recognized that the alternative explanation in terms of cosmological redshifts, large distances, and enormous luminosities and energies was very speculative but could find no strong arguments against it. The results for 3C 273 and 3C 48 were published six weeks later in four consecutive articles in *Nature* (Hazard, Mackey, and Shimmins 1963; Schmidt 1963; Oke 1963; Greenstein and Matthews 1963).

18.5 Comments

I might use this occasion to make a few technical and one or two philosophical remarks.

1 The argument that objects that are variable on a time scale of x years cannot be much larger than x light years, was used – I believe for the first time – on quasars with great effect. Since the time scale of variability for 3C 48 (Sandage and Matthews 1963) and 3C 273 (Smith 1963) was of the order of months or years, their size could not be much larger than one light year.

2 Even though the luminosities of quasars based on the cosmological interpretation of the redshifts were large (say, 10 or 100 times that of the brightest galaxies), it was – I believe – their small sizes, of the order of a light year or less, that caused serious doubts about their distances in many astronomers' minds in the late 1960s.

3 These doubts would probably not have arisen if the variability of nuclei of Seyfert galaxies had been discovered earlier. If that had happened, doubts about their distances would have been soon resolved since they show many characteristics of ordinary galaxies. On the contrary, the point-like appearance of quasars allowed no such deductions to be made.

4 It is remarkable that after 25 years our basic understanding of quasars is still so unsettled. We think that they are black holes feeding on stars and gas in the nucleus of a galaxy, but the detailed evidence for this remains vague. In contrast, it was clear within one or two years after their discovery that pulsars were highly magnetized, rotating neutron stars – a much more precise description of the physics of the situation.

5 The discovery of quasars signaled a revolution in astronomy that included the discovery of molecules of space, the microwave background, pulsars, and many others. This had, I believe, a profound impact on the conduct of those practicing astronomy. Before the 1960s, there was much authoritarianism in the field. New ideas expressed at meetings would be instantly judged by senior astronomers and rejected if too far out. The discoveries of the 1960s were an embarrassment in the sense that they were totally unexpected and could not be evaluated immediately. In reaction to these developments, an attitude has evolved where even outlandish ideas in astronomy are taken seriously. Given our lack of solid knowledge in extragalactic astronomy, this is probably to be preferred over authoritarianism.

References

Greenstein, J. L. and Matthews, T. A. (1963). Red-Shift of the Unusual Radio Source: 3C 48. *Nature*, **197**, 1041.

Hazard, C., Mackey, M. B., and Shimmins, A. J. (1963). Investigation of the Radio Source 3C 273 by the Method of Lunar Occultations. *Nature*, **197**, 1037.

Matthews, T. A. and Sandage, A. R. (1963). Optical Identification of 3C 48, 3C 196 and 3C 286 with Stellar Objects. *Astrophysical Journal*, **138**, 30.
Matthews, T. A., Bolton, J. G., Greenstein, J. L., Münch, G., and Sandage, A. R. (1961). Reported in *Sky and Telescope*, **21**, 148.
Oke, J. B. (1963). Absolute Energy Distribution in the Optical Spectrum of 3C 273. *Nature*, **197**, 1040.
Schmidt, M. (1962). Spectrum of a Stellar Object Identified with the Radio Source 3C 286. *Astrophysical Journal*, **136**, 684.
(1963). 3C 273: A Star-Like Object with Large Red-Shift. *Nature*, **197**, 1040.
Smith, H. J. (1965). Light Variations in 3C 273. In *Quasi-Stellar Sources and Gravitational Collapse*, ed. I. Robinson, A. Schild, and E. L. Schucking, p. 221. Chicago: University of Chicago Press.

Discussion

McCrea:

Dr. Schmidt seems through modesty not to have mentioned the first Texas Symposium in December 1963. I think it was he who most firmly supported the cosmological interpretation of the redshift at that meeting.

Dr. Schmidt mentioned the work at Jodrell Bank. At the Tercentenary celebration of the Royal Society in 1960 Sir Bernard Lovell spoke of sources of very small diameter. Could Dr. Schmidt say if these included some quasars?

Schmidt:

I believe I did not speak at the first Texas Symposiumin Dallas in December 1963. Jesse Greenstein and I had been working on the interpretation of the spectra of 3C 273 and 3C 48, and he talked about our work in Dallas.

I imagine some of the small diameter sources mentioned by Sir Bernard Lovell were quasars – see the comment by Woodruff Sullivan.

Bill, I recall the first big event at which I did talk about quasars was a Joint Discussion at the IAU General Assembly in Hamburg in 1964. I presented data on the first nine quasars, four of which had redshifts at the time. At the end, there was only one question, by you – and you asked whether I was aware that the quasars I had discussed were all within 5 degrees of a large circle on the sky. I was very surprised, and probably did not reply. In a sense, this question was the first of many to come that queried the nature of quasars.

Marcaide:

Once you found out about the more compact of Hazard's radio sources, what was your attitude with respect to the other component?

Schmidt:
The jet, which was identified with the extended radio component, was very faint. No good spectra were obtained for many years. We suspected that its nature might be like that of the jet of M87.

Osterbrock:
I remember very well the period in 1961 of which you spoke, in which 3C 48 was believed to be a radio star. I heard the late paper on it presented at the AAS meeting in New York. Woltjer and I, then both at the Institute for Advanced Study in Princeton, worked very hard at trying to identify the available wavelengths of the emission lines in it – always in the context of a star with essentially 0 velocity but anomalous abundances, ionization, and physical conditions. Also I remember, several years later, advising a graduate student (Joe Miller) to use a Seyfert galaxy nucleus as a standard object for photometry which we *knew* would not vary in light!

Sullivan:
(1) You have correctly emphasized the importance of Hazard's lunar occultation work in the discovery of the quasars, but I would like to take it one step further back. Interferometric observations at Jodrell Bank continued on ever-longer baselines across the Cheshire countryside throughout the 1950s. These represented a long-term commitment to better and better source sizes that was vital in demonstrating the remarkable physics of these sources. This work began with Hanbury Brown's intensity interferometer and then was carried on by the conventional interferometry of Jennison and in particular Palmer's group. A milestone along their path that makes very interesting reading even today was "Five Radio Sources of Small Angular Diameter" in *Observatory*, **77**, 103–6 (1957) by Morris, Palmer, and Thompson. Three of these sources were shown to have sizes <12″ and two of these were the quasars 3C 147 and 3C 196, then called HBH 10 and HBH 11.

(2) You mentioned that the variations in optical intensity of the quasars were the first application you knew of the technique of inferring an upper limit for the linear size of an object by the time scale on which it varies in intensity. There is, however, an application of this principle a good bit earlier during the late 1940s. (a) When in 1946 Hey discovered a region (less than 2 degrees in size) in the constellation of Cygnus varying in intensity on a scale of minutes, he deduced that it had to be a small discrete source in order to vary so. (The reasoning was found on the assumption that the variations were intrinsic to Cygnus A, but we now know they are solely due to the ionosphere.) (b) A few years later, even after it had become established that most or all of these scintillations were ionospheric

in origin, Ryle frequently argued (1948–52) that a certain subset of the intensity variations of Cygnus A were indeed intrinsic and therefore constituted evidence that the radio stars were a new kind of star, no larger than 20 light-seconds in size.

19

History of dark matter in the universe (1922–1974)

Virginia Trimble

19.1 Introduction

Dark matter, both as an idea and as a phrase, has a surprisingly long record, if the words are taken to mean any sort of non-luminous or sub-luminous material whose existence we deduce from its gravitational effects on brighter things. This definition reserves to Professors Harrison (1990) and Bondi (1990) the realm of material whose only effect is to obscure light.

The history of dark matter in our restricted sense can then be divided into five periods, roughly analogous to the periods of human history whose relics we saw around us during the Bologna meeting, though the dates are somewhat different. These are the Neolithic (before 1900), the Classical (1922–37), the Mediaeval (1939–60), the Renaissance (1961–73), and the Soviet and American Revolutions (1974).

19.2 Neolithic dark matter (before 1900)

From the eighteenth and nineteenth centuries come three kinds of studies, one theoretical and two observational, concerned with evidence for non-luminous objects. First, John Michell (1784), followed by Pierre Simon, Marquis de Laplace (1799), considered objects sufficiently compact to have escape velocities in excess of the speed of light. Michell noted that "all light emitted from such a body could be made to return to it by its own proper gravity," and Laplace concluded that "the largest luminous bodies in the universe may, through this cause, be invisible." These remarks are generally regarded as the first predictions of black holes.

On the observational front, Wilhelm Friedrich Bessel (1845a and b) analyzed irregularities in the motions of Sirius and Procyon shortly before John Couch Adams (1847) and Urbain J. J. Leverrier (1845) looked at the orbit of Uranus. Bessel concluded that each star had an invisible companion of roughly its own mass, while Adams and Leverrier predicted a new planet. The latter (Neptune) remained unseen for less than a year, but the stars counted as dark until 1862 and 1896 respectively, when they were seen by Alvan G. Clark, testing an 18½″ objective, and J. M. Schaeberle (1896), using the Lick 36″. Schaeberle seems to have been rather a cantankerous character, given to discussing the "worthless telescopes" and "strange articles" of his contemporaries.

Curiously, all three classes of object implicated in these early studies are still viable dark matter candidates on one scale or another, under the names of primordial (10^{15-30} g) or pregalactic (10^{4-8} $M_\odot$) black holes, white dwarfs (left from bimodal star formation), and gas giants or brown dwarfs. Incidentally the "brown giants" whose possible existence was noted by Bondi and Bondi (1950) are not viable candidates, being intrinsically very bright, though very (infra)red.

19.3 Classical dark matter (1922–1937)

Moving into the twentieth century, we find both Sir James H. Jeans (1922) and Johannes C. Kapteyn (1922) determining the local mass density from stellar velocity dispersions and distributions above the galactic plane. Their results (0.143 and 0.099 $M_\odot/pc^3$) roughly bracket the modern range of values, as do their conclusions, based on comparing the mass per luminous star implied by these numbers with the average mass of known binary systems (1.6 $M_\odot$). Jeans opined that "there must be about three dark stars in the universe to every bright star," while Kapteyn, noting that "we therefore have the means of estimating the mass of dark matter in the universe," concluded that "as matters stand at present it appears at once that this mass cannot be excessive." Their "universe" was, of course, our "galaxy."

The definitive early study of this kind by Jan H. Oort (1932) gives us the usual name, Oort limit, for the dynamically determined density. He found 0.08 to 0.11 $M_\odot/pc^3$. Stars brighter than $M_v = 13.5$ contributed 0.038 $M_\odot/pc^3$ to this. Oort's opinion that stars between 13.5 and 18.5 would account for most of the rest was based heavily on apparent gravitational redshifts near 240 km/s for the white dwarfs Procyon B and van Maanen 2, which (using R. H. Fowler's non-relativistic degenerate matter equations) would have required them to have masses much larger than 1 $M_\odot$. By the time of Oort's (1965) article in the Kuiper compendium, his estimate had risen to 0.150 $M_\odot/pc^3$, with "invisible stars and gas" responsible for 40 percent of the total.

The other historical pillar of dark matter studies is the investigation of velocity dispersions in rich clusters of galaxies. Fritz Zwicky (1933) looked especially at Coma and concluded "dass dunkle Materie in sehr viel groesserer Dichte vorhanden ist als leuchtende Materie." Incidentally, this sentence is not an illustration of the widespread story that Zwicky spoke seven languages, all badly – "dass" functions as a conjunction, not an article. A few years later, Sinclair Smith (1936) examined the velocities of Virgo cluster galaxies, and found a total mass of 10^{14} $M_\odot$ (for a distance of 2 Mpc, thus 5–10 × 10^{14} $M_\odot$ at the modern distance) and a mass per galaxy about 100 times that implied by the luminosities.

Holmberg (1937), in the only contemporary examination of binary galaxies, found an average mass per galaxy of 5 × 10^{11} h^{-1} $M_\odot$ (10^{11} $M_\odot$ on the then-current distance scale) from a handful of pairs. Remarkably, given the large errors and limited statistics of all determinations, this fits comfortably between the single-galaxy and the rich-cluster results, as we now think it should.

Zwicky, (1937a, 1937b, 1939, 1951, 1959, 1960, 1961) continued to take an active interest in the subject for a number of years, advocating a combination of dwarf galaxies and intergalactic material to make up the dominant dark component at the time of his *Handbuch* article (Zwicky 1959) and intergalactic stars and pygmy galaxies at the IAU Symposium (Zwicky 1961), by which time a 21 cm study of Coma (Muller 1959) had ruled out neutral hydrogen gas as a candidate. Dying in 1974, Zwicky just missed the chance to say "I told you so" to his converted colleagues on the subject of dark matter. Smith, whose 1936 paper explicitly cites Zwicky's work, passed from the scene in 1938 and does not, in any case, seem to have been the "I told you so" type (see chapter 22 below).

19.4 Mediaeval dark matter (1939–1960)

To this period belong a number of studies of velocity dispersions in clusters and groups of galaxies, hints that single galaxies might have larger masses than at first evident, and the advent of steady-state cosmology with its firm prediction of significant intergalactic matter. First, the cluster velocity of dispersions were persistently large, independent of the philosophical predispositions of the observers. Mass-to-luminosity ratios (in solar units, and assuming gravitationally bound systems) of 100 to 1000 were found for the clusters in Virgo (Ambartsumian 1958; Oort 1958; van den Bergh 1960), in Coma (Schwarzschild 1954; de Vaucouleurs 1959a; van den Bergh 1960), in Hercules (Burbidge and Burbidge 1959a), and in Canum Venaticorum (van den Bergh 1960) and for a number of small groups (Holmberg 1950; Zwicky 1956; Burbidge and Burbidge 1959b, 1960; de Vaucouleurs 1959b, 1960; Limber and Mathews 1960).

Next, several related lines of evidence gradually indicated that, in general, masses of galaxies got larger when measured on larger scales. Kunth (1952) and Lohman (1956) found, for instance, that the mass of the Milky Way implied by globular cluster velocities was two or three times larger than that calculated from the local circular velocity. For M31, the rotation curve failed to drop outside 15 kpc as predicted by Kepler's laws (Babcock 1939 provided optical data and van de Hulst et al. 1957 a 21 cm rotation curve). The implied global M/L was 20 or so, vs. two in the nucleus (Lalleman et al. 1960). Ellipticals also showed smaller nuclear M/L's than global ones (Osterbrock 1962). And galaxy pairs had smaller M/

L's than whole clusters (van den Bergh 1960) but larger M/L's than the galaxies considered separately (Kahn and Woltjer 1959 on the Local Group). Schwarzschild (1954) suggested that the dark matter, at least in elliptical galaxies, might be contributed by old white dwarfs that had had M_v near +1 on the main sequence, an interesting foreshadowing of the idea we now call bimodal star formation.

Finally, this was the era of the steady-state universe (Bondi and Gold 1948; Hoyle 1948), which necessarily has a good deal of pregalactic dark matter around at all epochs. In steady state with a creation field, the average density is exactly $3H^2/8\pi G$, a hundred times or so larger than the density contributed by luminous stars and galaxies. Hoyle (1959) predicted that the missing 99 percent would be largely in the form of intergalactic hydrogen at $T > 10^5$ K. Other members of the astronomical community thought this large preponderance of invisible stuff considerably less likely, in language gentle (Oort 1959; Bonnor 1960) to scornful (McVittie 1961b).

Despite the large number of observational results, this mediaeval period was marked by a profound lack of consensus on what the observations meant and even on whether they need to be taken terribly seriously.

19.5 Renaissance dark matter (1961–1973)

By the time of a pair of 1961 conferences in Santa Barbara, California, on the stability of clusters of galaxies (proceedings, edited by J. Newman, T. L. Page, and E. Scott in the *Astronomical Journal*, **66**) and problems of extragalactic research (IAU Symposium 15, proceedings, McVittie 1961a), opinion had crystallized around two strongly opposing views. The rich clusters must either be bound by dark matter, associated more with the clusters as a whole than with the individual galaxies, or the clusters must be short-lived and currently expanding out of some fairly violent explosion.

Binding was advocated by Zwicky (1961), Minkowski (1961), Limber (1961), Baum (1961, who, with Zwicky, believed there might be enough intergalactic star light to bind with luminous material), and Abell (1961, who thought it plausible that superclusters might be similarly bound at $\sim 10^{16} M_\odot$ each). Van Albada (1961) and von Hoerner (1961) presented results of analytic and numerical models of cluster dynamical evolution whose implication was that, for bound clusters, Virial masses were likely to be wrong, but not by factors of more than two or three. Today, von Hoerner's calculation would be described as an n-body simulation, with $n = 16$!

Unstable, expanding clusters were first proposed by Ambartsumian (1958, 1961) partially by analogy with known expanding clusters of young stars such as the Trapezium. Supporters included Vorontsov-Velyaminov (1961), Kalloghjian (1961), Markarian (1961), de Vaucouleurs (1961), and the Burbidges (1961).

The dissenters from the two main streams included Holmberg (1961), who believed that observational errors, substructure in clusters, and foreground/background interlopers could account for the large apparent velocity dispersions with no need for either dark matter or explosions, and Lemaître (1961), who proposed that rich clusters might be constantly exchanging galaxies with the field, so that the configurations were permanent, but the individual members not gravitationally bound. Van den Bergh (1961) considered, but rejected, a contribution from non-gravitational forces that might bind the clusters without requiring extra mass. Shortly after these meetings, Arigo Finzi (1963) summarized the evidence for M/L increasing with radius and suggested the closely related idea of non-Newtonian gravitation as an alternative to dark matter.

Virtually all these points of view have modern adherents. Ideas related to Ambartsumian's are explored by Halton C. Arp and Fred Hoyle elsewhere in this volume. Binney (1986) has revived Smith's (1936) suggestion of faint luminous extensions to galaxies, sufficient to keep M/L roughly constant with radius. The primacy of errors in the observations and their interpretation continues to be emphasized by Mauri Valtonen and his colleagues (Valtonen and Byrd 1986). Lemaître's suggestion seems to have no current proponents, but the density wave model of spiral structure has some similar features. Finally, Milgrom (1986) and Sanders (1986) among others have presented testable versions of non-Newtonian gravity, in the form either of a minimum gravitational acceleration or of an additional Yukawa-form gravitational potential.

19.6 The Soviet and American revolutions (1974)

The winner was, however, undoubtedly dark matter. Rather suddenly, at the time of the publication of two short summaries of the data indicative of M/L increasing monotonically with R (Einasto, Kaasik, and Saar 1974; Ostriker, Peebles, and Yahil 1974) most of the astronomical community decided, first, that the observations should be taken seriously, and, second, that non-luminous, gravitating material or dark matter was much the most likely explanation. Since then, it is the non-believers who have found themselves out of step and, sometimes, out of print. Thus this 1974 paradigm shift logically marks the division between history and current events in the field of dark matter research. Two recent reviews (Trimble 1987, 1988) provide an *entré* to the post-revolutionary literature.

References

Abell, G. O. (1961). *Astronomical Journal*, **66**, 607.
Adams, J. C. (1847). *Memoires of the Royal Astronomical Society*, **16**.
van Albada, C. B. (1961). *Astronomical Journal*, **66**, 590.

Ambartsumian, V. A. (1958). In *Solvay Conference: La Structure et l'évolution de l'universe*, p.163. Institute de Phys. Solvay.
(1961). *Astronomical Journal*, **66**, 536.
Babcock, H. W. (1939). *Lick Observatory Bulletin*, **19**, 41.
Baum, W. A. (1961). In *Problems of Extragalactic Research* (IAU Symposium No. 15), ed. G. C. McVittie, p. 255. New York: McMillan.
van den Bergh, S. (1960a). *Astrophysical Journal*, **131**, 558.
(1960b). *Monthly Notices of the Royal Astronomical Society*, **121**, 387.
(1961). *Astronomical Journal*, **55**, 566.
Bessel, F. W. (1845a). *Astronomische Nachrichten*, **22**, 145, 169, 185 (subm. 1844).
(1845b). *Monthly Notices of the Royal Astronomical Society*, **6**, 136.
Binney, J. (1986). *Philosophical Transactions of the Royal Society*, **A320**, 431.
Bondi, H. (1990). This volume.
Bondi, C. M. and Bondi, H. (1950). *Monthly Notices of the Royal Astronomical Society*, **110**, 287.
Bondi, C. M. and Gold, T. (1948). *Monthly Notices of the Royal Astronomical Society*, **108**, 252.
Bonnor, W. B. (1960). *Monthly Notices of the Royal Astronomical Society*, **121**, 475.
Burbidge, E. M. and Burbidge, G. R. (1959a). *Astrophysical Journal*, **130**, 629.
(1959b). *Astrophysical Journal*, **130**, 15 and 23.
(1960) *Astrophysical Journal*, **134**, 244.
(1961). *Astronomical Journal*, **66**, 541.
Einasto, J., Kaasik, A., and Saar, E. (1974). *Nature*, **250**, 309.
Harrison, E. R. (1990). This volume.
von Hoerner, S. (1961). *Astronomical Journal*, **66**, 580.
Holmberg, E. (1937). *Lund Annals*, **6**, 1–173.
(1950). *Meddeilung Lund Astronomical Observatory*, Ser. 2, No. 128.
Hoyle, F. (1948). *Monthly Notices of the Royal Astronomical Society*, **108**, 372.
(1959). In *Paris Symposium on Radio Astronomy* (IAU Symposium No. 9), ed. R. N. Bracewell, p. 529. Stanford: Stanford University Press.
van de Hulst, H. C., Raimond, E., and van Woerden, H. (1957). *Bulletin of the Astronomical Institute of the Netherlands*, **14**, 1.
Kahn, F. and Woltjer, L. (1959). *Astrophysical Journal*, **180**, 105.
Kalloghjian, A. (1961). *Astronomical Journal*, **66**, 554.
Kapteyn, J. C. (1922). *Astrophysical Journal*, **55**, 302.
Kunth, R. (1952). *Zeitschrift für Astrophysik*, **28**, 234.
Lalleman, A., Cuchesne, M., and Walker, M. (1960). *Publications of the Astronomical Society of the Pacific*, **72**, 76.
Laplace, P. S. de (1979). Quoted by S. W. Hawking and G. F. R. Ellis (1973), *The Large Scale Structure of Space Time*, p. 365. Cambridge: Cambridge University Press.
Lemaître, G. (1961). *Astronomical Journal*, **66**, 603.
Leverrier, U. J. J. (184). *Astronomische Nachrichten*, **25**, No. 580.
Limber, D. N. (1961). *Astronomical Journal*, **66**, 572.
Limber, D. N. and Mathews, W. G. (1960). *Astrophysical Journal*, **132**, 286.
Lohman, W. (1956). *Zeitschrift für Astrophysik*, **144**, 66.

Markarian, B. E. (1961). *Astronomical Journal*, **66**, 555.
McVittie, G. C. (ed.) (1961a). *Problems of Extragalactic Research* (IAU Symposium No. 15). New York: MacMillan.
(1961b). In McVittie (1961a), p. 449.
Michell, J. (1784). *Philosophical Transactions of the Royal Society*, **74**, 35.
Milgrom, M. (1986). *Astrophysical Journal*, **306**, 9.
Minkowski, R. M. (1961). *Astronomical Journal*, **66**, 558.
Muller, C. A. (1959). *Bulletin of the Astronomical Institute of the Netherlands*, **14**, 339.
Oort, J. H. (1932). *Bulletin of the Astronomical Institute of the Netherlands*, **6**, 249.
(1958). *Solvay Conference: La Structure et l'évolution de l'universe*, p. 163. Institute de Phys. Solvay.
(1959). In *Paris Symposium on Radio Astronomy* (IAU Symposium No. 9), ed. R. N. Bracewell, p. 353. Stanford: Stanford University Press.
(1965). In *Galactic Structure*, ed. A. Blaauw and M. Schmidt, p. 455. Chicago: University of Chicago Press.
Osterbrock, D. E. (1962). In *Interstellar Matter in Galaxies*, p. 117. New York: Benjamin.
Ostriker, J. P., Peebles, P. J. E., and Yahil, A. (1974). *Astrophysical Journal*, **193**, L1.
Sanders, R. H. (1986). *Monthly Notices of the Royal Astronomical Society*, **223**, 539.
Schaeberle, J. M. (1896). *Astronomical Journal*, **17**, 37.
Schwarzschild, M. (1954). *Astronomical Journal*, **59**, 273.
Smith, S. (1936). *Astrophysical Journal*, **83**, 23.
Trimble, V. (1987). *Annual Reviews of Astronomy and Astrophysics*, **25**, 425.
(1988). *Contemporary Physics*, **29**, 373.
Valtonen, M. J. and Byrd, G. C. (1986). *Astrophysical Journal*, **303**, 523.
de Vaucouleurs, G. (1959a). *Observatory*, **79**, 113.
(1959b). *Astrophysical Journal*, **130**, 718.
(1960). *Astrophysical Journal*, **131**, 585.
(1961). *Astronomical Journal*, **66**, 629.
Vorontsov-Velyaminov, B. (1961). *Astronomical Journal*, **66**, 551.
Zwicky, F. (1933). *Helvetica Physica Acta*, **6**, 110.
(1937a). *Astrophysical Journal*, **86**, 217.
(1937b). *Proceedings of the National Academy of Sciences*, **23**, 251.
(1939). *Proceedings of the National Academy of Sciences*, **25**, 605.
(1951). *Publications of the Astronomical Society of the Pacific*, **63**, 61.
(1956). *Ergebnisse der Exakten Naturwissenschaften* (Springer Tracts in Modern Physics), **29**, 344.
(1957). *Morphological Astronomy*. Berlin: Springer.
(1959). *Handbuch der Physik*, **53**, 373.
(1960). *Publications of the Astronomical Society of the Pacific*, **72**, 365.
(1961). In *Problems of Extragalactic Research* (IAU Symposium No. 15), ed. G. C. McVittie, p. 347. New York: MacMillan.

Discussion

McCrea:

Professor Trimble and other speakers at the meeting have made numerous mentions of J. H. Jeans. He was a very influential thinker in the years with which we are concerned.

Bondi:

I have three connections with dark matter:

1) In Gold's and my formulation of the steady-state theory the newly created (pregalactic) matter must fill the intergalactic spaces with considerable density, a deduction that was found uncomfortable by many astronomers (1948);

2) In my work on red giants I could not see why the radius of many such stars could not be much bigger, making them invisible in the optical domain (1952);

3) In some lecture I gave to the Royal Astronomical Society in 1968 I strongly advocated the need for short time constant astronomy especially to detect dark bodies crossing the line of sight suggesting that, for all we knew, there were many such bodies.

Hoyle:

In relation to what was said by George Ellis, there were many international meetings in the late 1950s and 1960s at which the possibility of dark matter was mentioned. The possibility was quite strongly opposed by the establishment of the day, largely because of the difficulties of observing it – a difficulty which still persists of course. The difference between then and now is that people today are far more ready to admit that unobserved particles may have an important influence in cosmology.

VI
Dramatis personae

20

Carl Wilhelm Wirtz – a pioneer in observational cosmology

Waltraut C. Seitter and Hilmar W. Duerbeck

20.1 Prologue

Science has provided us with two concepts which are now used to describe the progress of science itself. The first one goes back to astronomy, to the work of Copernicus, and is found in the title of his Book *De revolutionibus orbium coelestium* (1543). From the revolution of the heavenly bodies in their celestial orbs the meanings of a return to the former state and a new beginning, as well as a complete turn-over in power, were derived (Arendt 1963; see also Griewank 1952). The second one results from the life sciences of Darwin as presented in his book *On the Origin of Species by Means of Natural Selection, or the Preservation of Favoured Races in the Struggle for Life* (1859). Here the emphasis is on evolution, a process soon envisaged for the other sciences as well.

The history of science seems to us to bear the signs of *evolution* (see Einstein and Infeld 1938), rather than the signs of *revolution* (Kuhn 1967), as the following chapter from the history of cosmology may exemplify. While singular events of far-reaching influence do occur, their signature is not the breaking with the past, but the opening of the future. When the scientific bequest to their followers (to paraphrase Plinius writing about Hipparchos) surpasses to a high degree their own inheritance, the individuals thus gifted will have a dominating influence for a long time. The cause of science, however, is also furthered by the large flux of lesser contributions, which precede such a rise and follow in its wake. The story to be told here is that of the masterminds who set the scene – William Herschel in collaboration with his family on the observational, Einstein and some of his followers on the theoretical side – and the story of others, among them Carl Wirtz, whose more modest, yet brilliant work, contributed to the sequence of events that were essential for the institution of modern cosmology.

The acceptance of new ways of thinking and new results is the other part of evolution; its role in early cosmology will be discussed below.

20.2 The bequest of the Herschel family

Three contributions of William Herschel, assisted by his sister Caroline and followed by his son John, had the same paramount influence on the astronomy of the late nineteenth and early twentieth centuries as had the work of the great spectroscopists in creating the new branch of astrophysics. They are the solar motion (1783),[1] star gauging and the structure of the Milky Way (1785), the catalogues of nebulae (1786, 1789, 1802, 1864) and the analysis of nebulae (1811):

– The solar motion, first measured as it is reflected in the proper motion of stars, later also through their radial velocities, led by various steps to astronomically measurable parameters of space-time.

– The knowledge of the extent and structure of our own sidereal system became the prerequisite for our understanding of the distribution of matter in two basic ways. The changing role of our galaxy from the *sidereal universe* to the prototype of galaxies taught us about universal dimensions and the building blocks of luminous matter. The mathematical methods introduced in the process, statistical ones in particular, are still developing into increasingly more useful tools of observational cosmology.

– The observation of the first several thousand nebulae[2] – later on, with the help of spectroscopy, recognized either as gaseous emission line nebulae concentrated towards the plane of the Milky Way, or others, with continuous or absorption line spectra populating the higher galactic latitudes – is the third part of the Herschel heritage. Combined with the other two, it inspired astronomers to discover the direct route to modern cosmology. Among the investigators is Carl Wirtz, often taking first steps in a new field.

20.3 Carl Wilhelm Wirtz: his life

Wirtz was born in Krefeld, Rhineland, on August 24, 1876. He studied at Bonn University and obtained his Ph.D. in November 1898 with

[1] The 'Astronomisches Jahrbuch für 1786', Berlin 1783, contains a communication by Prévost concerning his own work on solar motion. In the last paragraph he writes: 'Though, according to this, the comparative list of Mayer gives the announcement of a progression of our solar system; nevertheless, the honor of its discovery belongs to Mr Herschel. The subject of his efforts, which has been made public by Mr de la Lande (1776) in the *Journal de Paris*, has inspired my investigation . . .'

[2] The terms *nebulae* and *nebular spots* remained somewhat ambiguous in the early years of our century up to roughly 1910. In most catalogues mentioned here, however, a clear dominance of extragalacitc nebulae is found, and in all cases where the spectra showed emission lines, the nebulae are called gaseous. The most distinctive physical property of the non-gaseous nebulae was their spiral structure. Some investigations of Wirtz concern the 'small nebulae' which were first suspected and then shown to be related to the spirals, but too small for resolving their spiral pattern. This terminology is characteristic for the period 1910–20. Spherical nebulae are now called ellipticals. The terms non-galactic (*nicht-galaktische*) and anagalactic nebulae which were used in the 1920s and '30s were subsequently replaced by extragalactic nebulae and galaxies.

the declination measurement of 487 stars. Besides other applications, his results provided improved coordinates for the evaluation of positions given in different zones of the first catalogue of the Astronomische Gesellschaft (Wirtz 1898). One of the seven propositions, which Wirtz debated in his doctoral examination, concerns the observation of 'nebular spots' and thus testifies to Wirtz's early interest in what was to become his field of major contributions. From the same period dates a triangulation of the Hyades stars with the Bonn heliometer, published four years later (Wirtz 1902). There is no doubt that Wirtz received a thorough education in data acquisition and reduction from his teacher Friedrich Küstner.

After short stays at the Kuffner Observatory in Wien-Ottakring (1899) and at the Nautical School in Hamburg (1901–2), Wirtz became observer and later professor at the Imperial Strasburg Observatory, Alsatia. He was appointed university lecturer in astronomy at the University of Strasburg in 1903. His introductory lecture on 'Recent Methods for the Determination of Stellar Parallaxes'[3] (1906) stressed the importance of Kapteyn's photographic and stereoscopic method. During 1916–18 Wirtz served in the German army and in 1919, after Alsatia had been returned to France, he became professor and observer at the Kiel Observatory. In 1937, he lost his teaching licence 'for political reasons' (Schmidt-Schönbeck 1965); after a prolonged illness he died in Hamburg on February 18, 1939, having just returned from a stay in the United States.

The *Astronomische Nachrichten*, the journal in which he had published his major articles, and which normally contains extensive obituaries, carried no more than the announcement of his death in a single line (Schaub 1939). The *Vierteljahrsschrift* to which he had contributed numerous reviews, and which is published by the Astronomische Gesellschaft, whose member Wirtz was, did not even mention his death. This is unusual since the deaths of members were always communicated, often accompanied by extensive necrologues with portraits of the deceased. Only the amateur astronomers remembered Wirtz and the many lectures and writings with which he supported public interest in astronomy (Kienle 1939).

One of the two known portraits of Wirtz (figure 20.1) is taken from the *Porträt-Gallerie* of the Astronomische Gesellschaft (Tass 1931) and shows Wirtz as he must have looked during his most productive years in cosmology.

Wirtz's astronomical œuvre covers a wide range of topics. His observational contributions concern planets and their satellites, minor planets, comets, eclipses and transits, variable stars, their brightness and colours, and, of course, nebulae. His work also includes topics which are equally important to astronomy, geodesy and nautics, especially investigations of

[3] Translations of titles and quotations from the German are by the present authors.

Fig. 20.1 Carl Wilhelm Wirtz (1876–1939), photographed 1930 or earlier (from Tass 1931).

atmospheric effects. He published extensively: articles, mostly in the *Astronomische Nachrichten*, a book *Tafeln und Formeln der Astronomie und Geodäsie* (1918a), contributions to encyclopaedias, numerous abstracts of scientific papers and articles in popular scientific journals.

20.4 Wirtz in Strasburg (1902–1916)

Two major tasks were carried out by the observer Wirtz in his early Strasburg days. He made regular and numerous observations with the 49-cm refractor of the observatory of all solar system events – his measurement of the diameter of Neptune in 1903 received widespread attention – and he participated in the major project of the observatory, the visual determination of exact nebular positions.

The project had been initiated by the first director of the Strasburg Observatory, Friedrich August Theodor Winnecke, who himself had contributed earlier measurements with a small telescope, but had retired prematurely by the time the large refractor became available. His successor, Ernst Becker, continued with the project and called Wirtz to Strasburg to share the workload of the extensive reductions (Wirtz 1911).

In 1911 when Becker's successor, Julius Bauschinger, took over it was apparent to all participants in the programme that photographic measurements of nebular positions were far superior and the project was terminated after 1257 nebular positions and magnitudes had been published in the bulky volumes of the *Annalen der Sternwarte Strassburg* 3 and 4 (1909–12).

20.5 Early apex determinations

At this point one might recall some details from the long history of apex determinations. Herschel's direct graphical solution on the celestial sphere was replaced, as more proper motions became available, by approximations starting with an assumed apex. Airy (1860) introduced rectangular coordinates and used correction terms in right ascension (determination of the precessional constant) and declination (correction of the first epoch measurements of Bradley). Airy's equations became the major tool in deriving the solar apex. Kapteyn (1901) pointed to the importance of including error terms for both epochs, permitting the use of stars with small proper motions and large measuring errors. This procedure helped to avoid selection effects in stellar positions, and thus systematic displacements of the apex.

While the classical apex determination had to rely on proper motions only, radial velocities were used as soon as they became available. Early attempts were made by Homann (1886) and Kövesligethy (1886), to be followed by others who were able to use increasingly more data. A *K*-term, taking care of systematic radial motions of certain groups of stars, was introduced by Campbell (1911).

20.6 Nebular parallaxes and solar apex determinations from proper motions

One of the early attempts to obtain quantitative data on the distances of *nebular spots* was the use of the antapex components of the proper motions of nebulae by Kapteyn (1906). He considered his result of $+0''.0046$ for the absolute parallax preliminary, and did not doubt that improved data would lead to more reliable conclusions.

Wirtz realized that the one quantitative study for which the Strasburg nebular data had a singular use and quality was the measurement of nebular proper motions and the corresponding determination of the solar apex. A paper in the *Strassburg Annalen* (1912) with the title 'An Attempt Concerning the Cosmical Role of Nebulae' shows, however, that his apex determination was the means towards another goal – finding the cosmic role of the nebulae. In this sense, it was used even in Hubble's (1929) paper.

Wirtz (1912) wrote that he found 241 nebulae in common with the earlier visual list of Schultz (1874), Uppsala, which have position measurements of small internal errors and no systematic deviations with respect to the Strasburg data. The purpose of his comparison was to check whether the 30 years of epoch difference were sufficient to give reliable results. Soon after conclusions had been reached, photographic observations by Lorenz (1911), Heidelberg, became available, and were used by Wirtz towards the same end. Wirtz defined 'proper motions' as the differences in nebular positions at the two epochs divided by the time interval, and found:

> – a large deviation of the nebular apex, RA = 27°, Decl. = −5° for Strasburg–Uppsala, and RA = 90°, Decl. = −3° for Heidelberg (corrected to Strasburg)–Uppsala, from the stellar apex at RA = 270°, Decl. = +30°.
> – differences of proper motions well within the measuring errors of the mean Strasburg parallax (−0.″0020 ± 0.″0024) which included Kapteyn's value (+0.″0046 ± 0.″012).

In Wirtz's own, as always, careful phrasing,

> Considering the uncertainties of the values accepted as proper motions, the non-uniform distribution of the material with respect to the normal [reference] positions and over the sky, and the differences in the nebular individuals used in the two cases, the supposition nevertheless gains some justification that the approximate agreement in several cases could be due to more than an accident. An extension of the Lorenz series of all 500 nebulae of Schultz may lead onward; in the present stage, a decision cannot be made as to whether the computed nebular drift is real or accidental.

Five years later, after more photographic material had become available, Wirtz continued with two more papers on 'The Drift of the Nebular Spots' (1916a, 1916b). His conclusions are:

> 1. The large mass of nebulae drifts towards an apparent vertex in RA = 245°, Decl. = −3°. The apparent direction of the dominant nebular motion is almost opposite to the parallactic motion of the fixed stars.
> 2. The *gaseous*[4] nebulae have the same parallactic motion as the fixed stars; they are located at the limit of the solar star cluster which surrounds our position in space.

In view of the opposition which arose immediately after his publications, one should mention the comment with which he concluded his nebular apex determination (1916b):

> One objection which may be raised against the general nebular drift shall not go unmentioned: let us assume that the nebulae remain at rest, while

[4] Italics by the present authors.

> the comparison stars move and we interpret the measured motion as nebular motion, then one sees that the result will lead us to the solar antapex, if we assume fixed stellar positions at any given epoch. Even when one tries to get stellar positions nearest to the epoch of nebular measurements, it is certain that this objection will always remain in all such investigations. This is not dangerous to the basic nature of our results, but one must expect that to this remainder of unrecognized proper motions of the comparison stars must be attributed some of the contradictions which we found and which one will find again in future investigations.

The first objection, raised by Kobold (1916) refers to this point when he writes:

> Only with a complete knowledge of the proper motions of the reference stars would one obtain, even with entirely error-free micrometer measurements, reliable values for the proper motions of the nebulae, and since it is not expected that this requirement can be fulfilled with the present data one will obtain systematic errors in the nebular motions due to the motions of the reference stars; this then has to be taken into account when the nebular motions are derived.

The second objection by Seeliger (1916) appears to be biased by the preconceived notion that all nebulae are members of our own stellar system. One point, however, is of interest. In order to account for a constant error term in the measuring series, Seeliger follows the procedures mentioned above which, instead of correcting the data independently, take care of the effects of random and systematic errors through additional unknowns introduced into the reduction process.

The answer to these objections was given by Wirtz (1917) in a further paper, written when he was already serving in the German army and stationed in Berlin. He adopted Seeliger's correction constants in Airy's equations, but the results still led to basic differences between the apex obtained from gaseous nebulae and the apex found from what he now called 'small nebulae'. He accepted the state of rest of the small nebulae as one possible result of such investigations. Not a useless one, however: the differences in kinematic behaviour of the two types of nebulae is one more piece of evidence for their basically different nature. Furthermore,

> Should the small nebulae have to be considered at rest relative to the system of fixed stars, we would even by this gain a valuable hint; because the small objects which can be centered reliably on the photographic plates would yield a new reference system and zero points for the studies of stellar motions, as would be apparent even from the few not very accurate micrometer measurements.

In spite of Wirtz's repeated references to this point (the last one on the occasion of the Fifth General Assembly of the International Astronomical

Union in 1935, see below), and subsequent suggestions by others, this was not taken up actively until 1940 (Zverev 1940 and further work in Russia), while the first results of the Lick programme, started by Wright in 1950, appeared in 1987 (Klemola, Jones and Hanson),[5] together with the most recent determination of the solar apex from proper motions (Hanson 1987).

20.7 Nebular motions and apex determinations from radial velocity measurements

In spite of the fact that the First World War was in its fourth year and Wirtz was still serving in the army, he published a new paper on the 'Motion of the Nebular Spots' (1918b). The first part is still devoted to the proper motions. His sentence, 'Then one would have with the nebulae an excellent reference system for stellar motions . . .' appears much more affirmative than his earlier suggestions. The most important section, however, is devoted to a discussion of the radial velocities of nebulae. Surprisingly, the *Publications of the Astronomical Society of the Pacific* had reached Wirtz by December 1917 with new and valuable information on nebular radial velocities (while de Sitter in July 1917 had lacked these data which were to become so important in the context of his relativistic model of the universe).

With the paper 'The Relation of the System of Stars to the Spiral Nebulae' by Paddock (1916) in the said *Publications*, Wirtz became aware not only of the then large number of 15 measured radial velocities of nebulae but also of the solutions for solar motion including a constant *K*-term in Airy's equations. Paddock (following Truman and others) wrote:

> These objects [nebulae], however, can hardly be considered to form a unitary system of associated objects, for it must be noted that the average velocity of each of the three groups [at different locations in the sky] is decisively positive, which means that they are receding not only from the observer or star system but from one another. Accordingly a solution for the motion of the observer thru [*sic*] space should doubtless contain a constant term to represent the expanding or systematic component whether there be actual expansion or a term in the spectroscopic line displacements not due to velocities.

Two statements of Wirtz in his paper (1918b) are of interest. The first paragraph concerns the solar velocity relative to the apex and the *K*-term, the second one relates to the philosophy of observational science:

> It is remarkable that our system of fixed stars should have such an

[5] A somewhat misleading historical review is given by Vasilevskis (1973) where it is claimed that Wright, who initiated the Lick proper motion programme in 1950, *conceived* the idea in 1916. A detailed account is given of how in 1951 he tried to confirm that at this early date and for many years afterwards he 'never heard of anyone else proposing that extragalactic nebulae be used for the purpose'.

incredibly large displacement of 820 km/sec, and equally strange is the interpretation of the systematic constant[6] k = +656 km. If we give this value a literal interpretation it means that the system of spiral nebulae relative to the momentary position of the solar system as center disperses with a velocity of 656 km ... One will criticize the attempt made here to understand the characteristic arrangement and motion of the nebulae, and argue that everything is built on material which is much too incomplete and much too uncertain. True! Against this, however, I hold two arguments which are actually one. On the one side, experience shows again and again that the law of large numbers emerges already at remarkably small quantities of things, and then, W. Herschel deduced his apex of 1783 (with RA = 262°, Decl. = +26° for 1900) with only 13 stars.

20.8 The velocity–magnitude relation and other interrelations

Wirtz's paper 'Some topics concerning the statistics of spiral nebulae and globular clusters' (1922a) appears to be the largest single step towards understanding the role of redshifts in cosmology. The paper begins as follows:

> At present the radial motions of 29 spiral nebulae are available in the literature. Though it is definitely expected that this material, being as scarce as it is valuable, will gradually grow, it does not seem premature to search for rules, even in these few data, which point to systematic interrelations. In other ways, and from different kinds of material, common properties have already been recognized and have been worked into a hypothesis of the structure of the universe.

The latter may be a reference to Shapley and Shapley (1919),[7] who were the first to point out a general connection between apparent magnitude and redshift. This early source is mentioned by Fernie (1970) who is one of the authors to give full credit to Wirtz.

Wirtz's determination of the solar apex motion and the *K*-term using the new radial velocities is extended into a discussion of the error distribution of the individual nebular motions:

> If one computes from *X*, *Y*, *Z*, *K* the radial motions of the 29 nebulae, then the error distribution agrees well with the theoretical frequency curve ... These errors may be taken as the individual radial motions of

[6] The unit km instead of km/sec is frequently used by Wirtz after he has once introduced the proper unit – an idiosyncrasy also found in German signs for speed limits!

[7] The Shapleys write:

> The speed of spiral nebulae is dependent to some extent upon apparent brightness, indicating a relation of speed to distance or, possibly, to mass. The six spirals with smallest radial velocity, including all of those with negative values, are not exceeded in brightness by any spirals in Holetschek's list of visual magnitudes...
>
> The arithmetical mean of these six velocities is ±188 km/sec; their algebraic mean is +49 km/sec, while for the other 19 spirals... it is fifteen times as large – that is +726 km/sec. Only three spirals besides these six bright ones are now known to have radial velocities of less than +500 km/sec.

> the spiral nebulae; they are free from solar motion and the redshift common to all these objects.

A search for correlations between the individual motions and other properties of the nebulae yielded one with galactic latitude (now known to be an artefact due to selection effects) and a strong one with magnitude:

> Very pronounced is the parallelity of v with the total magnitude *Mg* [the table given by Wirtz at this point is shown graphically as figure 20.2]. The sequence of absolute v-values means nothing. In the mean values obtained using the signs, however, an approximately linear relation is present in the sense that the spiral nebulae nearer to us have a tendency of approaching, the more distant ones a tendency of receding from our Milky Way system. . . One finds, of course, analogous behaviour also in the diameters of the structures, where a positive change of distance belongs to small diameters and a negative one to large diameters.
>
> All these statistical features superimpose on the most striking major process, which can be described as an expansion of the system of spiral nebulae with respect to our own position . . . and the dependence on magnitude indicates that the nearer or the more massive nebulae show less expansion than the distant nebulae or those of lesser masses.

Here we have the first explicit velocity–distance relation. The reference to nebular masses (as in Shapley and Shapley) implies the existence of a unique mass/luminosity relation.

Another important conclusion of Wirtz's paper is the close agreement between the apex positions obtained from spiral nebulae and from globular clusters, whose membership in our stellar system had been established (Shapley 1918). Furthermore, Wirtz determined the solar motion to be −712 and −373 km/sec, respectively, while the *K*-terms are +870 and −55 km/sec. These results are early observational indications (see also Lundmark 1920; Strömberg 1925) of the fact that the solar apex motion measured from the radial velocities of spiral nebulae and globular clusters reveals the galactic rotation at the location of the sun, as was subsequently shown by Oort (1930).

A brief note by Wirtz (1922b) refers to correlations between radial velocities and other nebular properties. For a suggested dependence on inclination he finds no support; a new suggestion concerning the winding of spirals had to be abandoned later on.

20.9 The velocity–log diameter relation

Wirtz (1924a) in his paper 'De Sitter's Cosmology and the Radial Motion of the Spiral Nebulae' discusses the correlation between the logarithms of nebular diameters and radial velocities in terms of cosmological theory. With references to de Sitter's (1917) original contribution, its technical discussion by Eddington (1923), from which Wirtz's basic

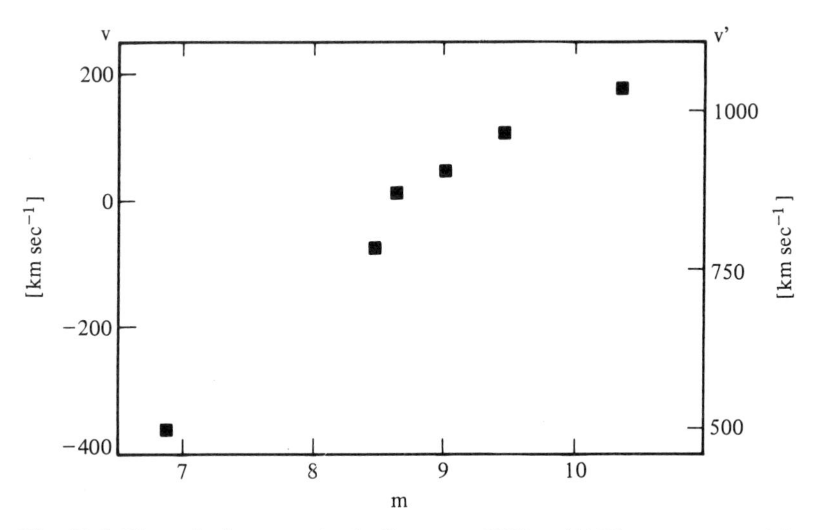

Fig. 20.2 The velocity–magnitude diagram of Wirtz (1922) reconstructed from his tabular data. Right, the observed average radial velocities; left, the velocities after subtraction of the mean K-value.

description of the de Sitter model is taken, and the more popular presentation by Weyl (1924), Wirtz points to the dual cause of expected nebular redshifts in de Sitter's case B universe (empty, stationary, hyperbolic):

> First the general scattering of matter towards the mass horizon, and then the spectral shift towards the red which appears in distant objects even when they are at rest with respect to the origin of coordinates, due to the slower course of time.

Here, neither origin nor horizon are fixed. According to Eddington (1923),

> we must recall that by the symmetry of the original formula . . . any point of space and time could be chosen as the origin with similar results. Thus there can be no actual difference in the natural phenomena at the horizon and at the origin.

Taking the redshift data assembled by Eddington and nebular diameters from various catalogues, Wirtz investigated the correlations between the measured quantities. He wrote:

> de Sitter pointed already to the well-founded observations of large redshifts in the spectra of three spiral nebulae, which, computed as Doppler effect, lead to average velocities of +600 km. From de Sitter's theory results, however, not only the fact of redshifts, but also their growth with increasing distance, and this for two reasons. It must become apparent from observations so taken that the Doppler radial velocities assume larger and more positive values with increasing distance of the

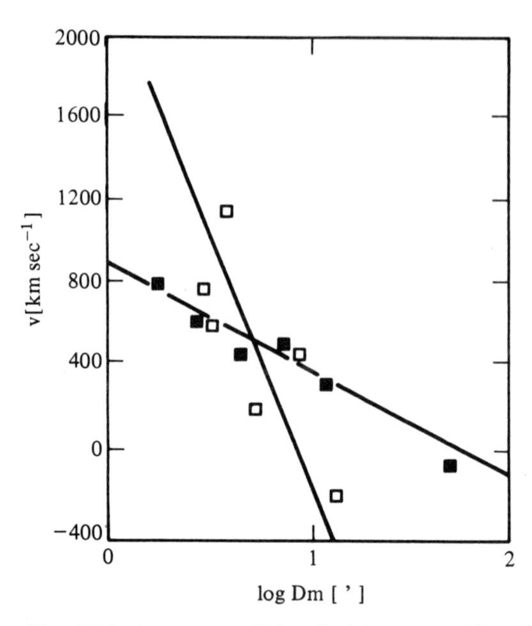

Fig. 20.3 Average radial velocities versus logarithms of the angular nebular diameters. Filled symbols: obtained with v as argument; open symbols: obtained with $\log Dm$ as argument. The two lines are the regression lines, $v = +479 \log Dm$ and $\log Dm = 0.96 - 0.000432v$, respectively. The data are given by Wirtz (1924a).

> spiral nebulae. Though nothing is known about the distances of the spiral nebulae, one can, under the assumption that on the average the linear diameters are equal, use their apparent diameters as a measure of distance. In de Sitter's cosmology the radial motion should increase with decreasing apparent diameter.

The table following in the text is presented here as figure 20.3; his regression lines are inserted. The tabulated mean values can be traced back to the original data in the references given. A diagram using these data had obviously been constructed by Wirtz, as is apparent from the following text, but it was not published. A reconstruction is shown in figure 20.4. (Figures obtained with the same data are also given by Priester and Schaaf 1987.) Wirtz described his graph as follows:

> When the different quantities v versus log Dm are represented graphically, the diagram shows a V-shaped or triangular form from which one deduces the following facts: for the apparently small nebulae one finds smallest and largest v, those with the apparently largest values have the smallest v, among the nebulae with small v one finds large and small objects, large nebulae with large v do not exist.
>
> From this one deduces that the dispersion of the linear dimensions of the nebulae fills the triangular plane in such a way that among the near nebulae absolutely small and large objects are visible while in the depth of space only the absolutely largest are subject to observing their radial

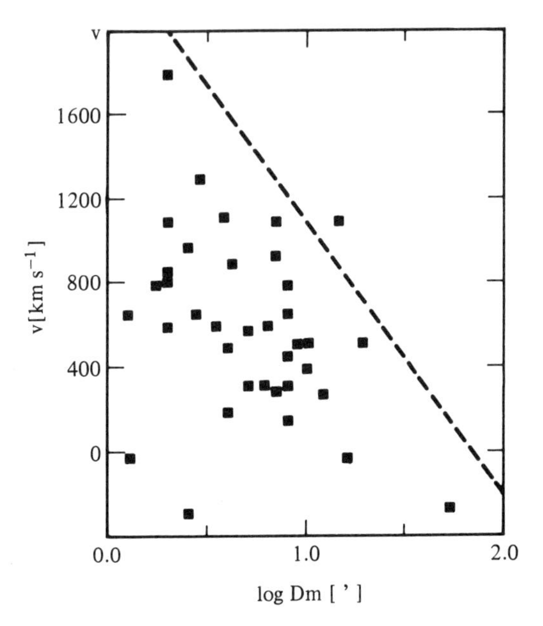

Fig. 20.4 The velocity–log diameter diagram, reconstructed from the data used by Wirtz (1924a).

> motions. The progression between v and apparent Dm is reproduced best through the hypotenuse of the triangle inclosing all observed points, which follows the absolutely largest nebulae under the assumption that the giants among the nebulae have the same average extent at all distances.

In this statistical analysis, Wirtz used the radial velocities v and the logarithms of the major axes of the nebulae, 'for obvious reasons not the numeri, but log *Dm*'.

This apparently refers to the use of diameters instead of magnitudes as distance indicators. Thus, in order to reproduce the apparently linear relation between radial velocities and magnitudes derived earlier (Wirtz 1922a), he had to use the logarithms of diameters. From the new data Wirtz found:

> For the [absolutely] largest nebulae the graphic presentation then yields the formula
>
> $$v(\text{km}) = 2200 - 1200 \cdot \log Dm \quad [Dm \text{ in arc minutes}].$$
>
> The coefficient 1200 indicates, that with a tenfold increase of distance the redshift computed as pure Doppler effect increases by 1200 km. This rule of thumb has one advantage: it does not come into conflict with the velocity of light, because the real and apparent radial motions which result from the properties of the de Sitter space then increase only very slowly with distance; and a redshift from which, in terms of the Doppler

effect, the velocity of light follows, would be reached only at distances which lie far beyond the opinions now held about the positions of the celestial bodies – approximately 10^{200} parsec, a quantity which exceeds also all estimates of the radius of curved space.

The correlation in terms of logarithms had the further advantage that it was well defined. A plot of v versus $1/Dm$, using the same data, appears more like a scatter diagram from which no conclusion can be drawn.

Summarizing the contents of his paper we come to the conclusion that Wirtz substituted the theoretical velocity–square distance law given by Eddington (1923) by an empirical velocity–logarithmic distance law of the observer. It is the best fit for the data available to Wirtz and guaranteed that the scale of the universe is large enough not to lead to a conflict between the special theory of relativity and possible future observations.

20.10 Wirtz's use of the linear velocity–distance law

The Festschrift dedicated to Hugo von Seeliger on the occasion of his 75th birthday concerns 'Problems in Astronomy'. One of the contributions, written by Wirtz (1924b), is devoted to 'Globular Nebulae, Spiral Nebulae and Surface Brightness'. Wirtz discusses the statistical behaviour of nebulae in the light of the evolutionary theory of Jeans (1919) with the following reservation:

> One can then check through the statistical behaviour of several directly observable characteristics, whether the sequences and transitions from one kind to another are those required by theory. No doubt, the conclusions from statistics are not conclusive . . . But these considerations apply to any cosmogonical investigation, apply to the whole of science. There exist neither compelling objections nor convincing proofs.

The first part of Wirtz's contribution concerns correlations of physical parameters of nebulae and will be discussed below. Here, his sections on nebular motion and distance determinations are of interest. Impressed by the suggested accuracy of van Maanen's proper motion measurements of nebulae, and obviously remembering his own unsuccessful attempts in this field, Wirtz accepts the resulting distance of spiral nebulae, 21,200 pc, with reserved confidence. Looking for support from other evidence, he writes:

> It concerns the important, most certainly observable, conclusion from the cosmology of de Sitter that an object must show a Doppler effect which depends on distance. Through this one can, in the reverse, determine the distance, when the radius of curvature of the de Sitter space-time world is known. L. Silberstein (*Nature* Vol. **113**, p. 350, 1924) has tried a derivation of the value of R using 7 globular clusters . . . he now obtains the radius of curvature of the world $R\ldots = 29 \times 10^6$ parsec.
>
> For 5 . . . spiral nebulae [introduced in the discussion of van Maanen's work] V. M. Slipher furthermore measured radial motions, whose mean

is |260| km. Thus one finds, using R, a mean distance of these spiral nebulae of 25,000 parsec.

Silberstein (1924) provided more than a numerical value for R. He wrote:

> I derived from de Sitter's cosmology a formula for the Doppler effect $D = d\lambda/\lambda$ of a star in purely inertial relative motion, radial and either from or to the observer.
>
> This formula practically amounts to
>
> $$D = \pm(1 - \cos^2 \sigma/\gamma^2)^{1/2}, \quad s = r/R,$$
>
> where R is the curvature radius of space time, r the distance of the star at the moment of observation, and γ an integration constant which is but little larger than unity ... For distant objects for which $\cos \sigma$ is at all distinguishable from 1, the constant γ may be replaced by its limited value 1, so that the formula becomes $D = \pm \sin \sigma$ or, to all purposes, for small or moderate σ
>
> $$D = \pm r/R.$$

This linear relation was used by Wirtz to obtain the nebular distances given above, though he does not state this explicitly. At the same time he suggests, that at this stage the near agreement between the results from proper motions and radial velocities is not yet more than a 'remarkable accident'.

20.11 From Wirtz to Hubble

A few months làter than Wirtz (1924a), Lundmark published the analysis of a slightly larger set of redshift data and their correlation with *actual* nebular distances, determined by himself. Here, in 'The Determination of the Curvature of Space-Time in de Sitter's World', Lundmark (1924) presents the first radial velocity–distance diagram (with radial velocities plotted irrespective of sign, following Silberstein). Contrary to Wirtz and to Hubble later on, he does not primarily try to verify a model but assumes the model in order to determine the radius of space curvature. He comes to the conclusion that the data do not yet yield a reliable numerical value.[8]

In 1927 Wirtz's student August Dose presented an almost unchanged

[8] The problem was first addressed by Schwarzschild (1900) in his paper 'On the Allowable Measure of the Curvature of Space'. His conclusions are:

> Summarizing one may, without contradiction to experience, conclude that the world is contained in a hyperbolical (pseudospherical) space with a curvature radius of more than 4 000 000 radii of the Earth's orbit, or in a finite, elliptical space with a radius of curvature more than 100 000 000 radii of the Earth's orbit, whereby, in the latter case, one assumes an absorption of light of 40 magnitudes during one orbit around space.
>
> With this we must now content ourselves.

It should be emphasized that Schwarzschild used the then available data to obtain *lower limits*.

version of the v–log D relation of 1924. It shows that little progress had been made during the intervening years. This conclusion is corroborated by the fact that only two papers of Lundmark (1924, 1925) bear the signs of orginality. The first one is discussed above, and in the second paper he presents the *direct* inclusion of a *variable* K-term in Airy's equation with three terms: independent of distance r, and dependent on (increasing with) r and r^2.

Wirtz's review (1928) of Lundmark's (1927) extensive paper and the comparison with his own work is of interest.

20.12 Hubble's apex determination and his velocity–distance relation

The results in Hubble's (1929) paper 'The Relation Between Distance and Radial Velocity Among Extra-Galactic Nebulae' are often considered to be the compelling observational contribution towards verification of the Friedmann–Lemaître models. The original interpretation of the data by Hubble, however, was still in terms of the de Sitter universe.

Of the radial velocity data for the 24 nebulae used by Hubble most agree with those in the discussion by Lundmark (1924). The differences then between the 'Lundmark' and the 'Hubble' diagram rest only on the distance determinations.

Hubble determined the solar apex and velocity, giving the K-term a linear dependence on distance, Kr. The resulting value for K is 513 ± 60 km/sec per 10^6 parsecs (the 'Hubble constant'). A linear fit to Lundmark's (1924) data, with the line passing through the origin of coordinates, would have given a roughly 7 times smaller value with the low correlation coefficient 0.3. Hubble's paper concludes with the following:

> The outstanding feature, however, is the possibility that the velocity–distance relation may represent the de Sitter effect, and hence that numerical data may be introduced into the discussions of the general curvature of space. In the de Sitter cosmology, displacements of the spectra arise from two sources, an apparent slowing down of atomic vibrations and a general tendency of material particles to scatter. The latter involves an acceleration and hence introduces the element of time. The relative importance of these two effects should determine the form of the relation between distances and observed velocities; and in this connection it may be emphasized that the linear relation found in the present discussion is a first approximation representing a restricted range in distance.

In summary, the sequence of observational correlations between redshift and distance between 1920 and 1930 is:

> 1922 and 1924: Wirtz is first in looking systematically for correlations between the radial velocities of spiral nebulae and their

relative distances. The data can be interpreted in terms of the de Sitter model.

1924: Lundmark presents the first velocity–distance diagram and a solution of Airy's equation which includes a quadratic dependence of the K-term on distance. The data are insufficient for the attempted derivation of a numerical value for the curvature radius of the de Sitter universe.

1929: Hubble uses new distance determinations for the nebulae which lead to a linear dependence of the K-term on distance and the numerical value for K in units velocity per length with a small error. An interpretation in terms of the de Sitter universe is still envisaged.

20.13 Observations and theory in early relativistic cosmology

The de Sitter universe provided a characteristic property, which could be tested observationally: the redshift. This explains the success of the de Sitter over the Einstein model, in spite of the unrealistic requirement of the universe being massless. Also, the former could account for the disturbingly large redshifts of nebulae, the latter could not. Observational cosmology between 1917 and 1930 meant the interpretation of data in terms of the de Sitter model.

In this context it should be noted that an empty (dustless) universe can have the metric originally used by de Sitter, or a time-dependent metric but constant radius of curvature at all times (Lanczos 1922; Lemaître 1925; Robertson 1928), or a metric in which the radius of curvature itself is time-dependent (Friedmann 1922, 1924; Lanczos 1923; Lemaître 1927; Robertson 1929).[9]

[9] In the first case the universe is stationary (Friedmann's terminology) in the sense that all spatial distances remain constant. However, according to Eddington (1923), 'a particle at rest will not remain at rest unless it is at the origin; but will be repelled from the origin with an acceleration increasing with the distance. A number of particles initially at rest will tend to scatter unless their mutual gravitation is sufficient to overcome this tendency.' This constitutes one part of the observed wavelength shift; the other is due to the $\cos^2 x_1$ factor of x_4 in de Sitter's metric (x_1, x_2, x_3 are the space coordinates and x_4 is the time coordinate), which produces a slowing down of time between the origin and the horizon R and thus leads to a redshift. The universe, though hyperbolical, is of finite size, $v = c$ at the horizon.

In the second case, time is introduced into the factor accompanying the spatial coordinates by formally assuming a matter term whose influence is cancelled through the presence of negative pressure. A test particle remains at rest while the spatial distances grow; because it does not move towards the horizon it does not experience the slowing down of time. To reconcile the growth of spatial distances with the constancy of R, the curvature radius must be infinite while the observable horizon is again defined by $v = c$. Robertson (1928) described the model in the following way:

> That the properties of the manifold are, however, independent of t...
>
> A perhaps more serious objection is that the world is on this interpretation of infinite extent, but, as will be seen later, its closed character is maintained in the sense that the only events of which we can be aware must occur within a sphere of finite radius.

The third case is the Friedmann universe with $R = R(t)$ (i.e. *the properties of the manifold* are *time dependent*) for the limiting case of zero density. A test particle remains at rest while the metric

Observers in the 1920s generally referred to the original de Sitter model, thus accepting real particle motion and a finite universe with observable curvature radius. Even so, they were provided with a choice of relations between the nebular velocity v/c, the nebular distance r, and the curvature radius R.

Eddington (1923) first derived an approximate relation under the assumption that all particles are originally at rest with respect to the observer. He obtained $v/c \propto r^2/R^2$. The same assumption and proper integration over the particle acceleration yields $v/c \propto r/R$ (Tolman 1929; de Sitter's 1930a approximation; see also North 1965). It also agrees with the approximation in Silberstein's (1924) derivation. Tolman (1929), with no restriction concerning the original particle velocities and proper integration, found $v/c \geqslant r^2/R^2$. Thus, when Wirtz, Lundmark and Hubble wrote their papers, *both linear and quadratic relations* between velocity and distance were known within the framework of the original de Sitter model. Both were compared with observations. Wirtz was the only one to advocate the velocity–logarithmic distance relation obtained directly from data and independent of theoretical predictions.

Friedmann did not provide an explicit velocity–distance law for the observer. Lemaître's (1927) relation in the non-stationary universe is

$$v/c = (R_{\text{obs}}/R_{\text{emit}}) - 1,$$

and his approximate solution for small distances r

$$R'/R = v/rc$$

with $R' = dR/dt$. It should be noted that R' is measured in fractions of c. The corresponding equation for R' measured in units of velocity, i.e., according to Friedmann, is

$$R'/R = v/r = H_0$$

and defines the Hubble constant. The numerical value given by Lemaître of $R'/R = 0.68 \times 10^{-27}\,\text{cm}^{-1}$ becomes

$$H = 630\,\text{km/sec/Mpc}$$

in Friedmann's notation and Hubble's units and is the first observational value obtained for the Hubble constant.

Three years later de Sitter[10] accepted the Lemaître models as those most

distances as well as the curvature radius grow. There is no restriction to the size of R. This model, with negative vacuum energy (Lemaître 1934), has gained interest in connection with the inflationary universe, where it describes the transitory 'de Sitter phase' (see e.g. Kolb et al. 1986).

The confusion of definitions exists far beyond the acceptance of the Friedmann–Lemaître universe, and is well exemplified by Whitrow (1959): 'Such models [Einstein's and de Sitter's] are often referred to as 'static', although strictly speaking we now realize that the de Sitter model is best regarded as the limiting form of an expanding universe in which the mean density is everywhere zero'.

[10] Willem de Sitter is certainly one of the most remarkable cosmologists. His early papers followed

suitable for the interpretation of data. While working on the details of a paper of which he had presented a brief and preliminary version at a meeting of the Royal Astronomical Society in early 1930 (for details, see Hetherington 1973) de Sitter had been referred to Lemaître's publication by Eddington. Here, it suffices to mention that at the meeting de Sitter 'was unable to account for this coincidence', i.e., the equality of the numerical values for R when derived from the Einstein model (A) and the de Sitter model (B), while in his written version (de Sitter 1930a) he had found the answer:

> Notwithstanding the remaining uncertainty of the values [$R_B = 2 \cdot 10^{27}$ cm] and [$R_A = 2.3 \cdot 10^{27}$ cm] it is thus certain that R_A and R_B are of the same order of magnitude. Consequently, in the solution (B) $\varrho = 0$ is not an approximation, and we thus come to the conclusion that neither solution (A) nor (B) can correspond to the truth, (A) being excluded by the large positive systematic velocity V ... and (B) by the finite density.

De Sitter then introduces Lemaître's $R = R(t)$, starting with a finite value for the initial R.

Following this, de Sitter (1930b, 1931) showed graphically for both expanding and oscillating universes the relation between two quantities which are, except for an additive constant and a factor, identical with respectively redshift z and time t. Cosmic time became an observational quantity within the framework of a chosen model.

20.14 Evolution and the acceptance of the Friedmann–Lemaître universe

The acceptance of cosmic time as a *measurable quantity* seems to us to be the major step towards the acceptance of the expanding universe. Because of its importance for observational cosmology – although at a time when Wirtz had already withdrawn from it – a small digression into this problem may be permitted. Our brief remarks concentrate on time and evolution and do not attempt to rival the extensive accounts of the early days of modern cosmology, especially by North (1965) and Smith (1982), but also the reviews on relativistic cosmology by Robertson (1933), and the historical evaluation by Ellis (1988).

Einstein's (1915) introduction of the general theory of relativity within less than a year. (Karl Schwarzschild was even faster, but died in May 1916, a loss that appears to have severely influenced the course of cosmology, as might have the early death of Friedmann in 1925.) Two models of the universe carry de Sitter's name: the dustless de Sitter universe (see note 9) and the Euclidean Einstein–de Sitter universe (1932). His efforts to connect observations with his first model are as impressive as his immediate acceptance of the time-dependent radius of curvature soon after he had become acquainted with Lemaître's work. As it seems improbable that the earlier work of Friedmann, published in one of the major journals of physics, had escaped his attention (contrary to Lemaître's original paper which was hidden in the *Bruxelles Annales*) we must conclude that de Sitter in 1922 was not yet ready to accept changes in the scale of the universe.

In Minkowski's (1909) well-known lecture in Cologne, he stated: 'From now on space by itself and time by itself shall be reduced to shadows of themselves and only a kind of union of the two shall be kept in existence.'

Throughout the 1920s and beyond, however, the understanding of time as a physical quantity, measurable by evolutionary processes on the scale of the universe, itself went through an evolutionary phase. In order to trace this process we shall begin with Einstein (1917): 'The character of space curvature is, depending on the distribution of matter, variable locally and with time, but on a large scale it can be approximated by spherical space.' Expressed in the words of de Sitter (1917) this means that in model A (Einstein model), based on the 'material postulate of relativity', 'Time is the same everywhere and always'. It is apparent that time measured locally is different from universal time: the former progresses measurably, while universal time does not; the evolution of stars and nebulae (e.g. Jeans 1919) is possible, the evolution of the universe as a whole is not.

For model B (the de Sitter model), based on the 'mathematical postulate of relativity', de Sitter finds that 'there is no universal time, no difference between coordinates in 4-dimensional space, no physical meaning of the coordinates'. In this case, evolution of the universe does not occur, because evolution concerns a physical reality which the coordinates of space-time do not possess.

The growth of physical space-time and the associated change of the curvature radius, as first proposed by Friedmann (1922), was not acceptable. The neglect of Friedmann's work and the attribute *mathematical* appears to rest on his concept of time. He introduced a *beginning* (his monotonous world of the first kind) or ever-returning cycles (his monotonous world of the second kind); and the only test he offered was to look for the 'time scale of the universe'. The fact that he quotes his own surprisingly good estimate of ten billion years together with the modest remark 'but these numbers can, of course, only serve as an illustration for our computations' may have given an excuse for calling his work a purely mathematical exercise.

With respect to Friedmann's work it should also be noted that its rejection by the leader in the field could have been detrimental. The fact that Einstein (1922) claimed to have found an error in the derivation of Friedmann's equations may have throttled interest before it could actually build up, and Einstein's (1923) quick acknowledgement that *he* and not Friedmann had made the error had no noticeable effect. The brief literary discussion may indeed have strengthened the opinion, held until today, that Friedmann's work was a mathematical exercise which could be *computationally* refuted or accepted.

The introduction of time-dependent spatial dimensions in lieu of the original de Sitter metric (see note 9 above) replaced one unrealistic (empty) universe by another – but it avoided the notion of physical evolution of the universe. Eddington (1931), even while proclaiming Lemaître's expanding universe (but assuming the finite initial radius given by de Sitter 1930a) confessed that 'philosophically, the notion of a beginning of the present order of nature is repugnant to me'. Being aware of the fact that he was asked as a mathematical physicist, not as a philosopher, he disclosed 'the beginning' as an artefact of our laws of physics. Furthermore, he swept away all calculable evolution of the future with Heisenberg's uncertainty principle:

> A discussion of the properties of time would be incomplete without a reference to the principle of indeterminacy . . . This change of view seems to make the progress of time a much more genuine thing than it used to be in classical physics. Each passing moment brings into the world something new – something which is not merely a mathematical extrapolation of what was already there.

Lemaître (1931) responded with his note 'The Beginning of the World from the Point of View of Quantum Theory' where he introduced what was to become known as the primeval atom and where he described the beginning of the universe as incalculable:

> Now, in atomic processes, the notions of space and time are no more than statistical notions; they fade out when applied to individual phenomena involving but a small number of quanta. If the world has begun with a single quantum, the notions of space and time would altogether fail to have any meaning at the beginning; they would only begin to have sensible meaning when the original quantum had been divided into a sufficient number of quanta. If this suggestion is correct, the beginning of the world happened a little before the beginning of space and time.

Quantum theory came to the resuce of evolutionary cosmology in relieving cosmologists of the obligation to describe the beginning and to predict the future. Thus reduced, Friedmann's universe, by then supported through the work of Lemaître and Robertson (who was the first to quote Friedmann and did so with some criticism), became acceptable. In October 1931, a Supplement to *Nature* presented a discussion of the British Association on the topic 'The Evolution of the Universe'.

Observations became important for the acceptance of the expanding universe only *after* the notion of a time-dependent universe was no longer rejected. In hindsight, the work of Wirtz (1922a, 1924a) appears to have been fundamental. He was the first to use sufficient material to look for a correlation between distance and redshift and to formulate an empirical law within the general framework of theory. Hubble's analysis did not *induce* the acceptance of the expanding universe, because it was well within the predictions of the de Sitter universe, but it became a stepping stone for the evaluation of the Friedmann–Lemaître universe. De Sitter's main argument in favour of the expanding universe, the fact that the observed space density is too large for the approximation $\varrho = 0$, proved to be convincing. The excellent observations of Humason (1931), which extended the available distance scale by a large factor, contributed greatly to the cosmological relevance of data.

20.15 Surface brightness and apparent total magnitudes of nebulae

One field of Wirtz's investigations, even less known than his contributions to the statistics of galaxy positions and motions, is the statistics of the physical properties of galaxies.

In 1910, when the Strasburg project of measuring nebular positions had been concluded, Wirtz started the ambitious task of measuring the *surface brightness* of spiral nebulae, gaseous nebulae and globular clusters with the 49-cm refractor. His project superseded the earlier measurements of surface brightness for 41 nebulae (Pickering 1900). The resulting 'Catalogue of Surface Brightness of 566 Nebulae' was published in *Lund Meddelanden* (1923). The leading role of Wirtz in this field was acknowledged by Curtis (1933):

> Our data in this field are due mainly to Wirtz. The valuable researches of Wirtz on the areal brightness of the spirals seem not to have received sufficient attention as yet, and certain of his numerous correlations may fittingly be repeated at this point . . .

In a series of papers, Wirtz (1924b, 1924c, 1924d, 1925, 1926) analysed photometric data in order to illuminate the nature and role of the non-galactic nebulae. One basic feature of his approach is the statistical sampling of data. He hoped to reduce biases by combining data: 'Several series of observations of spiral nebulae were found in the literature [and combined with his own data] each of which aims at systematic completeness in a different way.'

Another one is *computational statistics*, the determination of correlation coefficients between two or, in pairs of two, three parameters: surface brightness, diameter and axial ratio. Among his results are the luminosity function (discussed below), the magnitude of the Milky Way seen at 1 megaparsec, $6\overset{\mathrm{m}}{.}4$, and the 'conclusion of the statistical investigation of the *m* [apparent magnitude] and m_0 [apparent magnitude at the standard distance of one megaparsec = absolute magnitude]: Probability of the coordination of the spiral nebulae with the Milky Way system and limitation of the Milky Way of second order; near approach to its limit'. The last statement refers to Charlier's (1922) hierarchical universe and implies that the Milky Way, together with the then observed nebulae, constitutes a cluster of galaxies.

In his contribution to the Seeliger Festschrift, with the subtitle 'A

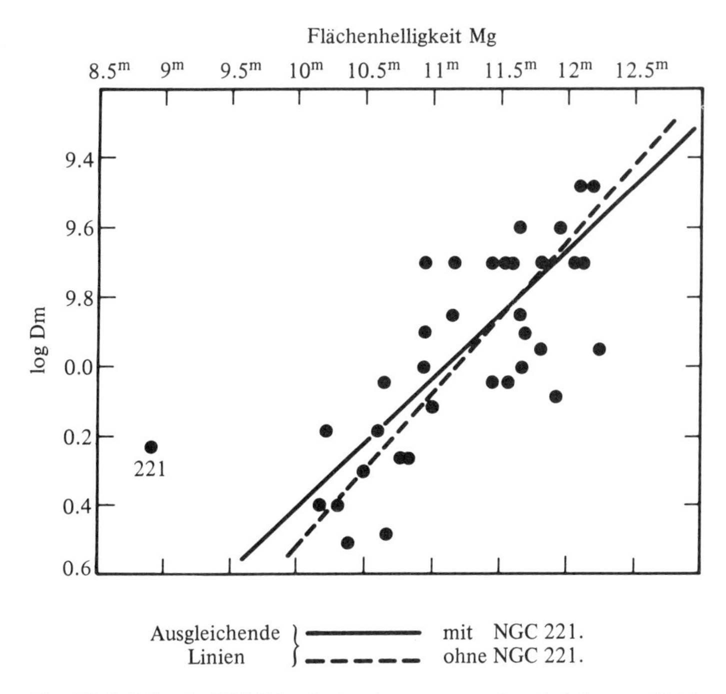

Fig. 20.5 Wirtz's (1924b) relation between surface brightness (*Mg*) and diameter (*Dm*) of globular nebulae (= elliptical galaxies).

Statistical Study', Wirtz (1924b) set out to find observational evidence for Jeans' theory of the evolution of non-galactic nebulae. Correlations of global parameters of the three classes of objects: globular nebulae (now called elliptical galaxies), spiral nebulae and globular clusters are obtained, and their comparison is checked against theory. From a modern point of view, the most interesting part is still (as already pointed out by Curtis) Wirtz's search for correlations between nebular parameters.

In figure 20.5 (one of the few figures actually published by Wirtz, and drawn not only for his own information), he compares surface brightness, which is independent of distance, with the logarithm of diameters, which is strongly affected by distance. The correlation between the two quantities cannot be attributed to distance but must relate to intrinsic parameters. With this investigation Wirtz introduced the *diameter versus surface brightness diagram* which is now considered important for the quantitative classification of galaxies (Kodaira, Okamura and Watanabe 1983): 'From the principal component analysis of photometric parameters, these two coordinates of the diagram are found to be most relevant to the study of characteristics of galaxies.' In view of this basic importance it appears less relevant that Wirtz found the wrong sign for ellipticals and the right one for spirals, and that he derived the confirmed fact (Kodaira et al. 1983) that the correlation for the latter is a much weaker one. The shortcomings of Wirtz's data as well as those obtained by others before the advent of two-

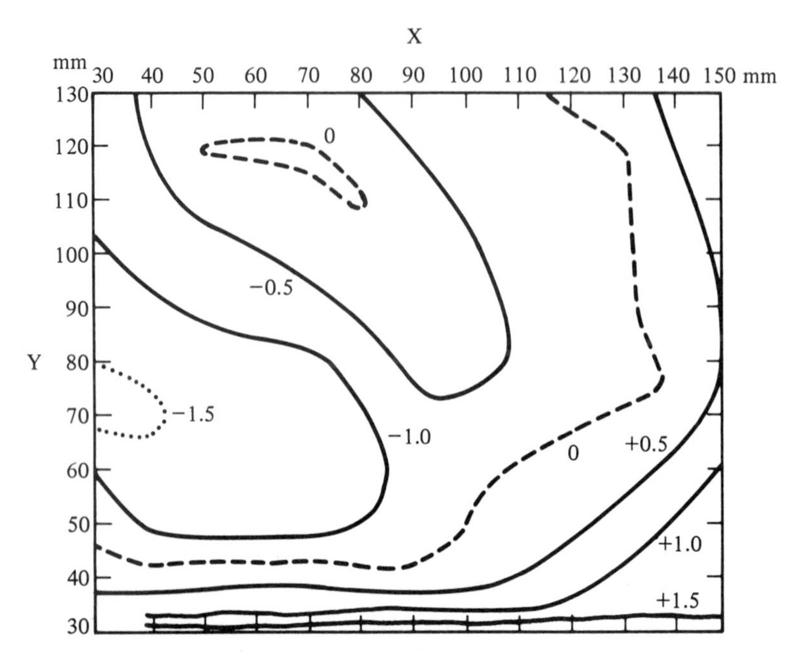

Fig. 20.6 Background isophotes of a photographic plate, taken from Wirtz's nebular photometry (1927).

dimensional electronic receivers are easily explained by the lack of suitable techniques for measuring surface brightness. Other correlations, such as the decreasing number of nebulae with decreasing galactic latitude, are now known to be due to interstellar absorption. Correlations under various aspects are discussed in more detail by Wirtz (1924c).

A surprisingly modern chapter is that of Wirtz's (1927) microphotometer measurements of photographic plates and the determination of total photographic magnitudes of nebulae. It reads not unlike the most recent microphotometer work, which, while now automated, is still confronted with the same basic problems. Figure 20.6 shows one of them, the uneven background density which has to be corrected before reliable magnitudes can be determined.

20.16 The luminosity function

In 1926, Wirtz published the first investigation concerning 'Total Brightness and Luminosity Curve of the Non-Galactic Nebulae, an Attempt'. He based his study on diameters and estimates of surface brightness of 1000 nebulae by Fath (1914), and on the assumption that the major axes of nebulae have, on the average, the same linear extent. The apparent diameters are then employed as distance indicators. They are calibrated with those of NGC6822, M31 and M33, whose absolute sizes are

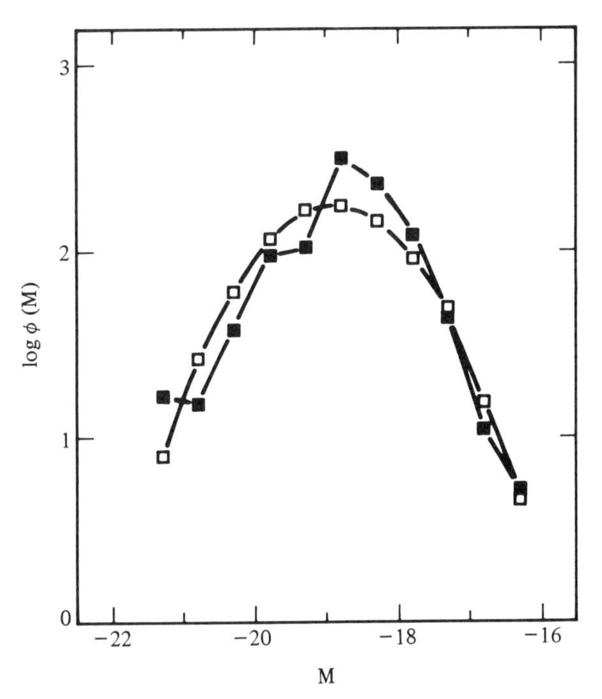

Fig. 20.7 The luminosity function reconstructed from the data given by Wirtz (1926).

known from their distances. These were determined by Hubble (1925a, 1925b) using the absolute magnitudes of Cepheid variables and other indicators. The luminosity function in figure 20.7 is plotted from the tabular values given by Wirtz, who wrote: 'These investigations include a portion of space which is made up from the 139 narrow pyramids of the Kapteyn areas. The luminosity function cannot yet be transposed to volume unit.'

In spite of this, the luminosity function, a near fit to a Gaussian distribution with a half width of a little more than 3 magnitudes, stands up well in comparison with the luminosity functions published five years later by Hubble and Humason (1931). The curve – by good luck or accidental cancellation of systematic errors – even holds up to modern values. It has been shown by Sandage, Binggeli and Tammann (1985) that the luminosity function of spiral galaxies (in the Virgo cluster) is indeed a Gaussian distribution with a half width of approximately 4 magnitudes.

20.17 Luminosity (mass) segregation in clusters of nebulae

In a photometric study of the Virgo-Coma group (the Virgo cluster in modern terminology) Wirtz (1927) noted:

> If the Virgo-Coma nebular group is a real physical cluster, then one may think that a peculiar distribution of magnitudes of the members must be

> revealed, like in the case of star clusters. In the manner, that the bright nebulae (the largest masses) are concentrated more towards the centre of the group than the faint ones ... The table teaches in three-fold repetition: there is really the tendency that the fainter nebulae press towards the border zones of the cluster, while the brighter nebulae dominate a broad central region.

The next statement to the same effect was made fifteen years later by Zwicky (1942). Since then it has been claimed (and contradicted) for a number of clusters. Recent numerical simulations support the process of mass segregation at different stages of cluster evolution.

20.18 Intergalactic absorption

References to intergalactic absorption are found in several of Wirtz's publications. In view of the fact that little is known, even today, his method and his results are of particular interest.

Wirtz (1924c) wrote:

> While all computations of the correlation coefficient for the nebular characteristics, surface brightness and diameter, give very low absolute values, uncertain in each individual case, one finds, however, from different material always the same negative sign, i.e. the intensity (surface brightness) increases with increasing apparent diameter. If one takes the apparent diameter as a measure of distance, one can interpret this effect as resulting from a general cosmic absorption. If one positions the nebulae, distant Milky Way systems, at distances which correspond to the parallaxes of K. Lundmark ... one can determine the amount of absorption. One finds, with the aid of the regression line, relating magnitude to the logarithm of the large axis, it to be extremely small, of the order of 10^{-5} magnitudes per 1000 parsec.

Around 1950, upper limits not much smaller than the value given by Wirtz in 1924 were quoted; the largest of them differs only by a factor 2. In the 1970s values of the order of 10^{-7} mag per 1000 parsecs were obtained by several groups (Margolis and Schramm 1977).

Following his computation of colour indices for 46 nebulae Wirtz (1927) addresses the problem of *selective* absorption in space. He had obtained from Holetschek's (1907) visual magnitude determinations (the same that were, corrected by Hopmann, used by Hubble 1929) and from his own photographic magnitudes, colour indices, which he considered individually unreliable, while the mean value appeared trustworthy within the given standard deviation:

> A notable fact would result from this. In the spectral classification of the Draper Catalogue FJ = $+0\overset{m}{.}94$ corresponds to the spectral type G8–K0. On the other hand, the few direct spectral determinations from the line character of the non-galactic nebulae yield a spectrum near G3. In this we

> would find new evidence for a weak absorption in space, which here appears to be selective in the sense that the nebulae appear redder by a smaller amount than corresponds to their spectra.

Further proof of the care executed by Wirtz in the analysis of his data is given by the subsequent discussion of an observational selection effect of the type now generally referred to as the Malmquist (1922) bias:

> If all nebulae up to the faintest [limit of detection] would be observed, the colour index would show a dependence on brightness. In the sense that it becomes smaller with decreasing magnitude. This would be necessary without a physical cause simply because of the selection forced by the visual and photographic detection limits.

20.19 Epilogue

In January 1936 Wirtz sent a brief note to the *Zeitschrift für Astrophysik*, published during the same year:

> A Literary Reference Concerning the Radial Motion of Spiral Nebulae
>
> The astronomical view of the world is in our days determined by the fact that the radial motions of non-galactic nebulae increase proportionally to their distance. It seems that this occurrence was first recognized by the author in 1921. From 29 radial motions, apex, velocity, redshift were derived, and in the remaining proper radial motions 'an approximately linear relation is present in the sense that the spiral nebulae nearer to us have a tendency of approaching, the more distant ones a tendency of receding from our Milky Way system' (*A.N.* **215**, 352). In *Meddel. Lund*, Ser. II, Nr. 29, S. 32, in 1922, the author could verify this result from different photometric material. The interpretation is also given explicitly 'that the nearer or the more massive nebulae show a lesser expansion than the distant nebulae or those of lesser mass'.

The emphasis is on *proportionality*, not on *linearity*. The latter was found by Wirtz to exist between the velocities and the *logarithms* of distance, not the distance itself. The same wording, though misleading for the modern reader, is, however, also used by Curtis (1933) in his discussion of Dose's (1927) work, and might have been understood at the time.

About a year before this note, Wirtz had written a letter to Harlow Shapley, then president of Commission 28 (Commission des Nébuleuses et des Amas Stellaires) of the International Astronomical Union, presenting some points for discussions at the Fifth General Assembly in 1935. Shapley (1936) introduced Wirtz's communication with the following words:

> Dr. Wirtz of Kiel has in the past paid much attention to the problem of surface brightness, of both galactic and extragalactic nebulae. He notes

> that we have not yet at our command sufficiently accurate material for the proper examination of these problems.

It is a recognition of the fundamental work of Wirtz in one of the two areas to which he refers in his letter: the measurement of nebular surface magnitudes, and the use of 500 spiral nebulae with well-defined nuclei as a reference system for position measurements. Reminding astronomers that observational *physical* cosmology has not yet achieved what he had set out to do, and that an inertial reference frame for proper motion measurements had not yet been realized, Wirtz closed the cycle of his work in observational cosmology.

Seen from the end of the century, whose early decades witnessed such progress in both theoretical and observational cosmology, Wirtz exemplifies the lone worker, who was denied ultimate success because his environment did not understand the signs of the time. Working with observations from rather outdated telescopes and from a body of data which he was not able to enlarge and improve, he suffered from the missed opportunities of German astronomy in the transition from the nineteenth to the twentieth century. It was not *actively* realized that astrophysics and cosmology needed institutions, sites, telescopes and auxiliary facilities of the new age, such as were available in the United States. (Plans for an observatory at an excellent site in Southwest Africa came too late, Schoenberg 1935.) Instead, astronomers with outstanding capabilities for sampling and analysing data were forced to spend much of their time working with outdated equipment, as Wirtz did with the faulty meridian circle at the insufficiently endowed Kiel Observatory.

Progress in theoretical cosmology was retarded in the 1920s perhaps to some degree by its prime mover, Einstein himself. The gifted physicist, who gave modern cosmology such a brilliant start, misinterpreted Friedmann and, in the case of the cosmological constant, even his own work. The interplay between theory and observations was largely dominated by the excellent writings of the *interpreter* Eddington, by Silberstein's abundant papers which advocated the *invariant curvature* of space-time (Silberstein 1930) and by the early model of the untiring cosmologist de Sitter. Only after the mental barrier against the expanding universe was broken down, was the theoretical work of Friedmann, Lanczos and Lemaître accepted and the investigations of Wirtz, Lundmark and Hubble valued as strong observational support.

The role of time in the Friedmann–Lemaître universe appears to have been the barrier which was removed only with help from another field of modern physics: quantum theory assured that an evolving universe does not mean that the creation becomes calculable. Lemaître, the priest, stated this quite openly (to himself) in an unpublished paragraph at the end of his short communication in *Nature* (Godart and Heller 1985): 'I think that everyone who believes in a supreme being . . . may be glad to see how present physics provides a veil hiding the creation.'

De Sitter, on the other side, was attracted by the large number of cosmological data available at about the same time when he became aware of Lemaître's work. He did not hesitate to replace his own model by the time-dependent universe. Both this action and the *dustless* de Sitter concept (resurrected as a special case of the Friedmann–Lemaître model in the context of inflationary theory) is a tribute to the great mind of de Sitter.

In the Germany of the 1930s and 1940s the political debasement prohibited further advances, just at the time when the full impact of Einstein's theory on cosmology became apparent. The only German astronomer whose contributions during this time merit mentioning is Otto Heckmann. But in spite of his early work on the general structure of world models (1931, 1932) and his excellent textbook *Theorien der Kosmologie* (1942) he never considered himself a cosmologist. It should be noted that one of his 1932 models, with a zero curvature parameter, attracted Einstein's and de Sitter's interest (1932) and became subsequently known as the 'Einstein–de Sitter universe'.

Wirtz observed, thought and wrote extensively during the transitory period in the first half of our century. His methods of data reduction and interpretation led to considerable improvements in the techniques of extragalactic research. His careful analysis of accidental and systematic errors is unsurpassed. Wirtz readily adopted the new ideas of his time, extended them and made his own contributions, such as the foundations of the velocity–distance relation, correlations between physical parameters of galaxies, the introduction of the luminosity function, the first recognition of mass segregation in clusters of galaxies, the possible membership of our galaxy in a cluster and the discussion of the still unsolved problem of intergalactic extinction. His writing is extremely clear, brief, yet full of information, and always as cautious as one would expect from an observer who is aware of the shortcomings of his data. Carl Wilhelm Wirtz was an outstanding pioneer in many branches of modern observational cosmology.

Acknowledgements

We are grateful to those who supported and helped us in tracing the work of Wirtz: B. Bertotti and S. Bergia for accepting the topic as an oral presentation and a written contribution to 'Cosmology in Retrospect'; P. A. G. Scheuer for his interest in Wirtz's work; W. McCrea and D. Osterbrock for very helpful comments; W. Becker (Basel), H. Brück (Edinburgh), J. Schmidt (Sachrang), H. Strassl (Münster) and A. Unsöld (Kiel) for personal recollections of the man and the times; H. E. Nager (University Library Münster) and S. Rucinski (York University, Toronto) for providing otherwise inaccessible references; the Zentrales Staatsarchiv (Merseburg), Bundesarchiv (Koblenz) and Geheimes Staatsarchiv Preußischer Kulturbesitz (Berlin) for archival investigations; R. Duemmler and H.-A. Ott (Münster) for extensive discussions and critical reading of the manuscript.

References

Airy, G. B. (1860). On the Movement of the Solar System in Space. *Memoirs of the Royal Astronomical Society*, **28**, 143.

Arendt, H. (1963). *On Revolution*. New York: Viking Press.

Campbell, W. W. (1911). On the Motions of the Brighter Class B Stars. *Lick Observatory Bulletin*, **6**, No. 195, 101.

Charlier, C. V. L. (1922). How an Infinite World may be Built Up. *Arkiv för Matematik, Astronomi och Fysik*, **16**, No. 22.

Copernicus, N. (1543). *De Revolutionibus Orbium Coelestium Libri VI*. Nürnberg: J. Petreius.

Curtis, H. D. (1933). The Nebulae. *In Handbuch der Astrophysik*, Bd. V/2, ed. G. Eberhard, A. Kohlschütter and H. Ludendorff, p. 774. Berlin: J. Springer.

Darwin, C. (1859). *On the Origin of Species by Means of Natural Selection*. London: J. Murray.

de Sitter, W. (1917). On Einstein's Theory of Gravitation, and its Astronomical Consequences. Third paper. *Monthly Notices of the Royal Astronomical Society*, **78**, 3.

(1930a). On the Magnitudes, Diameters and Distances of the Extragalactic Nebulae, and their Apparent Radial Velocities. *Bulletin of the Astronomical Institutes of the Netherlands*, **5**, No. 185, 157.

(1930b). The Expanding Universe. Discussion of Lemaître's Solution of the Equations of the Inertial Field. *Bulletin of the Astronomical Institutes of the Netherlands*, **5**, No. 193, 211.

(1931). Some Further Computations Regarding Non-Static Universes. *Bulletin of the Astronomical Institutes of the Netherlands*, **6**, No. 223, 141.

Dose, A. (1927). Zur Statistik der nichtgalaktischen Nebel auf Grund der Königstuhl-Nebellisten. *Astronomische Nachrichten*, **229**, 157.

Eddington, A. S. (1923). *The Mathematical Theory of Relativity*. Cambridge: Cambridge University Press.

(1931). The End of the World: From the Standpoint of Mathematical Physics. *Nature*, **127**, 447.

Einstein, A. (1915). Die Grundlage der allgemeinen Relativitätstheorie. *Sitzungsberichte Berlin = Annalen der Physik*, **49**, 767 (1916).

(1917). Kosmologische Betrachtungen zur allgemeinen Relativitätstheorie. *Sitzungsberichte Berlin = Annalen der Physik*, **53**, 130 (1918).

(1922). Bemerkung zur der Arbeit von A. Friedmann 'Über die Krümmung des Raumes'. *Zeitschrift für Physik*, **11**, 326.

(1923). Notiz zu der Arbeit von A. Friedmann 'Über die Krümmung des Raumes'. *Zeitschrift für Physik*, **16**, 228.

Einstein, A. and de Sitter, W. (1931). On the Relation between the Expansion

and the Mean Density of the Universe. *Proceedings of the National Academy of Sciences, Washington*, **18**, 213.
Einstein, A. and Infeld, L. (1938 and 1947). *The Evolution of Physics*. Cambridge: Cambridge University Press.
Ellis, G. F. R. (1988). In preparation.
Fath, E. A. (1914). A Study of Nebulae. *Astronomical Journal*, **28**, 75.
Fernie, J. D. (1970). The Historical Quest for the Nature of the Spiral Nebulae. *Publications of the Astronomical Society of the Pacific*, **82**, 1189.
Friedmann, A. (1922). Über die Krümmung des Raumes. *Zeitschrift für Physik*, **10**, 377.
Friedmann, A. (1924). Über die Möglichkeit einer Welt mit konstanter negativer Krümmung des Raumes. *Zeitschrift für Physik*, **12**, 326.
Godart, O. and Heller, M. (1985). *Cosmology of Lemaître*. Tucson: Pachart.
Griewank, K. (1952). Staatsumwälzung und Revolution in der Auffassung der Renaissance und Barockzeit. *Wissenschaftliche Zeitschrift der Friedrich Schiller Universität Jena (Gesellschaftliche und Sprachwissenschaftliche Reihe)*, **2**, 11.
Hanson, R. B. (1987). Lick NPMP II. Solar Motion and Galactic Rotation. *Astronomical Journal*, **94**, 409.
Heckmann, O. (1931). Über die Metrik des sich ausdehnenden Universums. *Veröffentlichungen der Universitäts-Sternwarte Göttingen*, Heft 17.
(1932). Die Ausdehnung der Welt in ihrer Abhängigkeit von der Zeit. *Veröffentlichungen der Universitäts-Sternwarte Göttingen*, Heft 23.
(1942). *Theorien der Kosmologie*. Berlin: J. Springer (2nd edn 1968).
Herschel, J. F. W. (1864). A New Catalogue of Nebulae and Clusters of Stars, Arranged in Order of Right Ascension and Reduced to the Common Epoch 1860.0. *Philosophical Transactions*, **154**, 1.
Herschel, W. (1783). On the Proper Motion of the Sun and Solar System: With an Account of Several Changes that have happened Among the Fixed Stars since the Time of Mr. Flamsteed. *Philosophical Transactions*, **73**, 247.
(1785). On the Construction of the Heavens. *Philosophical Transactions*, **75**, 213.
(1786). Catalogue of One Thousand New Nebulae and Clusters of Stars. *Philosophical Transactions*, **76**, 457.
(1789). Catalogue of a Second Thousand of New Nebulae and Clusters of Stars; With a Few Introductory Remarks on the Construction of the Heavens. *Philosophical Transactions*, **79**, 212.
(1802). Catalogue of 500 New Nebulae, Nebulous Stars, Planetary Nebulae, and Clusters of Stars; With Remarks on the Construction of the Heavens. *Philosophical Transactions*, **92**, 477.
(1811). Astronomical Observations Relating to the Construction of the Heavens, Arranged for the Purpose of a Critical Examination, the Result of Which Appears to Throw some New Light upon the Organization of the Celestial Bodies. *Philosophical Transactions*, **101**, 269.
Hetherington, N. (1973). The Delayed Response to Suggestions of an Expanding Universe. *Journal of the British Astronomical Association*, **84**, 22.
Holetschek, J. (1907). Beobachtungen über den Helligkeitseindruck von Nebelflecken und Sternhaufen. *Annalen der Universitäts-Sternwarte Wien*, **20**, p. 114.

Homann, H. (1886). Bestimmung der Bewegung des Sonnensytems [sic!] durch Spectral-Messungen. *Astronomische Nachrichten*, **114**, 25.
Hubble, E. (1925a). Cepheids in Spiral Nebulae. *Observatory*, **48**, 139.
(1925b). N. G. C.6822, a Remote Stellar System. *Astrophysical Journal*, **62**, 409.
(1929). A Relation between Distance and Radial Velocity among Extra-Galactic Nebulae. *Proceedings of the National Academy of Sciences, Washington*, **15**, 168.
Hubble, E. and Humason, M. L. (1931). The Velocity–Distance Relation among Extragalactic Nebulae. *Astrophysical Journal*, **74**, 43.
Humason, M. L. (1931). Apparent Velocity-Shifts in the Spectra of Faint Nebulae. *Astrophysical Journal*, **74**, 35.
Jeans, J. H. (1919). *Problems of Cosmogony and Stellar Dynamics*. Cambridge: Cambridge University Press.
Kapteyn, J. C. (1901). Der Apex der Sonnenbewegung, die Constante der Praecession und die Correctionen der Eigenbewegungen in Declination von Auwers-Bradley. *Astronomische Nachrichten*, **156**, 1.
(1906). On the Parallax of the Nebulae. *Proceedings of the Academy of Amsterdam*, **8**, 691. See also: *Astronomischer Jahresbericht*, **8**, 307 (1906).
Kienle, H. (ed.) (1939). Umschau. *Himmelswelt*, **49**, 147.
Klemola, A. R., Jones, B. F. and Hanson, R. B. (1987). Lick NPMP I. Goals, Organization, and Methods. *Astronomical Journal*, **94**, 501.
Kobold, H. (1916). Bemerkungen zur Untersuchung der Eigenbewegungen der Nebelflecke. *Astronomische Nachrichten*, **203**, 299.
Kodaira, K., Okamura, S. and Watanabe, M. (1983). Diameter versus Surface Brightness Diagram of Galaxies. *Astrophysical Journal*, **274**, L49.
Kolb, E. W., Turner, M. S., Lindley, D., Olive, K. and Seckel, D. (1986). *Inner Space – Outer Space*. Chicago: University of Chicago Press.
Kövesligethy, R.v. (1886). Bestimmung der Bewegung des Sonnensystems durch Spectral-Messungen. *Astronomische Nachrichten*, **114**, 327.
Kuhn, T. S. (1962). *The Structure of Scientific Revolutions*. Chicago: University of Chicago Press (2nd edn 1970).
Lanczos, K. (1922). Bemerkungen zur de Sitterschen Welt. *Physikalische Zeitschrift*, **23**, 539.
(1923). Über die Rotverschiebung in der de Sitterschen Welt. *Zeitschrift für Physik*, **17**, 168.
Lemaître, G. (1925). Note on De Sitter's Universe. *Journal of Mathematical Physics*, **4**, 37.
(1927). Un univers homogène de masse constante et de rayon croissant, rendant compte de la vitesse radiale des nébuleuses extra-galactiques. *Annales de la Société scientifique de Bruxelles*, **47**, 494; and (1931). A Homogeneous Universe of Constant Mass and Increasing Radius Accounting for the Radial Velocity of Extra-Galactic Nebulae. *Monthly Notices of the Royal Astronomical Society*, **91**, 483 (English translation).
(1931). The Beginning of the World from the Point of View of Quantum Theory. *Nature*, **127**, 706.
(1934). Evolution of the Expanding Universe. *Proceedings of the National Academy of Sciences, Washington,* **20**, 12.

Lorenz, W. (1911). Photographische Positionsbestimmungen von 178 Nebelflecken. *Veröffentlichungen der Grossherzoglichen Sternwarte in Heidelberg (Königstuhl)*, **6**, No. 4, 19.
Lundmark, K. (1920). The Relation of the Globular Clusters and Spiral Nebulae to the Stellar System. *Kungl. Svenska Vetenskapsakademiens Handlingar, Stockholm*, **60**, No. 8.
(1924). The Determination of the Curvature of Space-Time in de Sitter's World. *Monthly Notices of the Royal Astronomical Society*, **84**, 747.
(1925). The Motions and the Distances of Spiral Nebulae. *Monthly Notices of the Royal Astronomical Society*, **85**, 865.
(1927). Studies of Anagalactic Nebulae. *Meddelanden fran Astronomiska Observatorium Upsala*, No. 30.
Malmquist, K. G. (1922). On some Relations in Stellar Statistics. *Arkiv för Matematik, Astronomi och Fysik*, **16**, No. 23.
Margolis, S. H. and Schramm, D. N. (1977). Dust in the Universe? *Astrophysical Journal*, **214**, 339.
Minkowski, H. (1909). Raum und Zeit. *Physikalische Zeitschrift*, **10**, 104 (reprinted in: H. A. Lorentz, A. Einstein and H. Minkowski: *Das Relativitätsprinzip*. Leipzig: G. G. Teubner 1913 and later).
North, J. (1965). *The Measure of the Universe: A History of Modern Cosmology*. Oxford: Clarendon Press.
Oort, J. H. (1930). Note on the Velocities of Extragalactic Nebulae. *Bulletin of the Astronomical Institutes of the Netherlands*, **5**, No. 196, 239.
Paddock, G. F. (1916). The Relation of the System of Stars to the Nebulae. *Publications of the Astronomical Society of the Pacific*, **28**, 109.
Pickering, E. C. (1900). Observations of Nebulae. *Annals of the Astronomical Observatory of Harvard College*, **23**, No. 7, 134.
Prévost, P. (1783). Mémoire sur le mouvement progressif du centre de gravité de tout le système solaire. *Astronomisches Jahrbuch für 1786*, Berlin.
Priester, W. and Schaaf, N. (1987). Carl Wirtz und die Flucht der Spiralnebel. *Sterne und Weltraum*, **26**, 376.
Robertson, H. P. (1928). On Relativistic Cosmology. *Philosophical Magazine*, **5** (7), 835.
(1929). On the Foundations of Relativistic Cosmology. *Proceedings of the National Academy of Sciences, Washington*, **15**, 822.
(1933). Relativistic Cosmology. *Reviews of Modern Physics*, **5**, 62.
Sandage, A., Binggeli, B. and Tammann, G. A. (1985). Morphological and Physical Characteristics of the Virgo Cluster: First Results from the Las Campanas Photographic Survey. ESO Conference and Workshop Proceedings, No. 20, *The Virgo Cluster of Galaxies*, p. 239.
Schaub, W. (ed.) (1939). Todesanzeige. *Astronomische Nachrichten*, **268**, 195.
Schmidt-Schönbeck, Ch. (1965). *300 Jahre Physik und Astronomie an der Kieler Universität*. Kiel: F. Hirt.
Schoenberg, E. (1935). Astronomische Arbeiten in Südwestafrika. *Astronomische Nachrichten*, **257**, 201.
Schultz, H. (1874). Micrometrical Observations of 500 Nebulae. *Nova Acta Societas Scientiarum Upsaliensis*, Ser. 3, **9**, vol. 2.

Schwarzschild, K. (1900). Ueber das zulässige Krümmungsmaass des Raumes. *Vierteljahrsschrift der Astronomischen Gesellschaft*, **35**, 337.
Seeliger, H. (1916). Über die Eigenbewegungen der Nebelflecke. *Astronomische Nachrichten*, **203**, 305.
Shapley, H. (1918). Studies of Magnitudes in Star Clusters VIII. A Summary of Results Bearing on the Structure of the Sidereal Universe. *Proceedings of the National Academy of Sciences, Washington*, **4**, 224.
(1936). Commission des nébuleuses at des amas stellaires. *Transactions of the International Astronomical Union*, **5**, 173, ed. F. J. M. Stratton. Cambridge: Cambridge University Press.
Shapley, H. and Shapley, M. B. (1919). Studies Based on the Colors and Magnitudes in Stellar Clusters. *Astrophysical Journal*, **50**, 107.
Silberstein, L. (1924). The Curvature of de Sitter's Space-Time Derived from Globular Clusters. *Monthly Notices of the Royal Astronomical Society*, **84**, 363.
(1930). *The Size of the Universe.* Oxford: Oxford University Press.
Smith, R. W. (1982). *The Expanding Universe. Astronomy's 'Great Debate' 1900–1931.* Cambridge: Cambridge University Press.
Strömberg, G. (1925). Analysis of Radial Velocities of Globular Clusters and Non-Galactic Nebulae. *Astrophysical Journal*, **61**, 353.
Tass, A. (1931). *Porträt-Gallerie der Astronomischen Gesellschaft.* Budapest: Königlich-Ungarische Universitätsdruckerei.
Tolman, R. C. (1929). On the Astronomical Implications of the de Sitter Line Element for the Universe. *Astrophysical Journal*, **64**, 245.
Vasilevskis, S. (1973). Stellar Proper Motions with Reference to Galaxies. *Vistas in Astronomy*, **15**, 145.
Weyl, H. (1924). Massenträgheit und Kosmos. *Naturwissenschaften*, **12**, 197.
Whitrow, G. J. (1959). *The Structure and Evolution of the Universe.* London: Hutchinson.
Wirtz, C. (1902). Triangulation der Hyadengruppe. *Astronomische Nachrichten*, **160**, 17.
(1906). Über Kapteyn's Parallaxenbestimmung. *Astronomische Rundschau*, **8**, 11.
(1911). Beobachtung von Nebelflecken. *Annalen der kaiserlichen Universitäts-Sternwarte Strassburg*, **4**, Teil 1, 1.
(1912). Ein Versuch zur kosmischen Stellung der Nebel. *Annalen der kaiserlichen Universitäts-Sternwarte Strassburg*, **4**, Teil 2, 313.
(1916a). Die Trift der Nebelflecke. *Astronomische Nachrichten*, **203**, 197.
(1916b). Die Trift der Nebelflecke. Zweite Mitteilung. *Astronomische Nachrichten*, **203**, 293.
(1917). Über die Eigenbewegungen der Nebelflecke. *Astronomische Nachrichten*, **204**, 23.
(1918a). *Tafeln und Formeln der Astronomie und Geodäsie.* Berlin: J. Springer.
(1918b). Über die Bewegungen der Nebelflecke. Vierte Mitteilung. *Astronomische Nachrichten*, **206**, 109.
(1922a). Einiges zur Statistik der Radialgeschwindigkeiten von Spiralnebeln und Kugelsternhaufen. *Astronomische Nachrichten*, **215**, 349.
(1922b). Notiz zur Radialbewegung der Spiralnebel. *Astronomische Nachrichten*, **216**, 451.

(1923). Flächenhelligkeiten von 566 Nebelflecken und Sternhaufen. *Meddelanden fran Lunds Astronomiska Observatorium*, Ser. II, No. 29.
(1924a). De Sitters Kosmologie und die Radialbewegungen der Spiralnebel. *Astronomische Nachrichten*, **222**, 21.
(1924b). Kugelnebel, Spiralnebel und Flächenhelligkeit. Eine statistische Studie. In *Probleme der Astronomie. Festschrift für Hugo v. Seeliger*. Berlin: J. Springer, p. 66.
(1924c). Aus der Statistik der Spiralnebel. *Astronomische Nachrichten*, **222**, 33.
(1924d). Nebelstraße, Spiralnebel und Sterne. *Astronomische Nachrichten*, **223**, 123.
(1925). Die Korrelation von Himmelshelligkeit und Sternzahl. *Astronomische Nachrichten*, **225**, 299.
(1926). Totalhelligkeit und Leuchtkraftkurve der nichtgalaktischen Nebel, ein Versuch. *Astronomische Nachrichten*, **228**, 41.
(1927). Zur Photometrie der nichtgalaktischen Nebel. Die Virgo-Coma-Gruppe. *Publikation der Sternwarte in Kiel*, **15**, Nr. 3, 35.
(1928). Literarische Anzeigen. *Vierteljahrsschrift der Astronomischen Gesellschaft*, **63**, 34.
(1936). Ein literarischer Hinweis zur Radialbewegung der Spiralnebel. *Zeitschrift für Astrophysik*, **11**, 261.
Wirtz, C. W. (1898). Bestimmung der Declination von 487 Sternen und der Polhöhe der Bonner Sternwarte. *Veröffentlichungen der Königlichen Sternwarte zu Bonn*, No. 3.
Wright, W. H. (1950). On a Proposal to Use the Extragalactic Nebulae in Measuring the Proper Motion of Stars, and in Evaluating the Precessional Constant. *Proceedings of the American Philosophical Society*, **94**, 1.
Zverev, M. S. (1940). On the Catalogue of Faint Stars. *Astronomicheskiy Zhurnal*, **17**, 54.
Zwicky, F. (1942). On the Clustering of Nebulae I. *Astrophysical Journal*, **95**, 555.

21

Cosmic rays and cosmological speculations in the 1920s: the debate between Jeans and Millikan

M. De Maria and A. Russo

In the late 1920s cosmic rays offered scientists interested in astrophysical and cosmological problems a new window onto the universe. This mysterious radiation coming from the depths of space and time became, with starlight, a set of signals to be understood and eventually embodied in a coherent theory on the structure, origin, and evolution of the universe.

In this paper, which is a part of a wider study, we will discuss a controversy which emerged between the American physicist Robert A. Millikan and the English scientist James H. Jeans. Originally based on scientific arguments, the controversy was soon charged with ideological and religious significance, the views of Millikan and Jeans being publicly discussed, both in scientific journals and in daily newspapers, as conflicting philosophical visions about the fate of the universe and the destiny of mankind. This controversy has remained at the margin of the history of modern cosmology, but it is an interesting aspect of the lively debate about the cultural and social value of science which shook the scientific community, as well as the public, in the interwar period.

It was Robert Millikan who first used the evocative term "cosmic rays" to define the penetrating radiation which is responsible for the residual ionization of air. This occurred at a meeting of the American National Academy of Science held on November 9, 1925, where he presented experimental results on cosmic rays which definitely confirmed Rudolph Hess' original suggestion of the extra-terrestrial origin of what the German physicists called *Höhenstrahlung*. The address roused great excitement both in the scientific community and among the public and thus, before being printed in the *Proceedings* of the Academy, a shortened version of the paper was published in *Science* and in the December 5 issue of *Nature* (Millikan 1926).

The great penetrating power of the rays led to the obvious conclusion that they must consist of very high energy photons the origin of which, according to Millikan, was a "nuclear transformation of some sort" involving energies about fifty times greater than that involved in ordinary radioactive processes and taking place throughout space. In particular, as a possible mechanism capable of releasing the energies observed, the American physicist suggested the formation of helium out of hydrogen. In Millikan's image-making language, this process was supposed to go on "not in the stars but in the nebulous matter in space, i.e., throughout the depths of the universe" (Millikan and Cameron 1926: 868).

It took two years for Millikan's early speculations to become a consistent cosmological theory on the origin of cosmic rays; it required only one week for James Jeans to challenge these speculations. In a letter published in the December 12 issue of *Nature*, Jeans argued against the possibility of the radiation being caused by nuclear changes such as the formation of helium out of hydrogen: "An effect of the observed magnitude, which must presumably continue for millions of millions of years, would seem to require a quite incredible amount of hydrogen or similar matter scattered through space." He suggested rather that cosmic rays were produced in processes of annihilation of one electron and one proton occurring in spiral nebulae (Jeans 1925).

Jeans' hypothesis linked the problem of the origin of cosmic rays to that of the origin of stellar radiant energy. By 1925, in fact, it was generally accepted among astrophysicists that the main source of stellar energy was the gradual destruction of matter as protons and electrons annihilated one another in the interior of stars, a hypothesis suggested by Eddington's mass-luminosity relation which showed that stars lose considerable mass in their evolution (Jeans 1924; Eddington 1926).

In a lecture presented at University College, London, on November 9, 1926, and published both in England and in the United States, Jeans presented a general cosmological theory which embodied both the origin of cosmic rays and the evolution of stars (Jeans 1926a; *passim*; also Jeans 1926b). According to the English scientist, the matter of a star is continually annihilated and transformed into radiation; this radiation is softened during its journey through the star and eventually degenerates into ordinary temperature-radiation. "What is annihilating the matter of the stars," stressed Jeans, "is neither heat nor cold, neither high density nor low, but merely the passage of time." Matter, in fact, is transformed into radiation by a spontaneous process, which proceeds, like radioactive transformations, without any regard to temperature or density.

According to Jeans' theory, a typical star is originally made mainly of massive transuranic atoms. As the star ages, these transuranic elements disappear because of their "greater capacity for the spontaneous generation of radiation by annihilating themselves [while] the lightest elements survive for longest the disintegrating action of time." Moving back along the evolutionary sequence, one finds stars containing elements of higher and higher atomic weight and then, passing beyond the stars, one comes to the nebulae: "Here we ought to find the elements of highest atomic weight of all, and the matter of greatest radiating capacity." In these low density objects, radiation can escape into space without softening and eventually reach the Earth as cosmic rays.

The evolution of the universe, according to Jeans, "appears to be from complex to simple, and not, as in biology, from simple to complex." The terrestrial elements are but a non-transformable residue, dead ashes in which radioactive elements are "the last surviving vestiges of more vigorous primeval matter." It is in this inert ash that life can exist, Jeans concluded, and he then asked his listeners:

> What is then life? . . . Is it a mere accidental and possibly quite unimportant by-product of natural processes? Or . . . is it of the nature of a disease which affects matter in its old age?

In March 1928, Jeans gave his own answer to this question (Jeans 1928a: 467, 470):

> It does not at present look as though Nature had designed the universe primarily for life; . . . Life is the end of a chain of by-products; it seems to be an accident, and torrential deluges of life-destroying radiation the essential.

In the same address, presenting again his ideas about the evolution of the universe, he offered this image of the human condition in the cosmos:

> Observation and theory agree in indicating that the universe is melting away into radiation. Our position is that of polar bears on an iceberg that has broken loose from the icepack surrounding the pole, and is inexorably melting away as the iceberg drifts to warmer latitudes and ultimate extinction.

By the same date, having performed a series of new experiments, Millikan was ready to start a counterattack against this pessimistic vision. In fact, when processed by the use of the Dirac formula for the Compton absorption, Einstein's mass–energy relation, and Aston's data on atomic weights, Millikan's results showed evidence that the incoming photons had an energy spectrum consisting of three bands, corresponding to the energies released in the building up of the atoms of helium, oxygen, and silicon out of hydrogen. To these calculations Millikan added a "philosophical argument" (Millikan and Cameron 1928a: 402):

> In the radioactive process the heavier atoms are disintegrating into lighter ones. It is, therefore, to be expected that somewhere in the universe the building-up process is going on to replace the tearing-down process represented by radioactivity.

The conclusion was that cosmic rays were but "the signals broadcast through the heavens" of this creative process, "the announcement sent out through the ether of the birth of the elements," the "birth-cries of the infant atoms of helium, oxygen and silicon" (Millikan 1928: 281, 282).

A grand cosmological view sprang out of this conclusion, a perennial cycle of the birth and death of elements in a steady-state universe: protons and electrons existing in great abundance in interstellar space condense into atoms under the special conditions of low temperature and density existing there; the atoms then aggregate into stars under their gravitational force; in the interiors of stars, atom-annihilation processes take place, giving rise to radiant energy in the form of light and heat; finally, in order to close the cycle, Millikan had to suggest that "in the depths of space . . . positive and negative electrons are also being continually replenished through the conversion back into them, under the conditions of zero temperatures and densities existing there, of the radiation continually pouring out into space from the stars" (Millikan 1928: 283; also Millikan and Cameron 1928b).

In Millikan's eyes the most important aspect of this theory was that it provided an escape from the "the nihilistic doctrine" of the ultimate "heat death" imposed on the universe by the second law of thermodynamics, an escape which he had long sought in order to satisfy his deep religious beliefs, as well as the social demands on science as a factor of progress (Tobey 1971; Kargon 1981, 1982).

In the Old Continent, however, scientists were not inclined to give up such a fundamental physical principle in any cosmological theory. Arthur S. Eddington's arguments were presented in a book, *The Nature of the Physical World*, published in that same year 1928 (Eddington 1928: 74, 86):

> The second law of thermodynamics holds, I think, the supreme position among the laws of Nature . . . If your theory is found to be against the second law of thermodynamics I can give you no hope; there is nothing for it but to collapse in deepest humiliation . . . But since prejudice in favor of a never-ending cycle of rebirth of matter and worlds is often vocal, I may perhaps give a voice to the opposite prejudice. I would feel more content that the universe should accomplish some great scheme of evolution and, having achieved whatever may be achieved, lapse back into its chaotic changelessness than that its purpose should be banalized by continual repetition. I am an Evolutionist, not a Multiplicationist. It seems rather stupid to keep doing the same thing over and over again.

Jeans too did not fail to add his arguments (Jeans 1928b: 698–699). First he showed that all physical and astronomical evidence supported the unidirectionality of the evolution of the universe:

> The transformation "mass→radiation" occurs everywhere and the reverse tranformation nowhere. There can be no creation of matter out of radiation, and no reconstruction of radioactive atoms which have once

> broken up. The fabric of the universe weathers, crumbles, and dissolves with age, and no restoration or reconstruction is possible. The second law of thermodynamics compels the material universe to move ever in the same direction along the same road, a road which ends only in death and annihilation.

The second step was to challenge Millikan's theory. On the basis of experimental data, Jeans argued that the energy of cosmic radiation is too high to be explained by an atom-building process, an account which leaves the annihilation of matter as the only possible explanation of its origin. Then Jeans came to the central issue of the second law, according to which no macroscopic cyclic process can be possible in the universe:

> Indeed it is easy to find the exact spot at which Millikan's concept comes into conflict with the second law of thermodynamics: it is that we cannot have protons and electrons tranformed into radiation at high temperature and then have the process reversed at a lower temperature.

It is evident that the central point of the controversy was not so much the interpretation of experimental data, whose narrowness and inaccuracy hardly permitted such wide generalizations, as whether it was valid to extend the validity of the second law to the universe, an issue which inevitably called upon ideological tenets. For Millikan this constituted "an extravagant and illegitimate extrapolation from our very limited mundane experience," and he stressed that "modern philosophers and theologians have also objected on the ground that it overthrows the doctrine of Immanence and requires a return to the middle-age assumption of a *Deus ex machina*" (Millikan 1928: 284). Jeans, on the contrary, considered the second law "the simplest and the most fundamental of the physical laws," and maintained that "it is hard to imagine that such a wide law fails outside our laboratories." Thus he concluded by giving Millikan a lesson in scientific methodology (Jeans 1928b: 699):

> The obvious path for scientific progress would seem to lie in the direction of inquiring what consequences are involved in supposing these laws . . . to be of universal scope, and then testing these consequences against ascertained facts of observational astronomy.

By the end of 1930 Millikan could claim new experimental evidence in support of his theory: the use of the Klein–Nishina formula for the Compton effect confirmed the existence of the previously discovered three energy bands and a fourth band had been discovered, corresponding to the energy released in the building up of iron (Millikan and Cameron 1931). Thus, in his retiring presidential address at the meeting of the American Association for the Advancement of Science held in Cleveland on December 29, 1930, Millikan again presented his inspired vision of the eternal cyclic development of matter and energy in the universe and announced to the American people shattered by unemployment and economic depression that "the Creator is continually on his job" (Millikan 1931). An editorial in *The New York Times*, two days later, praised the prestigious Nobel prize winner's "cosmic optimism" (*New York Times*, 1930):

> Neither drought nor flood nor financial depression nor any other terrestrial ill can stay the cosmic optimism of the science that not only has such practical applications, but that has faith in a continuing creation and that cooperates with "a Creator continually on the job."

Millikan's latest "evidence for atom-building" was presented to the readers of *The Physical Review* in February 1931, and in May Jeans presented his ideas in the United States, lecturing at the universities of Princeton, Yale, and Harvard (Jeans 1931; Millikan and Cameron 1931). In September of that year it was the American to challenge the British astrophysicists in their own territory, namely at the centenary meeting of the British Association for the Advancement of Science. There a special symposium on "The Evolution of the Universe" was arranged, where the opening address by Jeans was followed by contributions by G. Lemaître, W. de Sitter, A. Eddington, R. Millikan, E. A. Milne, the president of the B.A.A.S. J. C. Smuts, the Bishop of Birmingham E. W. Barnes, and O. Lodge (*Symposium* 1931).

It was a "memorable discussion" before a crowd of 2,000, *The New York Times* reported (*New York Times* 1931a), in which Jeans' claim to discuss only the physical universe, where "strict determinism reigns" and whose ultimate fate is determined by the second law of thermodynamics, was countered by Millikan's lengthy list of "experimental facts" against the annihilation hypothesis as well as by the philosophical arguments presented by Smuts, Barnes, and Lodge. If someone asserts that the universe is not evolving towards the final state of maximum entropy, stressed Jeans, "he is entitled to his opinion either as a speculation or as a pious hope. All we can say is that his odds against his dream coming true involve a very high power of 10^{79} – in his disfavour." This Millikan countered by claiming the right of the experimentalist to reveal the proneness of what he calls "the thermodynamic theorist" to make "sweeping generalizations upon insufficient knowledge." Against anyone who "has gone too far in his dicta about the origin and destiny of the universe," Millikan could prove that "as an explanation of facts about cosmic rays the annihilation hypothesis fails at every point at which one can test it." Needless to say, atom-building in interstellar space was, in Millikan's scheme, quite a natural hypothesis.

Our historical account can stop here as the matter of the controversy, still comprehensible in the context of the cultural debate about science in the interwar period, was rapidly diverging from the actual development of research. In the cosmic-ray field the introduction of new experimental techniques (cloud chambers, Geiger counters, electronic coincidence circuits) and the discovery of new phenomena (geomagnetic effects, positron, electromagnetic showers, etc.) undermined the photon hypothesis about the nature of cosmic rays and caused a shift of interest from the "cosmic" problem of the origin of the primary relation to the "terrestrial" problems of properties and interactions of the radiation actually observed in the laboratory (De Maria and Russo 1985, 1989; Russo 1988). Even in the field of astrophysics the conceptual framework was radically changed after the discovery of the neutron opened new paths to the study of nuclear structure and nuclear interactions. On the other hand, Hubble's discovery of the redshift-distance relation transformed the very character of cosmological investigation, whose guiding line was no longer the study of the properties of stars but rather the analysis of the physical properties of relativistic space-time (Smith 1982).

As a matter of fact Millikan's theory, which was largely based on the cosmological speculations of his former colleague at the University of Chicago, the astronomer William D. MacMillan, was a rational attempt to reinstate a classical, Newtonian cosmology against emerging relativistic cosmologies (MacMillan 1920, 1923, 1925; Tobey 1971). Ideological and religious beliefs concurred with Millikan's scientific work in designing his unified perspective on nature, science, society, and progress. This typically American optimistic and progressive ideology of science clashed with Jeans and Eddington's claim of the cosmological validity of the second law of thermodynamics. This probably appeared to the Americans as European pessimism which, in the words of a reader of *The New York Times*, "would leave us with nothing on which to plant our feet and mighty little on which to base our faith" (*New York Times* 1931b).

References

De Maria, M. and Russo, A. (1985). The Discovery of the Positron. *Rivista di Storia della Scienza*, **2**, no. 2, 237–286.

(1989). "Cosmic Ray Romancing": The Discovery of the Latitude Effect and the Compton–Millikan Controversy. *Historical Studies in the Physical and Biological Sciences*, **19**, 2, 211–266.

Eddington, A. (1926). The Source of Stellar Energy. *Nature* (Supplement, May 1), **117**, 25–32.

(1928). *The Nature of the Physical World*. Cambridge: Cambridge University Press.

Jeans, J. H. (1924). The Ages and Masses of the Stars. *Nature*, **114**, 828–829.

(1925). Highly-Penetrating Radiation and Cosmical Physics. *Nature*, **116**, 861.

(1926a). Recent Developments of Cosmical Physics. *Nature* (Supplement, December 4), **118**, 29–40; reprinted in *Annual Report of the Smithsonian Institution, 1927*. Washington, D.C., 1928, 167–189.

(1926b). The Evolution of the Stars. *Nature*, **117**, 18–21.

(1928a). The Wider Aspects of Cosmogony. *Nature* (Supplement, March 24), **121**, 463–470.
(1928b). The Physics of the Universe. *Nature* (Supplement, November 3), **122**, 689–700.
(1931). The Annihilation of Matter. *Nature* (Supplement, July 18), **128**, 103–110.
Kargon, R. (1981). Birth Cries of the Elements: Theory and Experiment Along Millikan's Route to Cosmic Rays. In *The Analytic Spirit*, ed. H. Wulf, pp. 309–325. Ithaca: Cornell University Press.
(1982). *The Rise of Robert Millikan.* Ithaca: Cornell University Press.
MacMillan, W. D. (1920). The Structure of the Universe. *Science*, **52**, 67–74.
(1923). Cosmic Evolution. *Scientia*, **23**, 3–12, 103–112.
(1925). Some Mathematical Aspects of Cosmology. *Science*, **62**, 63–72, 96–99, 121–126.
Millikan, R. A. (1925). High Frequency Rays of Cosmic Origin. *Nature*, **116**, 823–825; *Science*, **62**, 445–448.
(1926). High Frequency Rays of Cosmic Origin. *Proc. Nat. Ac. Sci.*, **12**, 48–54.
(1928). Available Energy. *Science*, **67**, 279–284.
(1931). Present Status of the Theory and Experiment as to Atomic Disintegration and Atomic Synthesis. *Science*, **73**, 1–5; *Nature*, **127**, 167–170.
Millikan, R. A. and Cameron, G. H. (1926). High Frequency Rays of Cosmic Origin. III – Measurements in Snow-Fed Lakes at High Altitudes. *Physical Review*, **28**, 851–868.
(1928a). Direct Evidence of Atom Building. *Science*, **67**, 401–402.
(1928b). The Origin of Cosmic Rays. *Physical Review*, **32**, 533–557.
(1931). A More Accurate and More Extended Cosmic-Ray Ionization-Depth Curve, and the Present Evidence for Atom-Building. *Physical Review*, **37**, 235–252.
New York Times (1930). Cosmic Optimism (Editorial). December 31, p. 16.
(1931a). Scientists in Clash over Annihilation. September 30, p. 16.
(1931b). Millikan or Eddington (Letter to the Editor from A. O. Tate). January 11, p. III: 2.
Russo, A. (1988). I Raggi Cosmici e la Nascita della Fisica delle Particelle Elementari. In *Atti dell'VIII Congresso Nazionale di Storia della Fisica*, ed. F. Bevilacqua, pp. 429–471. Milano: GNCSF/CNR.
Smith, R. (1982). *The Expanding Universe. Astronomy's "Great Debate" 1900–1931.* Cambridge: Cambridge University Press.
Symposium (1931). The Evolution of the Universe. *Nature* (Supplement, October 24), **128**, 699–722; Jeans on pp. 701–704; Millikan on pp. 709–715.
Tobey, R. C. (1971). *The American Ideology of National Science.* Pittsburgh: University of Pittsburgh Press.

Discussion

A discussion took place about the role of J. Jeans as a precursor to the cosmological creation of matter and the theory of steady state. For example, in the paper "The Physics of the Universe" (*Nature*, Supplement **122**, 6, 89 [1928], p. 698), he says that at the beginning of the universe atoms must have begun to exist at some time and "this leads us to contemplate a definite event, or series of events, or continuous process, of creation of matter. If we want a naturalistic interpretation of this creation, we may imagine energy of any wavelengths less than 1.3×10^{-13} cm being poured into empty space; such radiation might conceivably crystallize into electrons and protons, and finally form atoms."

According to Arp the idea of creation of matter was very uncharacteristic in Jeans and does not justify the claim of his being a precursor to the steady-state theory. (The editors.)

22

Sinclair Smith (1899–1938)

Virginia Trimble

The names of most of the pioneers of dark matter astronomy are household words (at least in astronomical households). James H. Jeans is commemorated in his eponymous mass and length, Johannes C. Kapteyn in his heliocentric model of the "universe" (galaxy) and his selected areas. Fritz Zwicky is the subject of much folklore, and we all think of Jan H. Oort as a close friend (even those of us who have not been lucky enough actually to know him very well). Sinclair Smith is much less familiar for two reasons. First, his work was largely in the area we would now call instrumental physics, rather than observational astronomy, and, second, he died tragically early of cancer at the age of thirty-nine.

Sinclair Smith, born in Chicago in 1899, was a Californian from the age of thirteen, earning his Bachelor's degree at Caltech in 1921 and his PhD in 1924 (Smith 1925), working with John A. Anderson on electrically exploded wires as a method of obtaining laboratory spectra at high excitation and ionization energies. He remained in the physics laboratory of Mt. Wilson Observatory the rest of his life, apart from a year (1924–25) at the Cavendish Laboratory in Cambridge. A true Californian, if not quite native, Smith was an enthusiastic owner and sailor of small boats, in company with Tom Lauritsen and other CIT colleagues. His last years were spent largely on engineering design for the 200-inch telescope, especially its control system. An obituary by Anderson (1938) appeared in *Publications of the Astronomical Society of the Pacific.*

Smith was clearly a widgeteer par excellence. His outstanding contribution to exploding wire research was a spectrograph with microsecond temporal resolution (Smith 1924a, 1924b, 1925; Anderson and Smith 1926). It focussed the spectrograph slit on a rotating mirror, which, in turn, shined the light sequentially across photographic film, producing a record of the evolution of the spectrum of the wire as it vaporized. Rotating mirrors recur in several of his later devices.

Next came a vertical seismometer (Smith 1927) and an optical oscillograph (Smith 1928a, 1928b). This latter device provided a permanent record of a rapidly changing electric current by passing the current through a solenoid wrapped around a C_2S cell between crossed polarizers. A collimated beam of white light shined through the device, through a prism, onto a rotating mirror, and so to the film. As the current varied, changing the cell's rotation of the plane of polarization vs. wavelength, a varying pattern of light and dark fringes was recorded on the film.

Radiometers were Smith's next major love. He studied their sensitivity as a function of temperature (Smith 1930), using liquid air as a coolant, and adapted them for use as detectors in stellar photometry (Smith 1934) and for a laboratory registering microphotometer (Smith and Wilson 1932), also with a rotating mirror. This latter, his co-worker gently recalls, was "not competitive" with photoelectric cells and other technologies. Smith (1932, 1933) also developed a more conventional stellar photometer, with photoelectric cell and Hoffman electrometer. Attached to the 100-inch, it could produce a current of 500 electrons per second for a 14th magnitude star and determine its flux to 1 percent in 21 sec.

Smith's expertise in measurement of light intensity led to a collaboration with Richard C. Tolman (Tolman and Smith 1926) to consider experimental tests of the various possible interpretations of the wave-particle dualism. The device they considered appears never to have been built, perhaps because they concluded that it could not distinguish an early version of the absorber theory of radiation from more conventional interpretations. This appears to have been Smith's most nearly theoretical paper.

Perhaps the most productive widget was an f/1 quartz spectrograph, one of the very first to use a Schmidt camera (Smith 1935a). He turned this spectrograph toward M32 (Smith 1935b), along with a Wollaston prism (presumably also his own device), and the 100-inch telescope made into an interferometer with a dark strip across its tube. He concluded that the Galaxy was unpolarized, had a slightly resolved (0.8″) nucleus, and showed no gradient of spectral type (dG3 at all wavelengths) across its surface. These observations led him to rule out several now-forgotten models of elliptical galaxies in favor of the giant star clouds we all now accept.

The spectrograph was fast enough to record on the 60-inch, galaxies in Virgo fainter than the ones Humason was then studying with the 100-inch nebular spectrograph. Smith's (1936) analysis of his, Humason's, and Slipher's radial velocities of cluster members is the work that brings him into the history of dark matter. He concluded (using nine of his own velocities and about two dozen others) that Virgo had a roughly isotropic distribution of velocities, no significant equipartition of energy, and a mass of $5 \times 10^{14} h^{-1} M_\odot$ ($10^{14} M_\odot$ at his adopted distance of 2 Mpc). He noted the similarity to Zwicky's result for Coma, and concluded that there must be either large quantities of internebular material or enormous faint extensions beyond the visible galaxies (as just found for M31 by J. Stebbins and A. Whitford 1934, *Proceedings of the National Academy of Sciences*,

20, 93, doing photoelectric photometry down to 27 mags. per square arc sec.).

I am deeply indebted to Dr. Olin C. Wilson of Mt. Wilson and Las Campanas Observatories and his wife Kathy for an introduction to their friend and colleague Sinclair Smith. Dr. Wilson is the only one of Smith's close co-workers whose memories we can still tap first hand. Those memories are both vivid and fascinating.

Papers by and about Sinclair Smith

This is not a complete listing. It includes only papers I was able to consult first hand and serves to provide an overview of the topics that interested Smith during his brief but productive career. Parenthetical words indicate subjects rather than exact titles of the papers.

Anderson, J. A. (1938). *Publications of the Astronomical Society of the Pacific* (obituary).

Anderson, J. A. and Smith, S. (1926). *Astrophysical Journal*, **64**, 295 (characteristics of exploded wires).

Smith, S. (1924a). *Physical Review*, **24**, 206 (APS meeting abstract: spectrograph with high temporal resolution).

(1924b). *Proceedings of the National Academy of Sciences*, **10**, 4 (electrically exploded wires in high vacuum).

(1925). *Astrophysical Journal*, **61**, 186 (thesis: a study of electrically exploded wires).

(1927). *Physical Review*, **29**, 755 (APS meeting abstract: vertical seismometer).

(1928a). *Physical Review*, **32**, 319 (APS meeting abstract: optical oscillograph).

(1928b). *Astrophysical Journal*, **68**, 165 (optical oscillograph).

(1928c). *Nature*, **121**, 1 (high-frequency discharge spectroscopy).

(1930). *Proceedings of the National Academy of Sciences*, **16**, 373 (effect of low T on radiometer sensitivity).

(1932). *Astrophysical Journal*, **76**, 286 (photoelectric photometer and its sensitivity).

(1933). *Physical Review*, **43**, 211 (APS meeting abstract: stellar photoelectric photometer).

(1934). *Publications of the Astronomical Society of the Pacific* (ASP meeting abstract: spectral distribution of stellar energy determined with a radiometer).

(1935a). *Physical Review*, **47**, 257 (APS meeting abstract: an f/1 quartz spectrograph).

(1935b). *Astrophysical Journal*, **82**, 192 (properties of M32).

(1935c). *Publications of the Astronomical Society of the Pacific* (ASP meeting abstract: mass of Virgo; values given here are revised downward a factor of 10 in following paper).

(1936). *Astrophysical Journal*, **83**, 23 (mass of Virgo cluster).

Smith, S. and Wilson, O. C., Jr. (1932). *Astrophysical Journal*, **76**, 117 (radiometer as detector for laboratory microphotometer).

Tolman, R. C. and Smith, S. (1926). *Proceedings of the National Academy of Sciences*, **12**, 343 (test of quantum mechanical light propagation).

23

Revisiting Fritz Zwicky

Alessandro Braccesi

23.1 Zwicky involvements and scientific career

Fritz Zwicky was born in 1898 in Bulgaria, but of Swiss parentage, he rated the small Swiss town of Glarus as his home town (Zwicky 1962: 10). He received his Ph.D. in 1922 from the Federal Institute of Technology in Zurich. Three years later he joined the California Institute of Technology, where he was to remain until his retirement in 1969.

His thesis work involved the calculation of the tensile strength of NaCl crystals. He found by various methods 20,000 kg/cm^2, while the observed value was only 50 kg/cm^2. Nothing was wrong with the calculations, and indirect evidence showed that the result was correct, but, as he wrote in 1962, although "much work has been done . . . to elucidate the nature of the discrepancy . . . the problem of contradiction between simple theory and observation has not yet been clearly solved" (Zwicky 1962: 247).

In my opinion the problems raised by this work had a lasting influence on Zwicky.

In fact we find a scientist who undoubtedly was at ease with theories and with the related mathematical developments – he taught theoretical physics at Caltech from 1927 to 1942 – concerned with phenomena which, while at variance with the predictions, do not invalidate the theories on which the predictions themselves are based.

The investigation of such phenomena requires the highest degree of "open-mindedness." It also makes the investigator overcautious when a comparison between theoretical predictions and factual evidence has to be made.

If the world can be so rich as, it would seem, to override the very laws upon which it is made, than it follows that the primary task of the theoretician (not the theoretician in the technical sense, but every person in the process of figuring possibilities) is to investigate all possible alternatives, trying to be as comprehensive as possible.

This has to be the case with regard not only to the phenomena but to the theories themselves. In order to appreciate the more extended possible set of solutions, all the restrictions under which the theories had first been developed must progressively be dropped.

Most of Zwicky's contributions reflect this kind of attitude. We should remember his 1933 studies on the statistical mechanics of cooperative phenomena (Zwicky 1957: 250) as an example of the need to extend the domain of validity of theories and the 1934 suggestion that neutron matter could represent the most stable kind of matter in compact stellar nuclei (Baade and Zwicky 1934; Zwicky 1957: 255) as an example of open-mindedness in considering exotic physical possibilities.

Other examples of these attitudes can be found through analyzing Zwicky's pre-war activities. During World War II Zwicky became the director of research of the Aerojet Engineering Corporation, a position he held until 1949 when he was forced to resign [Zwicky 1962:25].

Between 1945 and 1955, summing up his scientific and managerial experiences Zwicky became more and more persuaded that method is the most important thing of all. His method, a cultural inheritance of which he vindicated only the formalization, he called "morphological analysis."

From this very moment he rated the formalization of this method as his most important contribution, and tried to ascribe to the use of this method, more than to his peculiar genius, his many achievements.

In his peculiarly provocative way he wrote: "it will be well to prove the value of the morphological method first in some . . . fields . . . where he [Zwicky] has no previous detailed knowledge or experience. Being originally a physicist by conventional standards, he thus has stepped more or less far out of his field and has occupied himself for a hobby with the fundamental pattern of things (morphology) in the problem of total war, astronomy, jet propulsion, education and epistemology" (Zwicky 1957: 11).

People who have not been acquainted with Zwicky may be disconcerted by his examples. Astronomers do not like to hear their art referred to as "hobby"; even more intriguing may appear the question of "total war."

Nevertheless, one should remember that Zwicky was charged with making military missions to Germany and Japan in 1945 and 1946 (Zwicky 1962: 134). In Germany he interrogated 600 of the Peenemunde personnel who had fallen into the hands of the Allies and, his mission accomplished, returned to the States with the first supersonic wind tunnel ever to reach that country. In Japan he had to evaluate the damage produced by the atomic bombs on the towns of Hiroshima and Nagasaki.

Incidentally he concluded that much could be done in the way of defense if the technology needed to produce materials having a strength approaching the theoretical predictions could ever have been developed. This idea is at the basis of the development of materials with peculiar structural properties such as carbon fibers, materials which have now been found to have wide application in the aerospace industry.

23.2 Zwicky's *Morphological Astronomy*

I think that most people who approached cosmology in the late 50s and early 60s are in debt to some outstanding books. Among these books are Bondi's *Cosmology*, Zwicky's *Morphological Astronomy*, the review papers published in the *Handbook of Physics*, and some published symposia such as *The Paris Symposium on Radio Astronomy* or the one on the *Problems of Extra-Galactic Research*, published in 1961.

Zwicky's book, however, has a peculiarity. Books devoted to personal research, mainly comprising original contributions, are very seldom found in the post-war astronomical literature, and a book from an observational astronomer is an absolute rarity. In fact, Zwicky's *Morphological Astronomy* is the only one I know.

A book such as this is a beautiful, perhaps a little old-fashioned, stimulating thing. It allows the reader to become familiar not only with a certain number of topics but, to an extent, with the author himself.

This is particularly true in Zwicky's case because he avoids the habit of presenting research as a series of achievements – the science in pills of the present day – but instead explains the problems confronted, the methods he will employ to deal with them, the results, and the new problems which may arise from the results themselves – all in a very lively way.

Most of Zwicky's book is devoted to the problem of the large-scale distribution of galaxies, their clustering, the evidence for superclustering, or for the existence of patchy intergalactic absorption etc.

This is probably the area where Zwicky contributed most to extragalactic astronomy, from his pioneering determination of the mass to light ratio in Coma made in 1937 (Zwicky 1937), to the six volumes of the *Catalog of Galaxies and of Clusters of Galaxies* (Zwicky et al. 1961–68), which really updated Herschel's list of nebulae extending catalogued galaxies from the 12th to the 15th magnitude.

The list and the atlas of clusters of galaxies included in the catalogue represents Zwicky's personal contribution. In order to find the clusters he personally scrutinized the 559, E-emulsion, $\lambda > 15°$, original P.S.S. plates, marking with dots of red ink all the galaxies he could identify, later to regroup them in clusters. We now know that some of his looser groups are spurious; nevertheless, he seldom dismissed a real one.

Also well known is the cooperative program for supernova search he promoted and to which he and his associates gave the major contribution first with the 18-inch, then with the 48-inch Palomar Schmidt telescopes, as well as his early 1934 suggestion that the energy of supernova explosions could originate from the collapse of ordinary matter to neutron matter. Moreover, his suggestion of using $L\gamma\alpha$ emission to study distant galaxies (Zwicky 1962: 365) ultimately proved to be a success.

Let us now, however, come back to *Morphological Astronomy*. Apart from the topics related to the distribution of galaxies which I have already mentioned, many other subjects are discussed in Zwicky's book, including: intergalactic matter, either luminous or obscure (Zwicky 1957: 3); the best possible operative definition of an inertial reference system: "A coordinate system fixed to the principal axes of the ellipsoid of inertia of all the galaxies and of the intergalactic matter and radiation within a large sphere of radius R" (Zwicky 1957: 114); the problem of the possible divergence of the galaxy luminosity function at the faint end (Zwicky 1957: 220); galaxies as gravitational lenses (Zwicky 1957: 215); stationary and non-stationary galaxies (Zwicky 1957: 214); Einstein redshift from galaxies' gravitational potential (Zwicky 1957: 188); the possibility of getting accelerating potentials up to 10^{19} eV from charge separation phenomena produced by radiation pressure in supernova outbursts to accelerate cosmic rays (Zwicky 1957: 253), as well as the pre-discovery prediction that very high energy cosmic rays could contain a sizable fraction of heavy nuclei (Zwicky 1957: 254); the possible disruption of associations because of supernovae blowing out the interstellar matter which made the system gravitationally stable (Zwicky 1957: 258); the need to improve the theory of statistical mechanics in the fields where "local conditions do not only depend on local parameters but are vitally influenced by the integral characteristics of the whole system" because "the principles and the methods which will allow us to deal with the statistics of large-scale gravitating systems such as clusters of galaxies still remain to be developed"; the need to extend statistical mechanics to the stationary systems which are not in local thermodynamic equilibrium, a discussion which includes a study of the perturbations growing in such systems; the equilibrium condition in very rarefied gas, i.e. in gases where the mean free path is comparable with other, macroscopic, scale lengths such as the scale height parameter (Zwicky 1957: 250–52), and so on.

Some other topics he introduces at first glance seem a little odd, such as the suggestion of possible extra terms in specific heats (Zwicky 1957: 256) or what he calls the "gravitational drag of light" (Zwicky 1957: 180). On closer inspection, however, one realizes that even here Zwicky is making a sound use of classical physics.

In the introductory section epistemological problems are dealt with at length. Zwicky did not consider the accumulation of knowledge as a linear process, but as a process which encounters many difficulties. He also thought that awareness of the difficulties of this process would help in finding more rational – i.e. more free – and productive approaches.

Let us quote from the *Morphological Astronomy* what he wrote on this subject:

> If rain begins to fall on previously dry areas of the earth, the water on the ground will make its way from high levels to low levels in a variety of ways

> ... Whatever courses are being followed by the first waters, their existence will largely prejudice those chosen by later floods. A system of ruts will consequently be established which has a high degree of permanence ...
>
> Just as rains open up the earth here and there, ideas unlock the doors to various aspects of life, fixing the attention of men on some aspects while partly or entirely ignoring others. Once man is in a rut he seems to have the urge to dig ever deeper. And what often is more unfortunate, he does not take the excavated debris with him like the waters, but throws it over the edge, thus covering up the unexplored territory and making it impossible for him to see neighbours in the eyes, intentionally or unintentionally and make it difficult for them to see anything at all.
>
> Thus, although inventions, discoveries, and research open up ever new fields and often illuminate them in many details, these fields, local in character, are chosen more or less at random. Investigators among our ancestors and up to the present were usually fully occupied with local successes. They often failed to incorporate these successes into, or even perceive their relation to, the totality of possible aspects of human life. The totality of these aspects is, for practical purposes, infinite in extent; it is therefore inexhaustible, but it is not without structure. There are mountains and valleys in this structure, just as there is a topography on the earth. And these mountains and valleys apparently determine the easiest subconscious flows of ideas, of inventions and discoveries much as the mountains on the earth roughly determine where the great water courses will be.
>
> If we, therefore, knew the general intrinsic structure of these predetermined valleys of thought we should be able to determine more easily the directions in which to search for new truths.
>
> (Zwicky 1957: 6–7)

I think that even today Zwicky's analysis is worth considering.

There is a last class of problems, among those envisaged by Zwicky, which are ordinarily regarded as lying outside the domain of exact sciences.

Nevertheless, one is forced to admit that Zwicky's explicitly stated Kantian approach to the problems of knowledge – his "communicable truth" appears to play the role of an historically and culturally defined "*a priori* of the reason" – offers a ground for the discussion of such odd questions as the three-dimensionality of space (Zwicky 1957: 279), the reason why out of "the non-denumerable infinity of transcendental numbers there are just two, namely "e" and "π" which appear in most of fields of mathematics and physics with obstinate persistence, while all others ... seem to be relegated to an anonymous background" (Zwicky 1957: 274), or the fact that "often the whole structure of [physically significant equations] can be derived almost *a priori*" (Zwicky 1957: 281).

I have described in some detail the content of Zwicky's book in the belief that his major legacy to our generation is probably not to be found in his specific, although outstanding, scientific contributions, but in the broad cultural influence his book exerted. In fact, the peculiar and unique combination of soundness and open-mindedness, of hard work and wild ideas, with which the readers of the book become familiar, probably had a deep influence on many of us, an influence of which as yet we are, perhaps, not completely aware. When, in order to prepare the present account, I went through Zwicky's book again, I was quite surprised to discover how many of the ideas and problems which have been on my mind through the years are found there. No doubt, I had forgotten my primary source.

23.3 The "lone wolf"

It would be unfair not to recall that Zwicky was not an easy person. Perhaps the astronomical community is essentially aware of the rather nominalistic, late 60s, dispute about blue stars, quasi-stars, and compact galaxies. But this was only the occasional flaring of a much more deeply rooted tension.

There is a second book by Zwicky which is seldom found in libraries because it was printed for private circulation. In this book he collected in 1962, under the heading of *Morphology of Propulsive Power*, many of the papers and technical reports referring to his war-time work on jet propulsion.

In this book, as well as in *Morphological Astronomy*, one finds many pages in which the discussion resembles a monologue of someone who is thinking aloud. A collage of these pages does indeed constitute an intellectual autobiography of the writer.

Messages of this kind are quite rare in present scientific literature; furthermore, they usually come only from the very top personalities. Thus Zwicky's pages present a double interest: apart from the personal adventure of the writer himself, they allow the reader to appreciate the influence which has been exerted on an average – even if, in this case, much better than average – scientist of his generation.

Zwicky's original background was undoubtedly that of a German-speaking middle-European born before World War I. This background, however, was colored by Zwicky's Swiss connections.

Immanuel Kant remained throughout Zwicky's life as firm a reference point for ethics as for epistemology. As far as ethics is concerned Zwicky wrote: "Within the fields of applied ethics . . . great tragedies have often resulted because of the conviction of many that there exists absolute good and absolute evil (Kant's categorical imperative)" (Zwicky 1957: 271). The reference is as clear as the criticism. In a peculiar way this also applied to the United States. Concerning epistemology, we find: "Kant . . . stated in his 'Kritik der reinen Vernunft' that the space in which we live is Euclidean, there being actually no other space thinkable. He furthermore postulated that all natural phenomena obey the law of strict *casuality*, and that no alternative to this law exists at all." He continues: "During the last two centuries of scientific progress all the mentioned axiomatic convictions of Kant's have been found incorrect or incomplete" (Zwicky 1957: 271).

The problem of the "*a prioris* of the mind" or more precisely, in Zwicky's words, of the "*structure of our modes of thought*" (Zwicky 1957: 272) is connected with what Zwicky identifies with the "*communicable truth*." Then the "*flexibility of communicable truth*" (Zwicky 1957: 12) should reflect a flexibility of the *a priori* of the mind, a flexibility which is rated by Zwicky as "*inexhaustible*" (Zwicky 1957: 12). In other words, real progress in knowledge means, if I rightly understand him, a continuous shift of the line of "separation of the subject and the object", i.e. in the possibility of extending the range of the things we can speak of.

For a scientist who saw the rise of relativity and quantum mechanics and who himself predicted the existence of neutron stars, this way of thinking seems to stem in a rather natural way from his own personal experience.

Let me now come to the peculiarities of Zwicky's human adventure. When the pages we have referred to were written, World War II had just finished and Cold War was still at its apex.

Zwicky in fact was an idealist, even if he recognized the deadly consequences of "Kant's categorical imperatives." Even more, he was inclined to consider himself in a rather heroic mold.

He wrote "history is made by individual men and groups of men" (Zwicky 1957: 5). He, as a man, had his own wars to fight. After all, "The Swiss are not pacifists. They are a nation of warriors, fortified, provisioned for a siege, armed to the teeth" (Zwicky 1962: 17).

In these wars the citizen was rated as much more important than the scientist: "those who believe in science and technology *per se* are on the level of a man whose ultimate goal in life it is to split with deadly accuracy through a keyhole from a distance of seven meters" (Zwicky 1962: 13). This is a further example of Zwicky's tendency toward extremely provocative judgements.

The intellectual autobiography of which we were speaking starts with his pre-war concern for the rise of the totalitarian regimes of Mussolini and Hitler. He, like many other Europeans, had a presentiment of the hard times which were to come.

In 1940, "having surveyed the fields in which he could himself contribute" (Zwicky 1962: 17), he decided to act. What he had in mind is not detailed in the documents I have at my disposal, but they were discussed with the Vice President of the United States and the top presidential adviser (Zwicky 1962: 19). Apparently he discussed the problem of nuclear bombs, without being aware of what had already gone on, and he got no audience. When he went back to Pasadena from his "lone wolf campaign . . . many of his colleagues no doubt felt that he was slightly off his rocker" (Zwicky 1962: 20).

In Pasadena at first he could work only in a civil defense project, then in 1943 he was appointed director of research of the Aerojet Company, becoming one of the top experts in jet propulsion in the United States. This is the reason for the mission to Germany mentioned above.

We have also mentioned his missions to Japan. Probably he was one of the few scientists not directly involved with the bomb project to have a sufficiently broad knowledge and the necessary stomach for a cool appraisal of the bombs' effects.

His judgement of the whole bomb affair is characteristic. He wrote: "Dr. Oppenheimer and other scientists ... have advised against the development of hydrogen bombs on moral and technical grounds, a position which, had it resulted in a policy of non-support of the United States government would have been perhaps more disastrous than the original decision to use atomic bombs on cities inhabited largely by civilians. It is one thing to decide to use a new and horrible weapon in this manner and quite another to be reluctant to develop such weapons." Then follow some words characteristic of the Cold War times and Zwicky's horror of the "Bolshevik" "so that free men be prepared to meet whatever may be brought against them in their struggle to preserve their freedom at all times and by all means" (Zwicky 1957: 268).

Knowing these attitudes one can understand Zwicky's dismay over subsequent developments in his relations with the U.S. administration. He had been enrolled in the military missions "in spite of the fact that he was a Swiss and had never acquired American citizenship" (Zwicky 1962: 135). Because his application for U.S. citizenship in 1949 was rejected, he was forced to resign from his position as research director of the Aerojet General Corporation and in 1955 lost his clearance for classified projects. His personal war was over.

The paragraph in *The Morphology of Propulsive Power* in which he recounts these happenings rather unexpectedly is subtitled "Back to Science and Life." He remarks: "It is an immense relief not to have to work behind the bars of security any more, to pursue at one's leisure some of the fascinating problems of science, to enjoy life" (Zwicky 1962: 191).

However, his bitterness with the "establishment" – and here he included the scientific establishment – did not dissipate. The final sentence of *Morphological Astronomy* reads: "after Pythagoras had discovered his famous theorem, the Greeks slaughtered 150 oxen and arranged for a feast. Ever since that happy time, however, whenever anybody proposed something drastically new, the oxen have bellowed" (Zwicky 1957: 290). Happily enough in the last thirty years many things "drastically new" have been accepted into the body of science without oxen bellowing, many of them before his death in 1974. Let us hope he could appreciate this.

Probably other people could have written more eloquently and with greater knowledge about Zwicky than I. My hope, then, is that this account might trigger further studies.

References

Baade, W. and Zwicky, F. (1934). On Super-Novae. *Proceedings of the National Academy of Sciences*, **20**, 254–259.

Zwicky, F. (1937). On the Masses of Nebulae and of Clusters of Nebulae. *Astrophysical Journal*, **86**, 217–246.

(1957). *Morphological Astronomy*. Berlin: Springer.

(1962). *The Morphology of Propulsive Power*. Pasadena: Society for Morphological Research.

Zwicky, F., Herzog, E., Wild, P., Karpowicz, M., and Kowal, C. T. (1961–68). *Catalog of Galaxies and of Clusters of Galaxies*, vols. 1–6. Pasadena: California Institute of Technology.

Index

absolute space and time, 52
accelerated frames, 53
α,β,γ theory, 136
Aristotle's spherical cosmos, 51
arrow of time, 234ff

Baryon–antibaryon asymmetry, 144
big bang, 27, 41, 97, 135, 162f, 170, 302
  in extenso, chapter 8
  little big bangs, 222
  see also cosmology
black holes, 355f
  see also quasars
Brans–Dicke theory, 84

Cepheid variables, 68
constants of physics, 68
Copernicus, xiif
cosmic abundancies and nucleosynthesis, 130
  in extenso, chapter 8 and 9
  abundancies (of elements), 160, 170f, 177ff
  nucleosynthesis in stellar interiors, 131, 179
  primordial (primeval, cosmological) nucleosynthesis, 42, 159, 344
  by neutron-capture, 131ff
  primordial Helium, 142
  primordial synthesis of D, ^{4}He, ^{7}Li, 244ff: of light elements, 167f, 180; of heavy elements, 175f; in little big bangs, 226
cosmic microwave background radiation, 170f, 194, 340, 343f
  in extenso, chapter 15
  prediction of, 137, 139, 165
  blackbody distribution of, 303f
  vs hypothesis of thermalization of e.m. energy produced by astrophysical objects, 244ff
cosmic rays
  in extenso, chapter 21
  see also radio
cosmic time, 22, 283
cosmology
  Euclidean vs steady-state, 340
  big bang vs steady-state, 324, 344
  steady-state vs Friedman or relativistic, 227, 321
  Newtonian (kinematic relativity, Milne's theory), 28, 78, 112
  Relativistic: *in extenso*, chapter 6
  time scale problem of, 191, 194
  see also universe
cosmological constant, 17, 59, 97, 220ff, 214
  and large numbers, 77, 82
cosmological models
  inhomogeneous
  in extenso, chapter 7
  principle, 39, 98: perfect, 39, 84
  see also under universe
cross sections (of thermonuclear reactions), 135, 141

dark matter
  in extenso, chapter 19
  astronomy, 362, 411
  invisible stars, 356f
  pregalactic dark matter (and steady-state universe), 358, 362
  intergalactic (internebular) matter, 412, 418
deceleration parameter, 224f
de Sitter space–time (including references to de Sitter metric, solution, model, universe, etc.), xv, 17, 61ff, 98ff, 221, 375ff, 381ff, 393
  see also Einstein, inflation, horizon
Dirac's Large Number hypothesis, 79f

Eddington number, 75
Einstein
  static or 'cylindrical' model, 11ff, 59ff, 97ff, 381ff
  Einstein–de Sitter universe, 28, 383
  equations, 59
  tensor, 213
entropy per baryon, 149
equivalence principle, 53

fine structure constant, 75

Friedmann, Lemaître, Robertson, Walker
solution (including references to 'Friedmann evolving models', 'Lemaître's expanding universe', 'Robertson–Walker metric', 'Lemaître-Eddington model', etc.(, 22f, 41, 101f, 130, 212f, 343, 380f

G variation, 80, 83
galaxies
formation of, 138
classification of, 278, 387
large scale distribution of, 417
clusters of, 222, 356ff, 386, 412, 417ff
superclusters, 358, 417
radio, 339f, 347f
Seyfert, 351, 353
recession of, 201f, 208, 261ff, 369ff
relation between distances and redshifts of, *see under* Hubble
general covariance, 55
general relativity, *see under* relativity
geodesic principle, 55

heat death, 79, 404
hierarchical universe, *see under* universe
horizon, 101
in de Sitter universe, 17, 101
Hubble
law, 20
redshift–distance relation (various forms of), 21, 226ff, 373ff
constant, 39, 202, 206, 280, 380, 382
time, *see under* universe (age of)
hydrogen, 21 cm line of, 295, 312, 330

inertia, 49, 58
inertial frames, 50
inflation
inflationary universe, 382
inflationary theory, 393
inflationary solution of Einstein's field equations, 221
'de Sitter phase', 382
inhomogeneous models
in extenso, chapter 7
see also universe
initial singularity, 23
interstellar absorption, 34

kinematic relativity, *see under* cosmology

Large Numbers, 79, 210ff
in extenso, chapter 5
hypothesis, 80
logN–logS plot, 315ff, 334
quasar and radio galaxy components of, 342

Mach's Principle, 6, 64
in extenso, chapter 4
relativity of inertia, 49, 61
local laws and dynamics of the universe, 7, 51
Mercury's perihelion, 14
Milne's theory, *see under* cosmology
missing mass, 149

Natural units, 70
nebulae, 5, 343, 366f, 386
as island universes (or sidereal systems), 5, 16, 20, 208, 252,
spiral nebulae, 206, 250ff
distances and motions of, 5, 369
see also galaxies
neutrinos
background of, 145
neutrino families, 170
neutron
capture reactions, 131, 135
to proton ratio, 144
neutron stars, 421
netron mater (in compact stellar nuclei), 416
non-Euclidean geometry, 16, 54
nucleosynthesis, *see under* cosmic abundances

Olbers' paradox, 312
in extenso, Chapter 3
origin of the elements, 161
see also cosmic abundances and nucleosynthesis

primeval atom, 26, 130, 385
pulsars: as neutron stars, 421

quasars
discovery of: *in extenso*, chapter 18
nature of, 6, 229
redshifts of, 340, 343, 350
cosmological interpretation of, 350
as black holes, 351
see also logN-logS plot, radio

radio
stars, 309ff: as powerhouses for the galactic cosmic rays, 311
3C 48 as a radio star, 348, 353
sources (nature of), 313, 331ff: optical identification of, 316, 347
galactic vs extragalactic radio sources, 320ff, 331, 333, 340
radio galaxies, *see under* galaxies
quasi-stellar radio sources, 349
source counts (tests), 228ff, 321ff, 343f
in extenso, chapter 17: and steady-state theory, 228ff, 324, 331f, 336, 339, 343; *see also* logN–logS plot
redshift
gravitational, 350, 356
cosmological (redshift of distant sources in an expanding universe), 39, 101
in the de Sitter universe, 99, 381
of galaxies (spiral nebulae, extragalactic nebulae), 21, 130, 206, 375, 391
of radio sources, 347ff
of quasars, 340, 348, 351
of radio galaxies, 348
redshift–distance relations (various forms of), *see under* Hubble

relativity
  principle, 56
  general theory of, xiv, 11, 324, 383
  first relativistic model of the universe, 11
  of inertia, *see under* Mach's Principle
  see also cosmology (relativistic)

stars
  evolution of, 402
  radio stars, *see under* radio
  brown giants, invisible, dark, *see under* dark matter
  white dwarfs, 348, 356, 358
steady-state cosmology (including references to steady-state model, theory, etc.), 7, 40, 84, 140, 170, 191ff, 324, 344
  in extenso, chapters 12, 13
  Jeans, as precursor of steady-state theory, 409
  see also dark matter, radio (source counts), cosmology
supernova
  remnants, 313, 332
  search, 417
  explosions, 417
  outbursts, 418

universe
  age of, *see under* Hubble
  models of: *in extenso*, chapter 6
  inhomogeneous: *in extenso*, chapter 7
  inflationary, 381, *see also* inflation
  hierarchical, 7, 12f, 36, 43
  static (static models of), 13, 200, 382, *see also* Einstein
  instability of Einstein static model, 201
  non static cosmological models, 206
  stationary, 381, *see also* Steady-state cosmology
  expanding, 39f, 99f, 134, 383f, 392, *see also* Friedmann, Lemaître, Robertson, Walker solution
  evolving (evolutionary), 97ff, 324, 384ff, 404
  see also de Sitter space-time

weak anthropic principle, 86
weak interactions, 167
Wheeler and Feynman's action-at-a-distance electrodynamics, 233ff
  Hogarth's elaboration of, 233ff
Wiggish (history), xvif